RYPINS' INTENSIVE REVIEWS

Series Editor

Edward D. Frohlich, MD, MACP, FACC

Alton Ochsner Distinguished Scientist
Vice President for Academic Affairs
Alton Ochsner Medical Foundation
Staff Member, Ochsner Clinic
Professor of Medicine and of Physiology
Louisiana State University of Medicine
Adjunct Professor of Pharmacology and
Clinical Professor of Medicine
Tulane University School of Medicine
New Orleans, Louisiana

RYPINS' INTENSIVE REVIEWS

Kent E. Vrana, PhD
Associate Professor and Director of Graduate Studies
Department of Physiology and Pharmacology
Bowman Gray School of Medicine of Wake Forest University
Winston-Salem, North Carolina

LIPPINCOTT WILLIAMS & WILKINS
A **Wolters Kluwer** Company
Philadelphia • Baltimore • New York • London
Buenos Aires • Hong Kong • Sydney • Tokyo

Acquisitions Editor: Richard Winters
Developmental Editor: Mary Beth Murphy
Managing Editor: Susan E. Kelly
Manufacturing Manager: Kevin Watt
Supervising Editor: Mary Ann McLaughlin
Production Editor: Jane Bangley McQueen, Silverchair Science + Communications
Cover Designer: William T. Donnelly
Interior Designer: Susan Blaker
Design Coordinator: Melissa Olson
Indexer: Linda Hallinger
Compositor: Silverchair Science + Communications
Printer: Courier/Kendallville

Printed in the United States of America

9 8 7 6 5 4 3 2 1

Library of Congress Cataloging-in-Publication Data

Vrana, Kent E.
Biochemistry / Kent E. Vrana.
p. cm. - - (Rypins' intensive reviews)
Includes index.
ISBN 0-397-51546-4
1. Clinical biochemistry--Outlines, syllabi, etc. 2. Clinical biochemistry--Examinations, questions, etc. 3. Biochemistry--Outlines, syllabi, etc. 4. Biochemistry--Examinations, questions, etc. 5. Molecular biology--Outlines, syllabi, etc. 6. Molecular biology--Examinations, questions, etc. I. Title. II. Series.
RB112.5 .V73 1998
612'.015'076--dc21

98-40990
CIP

Who Was "Rypins"?

Dr. Harold Rypins (1892–1939) was the founding editor of what is now known as the RYPINS' series of review books. Originally published under the title *Medical State Board Examinations,* the first edition was published by J. B. Lippincott Company in 1933. Dr. Rypins edited subsequent editions of the book in 1935, 1937, and 1939 before his death that year. The series that he began has since become the longest-running and most successful publication of its kind, having served as an invaluable tool in the training of generations of medical students. Dr. Rypins was a member of the faculty of Albany Medical College in Albany, New York, and also served as Secretary of the New York State Board of Medical Examiners. His legacy to medical education flourishes today in the highly successful *Rypins' Basic Sciences Review* and *Rypins' Clinical Sciences Review,* now in their 17th editions, and in the *Rypins' Intensive Reviews* series of subject review volumes. We at Lippincott Williams & Wilkins take pride in this continuing success.

—The Publisher

Series Preface

These are indeed very exciting times in medicine. Having made this statement, one's thoughts immediately reflect about the major changes that are occurring in our overall health care delivery system, utilization-review and shortened hospitalizations, issues concerning quality assurance, ambulatory surgical procedures and medical clearances, and the impact of managed care on the practice of internal medicine and primary care. Each of these issues has had a considerable impact on the approach to the patient and on the practice of medicine.

But even more mind-boggling than the foregoing changes are the dramatic changes imposed on the practice of medicine by fundamental conceptual scientific innovations engendered by advances in basic science that no doubt will affect medical practice of the immediate future. Indeed, much of what we thought of as having a potential impact on the practice of medicine of the future has already been perceived. One need only take a cursory look at our weekly medical journals to realize that we are practicing "tomorrow's medicine today." And consider that the goal a few years ago of actually describing the human genome is now near reality.

Reflect, then, for a moment on our current thinking about genetics, molecular biology, cellular immunology, and other areas that have impacted upon our current understanding of the underlying mechanisms of the pathophysiological concepts of disease. Moreover, paralleling these innovations have been remarkable advances in the so-called "high tech" and "gee-whiz" aspects of how we diagnose disease and treat patients. We can now think with much greater perspective about the dimensions of more specific biologic diagnoses concerned with molecular perturbations; gene therapy not only affecting genetic but oncological diseases; more specific pharmacotherapy involving highly specific receptor inhibition, alterations of intracellular signal transduction, manipulations of cellular protein synthesis; immunosuppresive therapy not only with respect to organ transplantations but also of autoimmune and other immune-related diseases; and therapeutic means for manipulating organ remodeling or the intravascular placement of stents. Each of these concepts has become inculcated into our everyday medical practice within the past decade. The reasons why these changes have so rapidly promoted an upheaval in medical practice are continuing medical education, a constant awareness of the current medical literature, and a thirst for new knowledge.

To assist the student and practitioner in the review process, the publisher and I have initiated a new approach in the publication of *Rypins' Basic Sciences Review* and *Rypins' Clinical Sciences Review.* Thus, when I assumed responsibility to edit this long-standing board review series with the 13th edition of the textbook (first published in 1931), it was with a feeling of great excitement. I perceived that great changes would be coming to medicine, and I believed that this would be one ideal means of not only facing these changes head on but also for me personally to cope and keep up with these changes. Over the subsequent editions, this confidence was reassured and rewarded. The presentation for the updating of medical information was tremendously enhanced by the substitution of new authors, as the former authority "standbys" stepped down or retired from our faculty. Each of the authors who continue to be selected for maintaining the character of our textbook is an authority in his or her respective area and has had considerable pedagogic and formal examination experience. One dramatic recent example of the changes in author replacement just came about with the 17th edition. When I invited Dr. Peter Goldblatt to participate in the authorship of the pathology chapter of the textbook, his answer was "what goes around, comes around." You see, Dr. Goldblatt's father, Dr. Harry Goldblatt, a major contributor to the history of hypertensive disease, was the first author of the pathology chapter in 1931. What a satisfying experience for me personally. Other less human changes in our format came with the establishment of two soft cover volumes, the current basic and clinical sciences review volumes, replacing the single volume text of earlier years. Soon, a third supplementary volume concerned with questions and answers for the basic science volume appeared. Accompanying these more obvious changes was the constant updating of the knowledge base of each of the chapters, and this continues on into the present 17th edition.

And now we have introduced another major innovation in our presentation of the basic and clinical sciences reviews. This change is evidenced by the introduction of the *Rypins' Intensive Reviews* series, along with the 17th edition of *Rypins' Basic Sciences Review, Rypins' Clinical Sciences Review,* and the *Questions and Answers* third volume. These volumes are written to be used separately from the parent textbook. Each not only contains the material published in their respective chapters of the textbook, but is considerably "fleshed out" in the discussions, tables, figures, and questions and answers. Thus, the *Rypins' Intensive Reviews* series serves as an important supplement to the overall review process and also provides a study guide for those already in practice, in preparing for specific specialty board certification and recertification examinations.

Therefore, with continued confidence and excitement, I am pleased to present these innovations in review experience for your consideration. As in the past, I look forward to learning of your comments and suggestions. In doing so, we continue to look forward to our continued growth and acceptance of the *Rypins'* review experience.

Edward D. Frohlich, MD, MACP, FACC

Preface

Biochemistry and molecular biology are disciplines that have blossomed during the twentieth century, and discoveries continue to occur at an accelerating pace. Biochemical research during the twentieth century has been performed by the vitamin hunters, the enzyme hunters, the pathfinders, and—most recently—by the gene hunters. By 2003, the complete DNA sequence of the human genome (3×10^9 nucleotides) will be known, and the sequence of approximately 100,000 genes contained in this genome will be available. The structures of new disease-associated genes are currently reported weekly, and the rate of discovery of such genes promises to increase exponentially. Moreover, it is likely that many new families of genes will be discovered, adding to the foundations of biochemistry and molecular biology. These studies will serve as the foundation for a new molecular medicine.

This volume in the *Rypins' Intensive Reviews* series encompasses traditional medical biochemistry and the beginnings of the new molecular medicine. The text, figures, and tables contain the essentials of contemporary biochemistry. The comprehensive examination comprises numerous questions about disease processes, and the answers provide information that aids in the biochemical reasoning required to understand various pathophysiologic states.

Students will find this review useful in preparing for course examinations and for the United States Medical Licensure Examination (USMLE) Step 1. There are approximately 300 questions (all presented in the format of the current Step 1 exam) with answers and discussions. Must-Know Topics, or key concepts, presented at the end of the book serve as a quick assessment of a student's overall understanding of biochemistry.

I thank Susan Kelly for her editorial assistance and for all of the work she did to bring this book to publication. On a personal note, I thank Professor Robert Roskoski, Jr., for providing the text and tables from his chapter in *Rypins' Basic Sciences Reviews*, on which this book is based. I have had the privilege of working with him for more than 20 years on both educational and research projects. I appreciate his adage that "biochemistry is fun."

Kent E. Vrana, PhD

Introduction

Preparing for the USMLE

UNITED STATES MEDICAL LICENSING EXAMINATION (USMLE)

In August 1991 the Federation of State Medical Boards (FSMB) and the National Board of Medical Examiners (NBME) agreed to replace their respective examinations, the FLEX and NBME, with a new examination, the United States Medical Licensing Examination (USMLE). This examination will provide a common means for evaluating all applicants for medical licensure. It appears that this development in medical licensure will at last satisfy the needs for state medical boards licensure, the national medical board licensure, and licensure examinations for foreign medical graduates. This is because the 1991 agreement provides for a composite committee that equally represents both organizations (the FSMB and NBME) as well as a jointly appointed public member and a representative of the Educational Council for Foreign Medical Graduates (ECFMG).

As indicated in the USMLE announcement, "It is expected that students who enrolled in U.S. medical schools in the fall of 1990 or later and foreign medical graduates applying for ECFMG examinations beginning in 1993 will have access only to USMLE for purposes of licensure." The phaseout of the last regular examinations for licensure was completed in December 1994.

The new USMLE is administered in three steps. Step 1 focuses on fundamental basic biomedical science concepts, with particular emphasis on "principles and mechanisms underlying disease and modes of therapy." Step 2 is related to the clinical sciences, with examination on material necessary to practice medicine in a supervised setting. Step 3 is designed to focus on "aspects of biomedical and clinical science essential for the unsupervised practice of medicine."

Today Step 1 and Step 2 examinations are set up and scored as total comprehensive objective tests in the basic sciences and clinical sciences, respectively. The format of each part is no longer subject-oriented, that is, separated into sections specifically labeled Anatomy, Pathology, Medicine, Surgery, and so forth. Subject labels are therefore missing, and in each part questions from the different fields are intermixed or integrated so that the subject origin of any individual question is not immediately apparent, although it is known by the National Board office. Therefore, if necessary, individual subject grades can be extracted.

Step 1 is a two-day written test including questions in anatomy, biochemistry, microbiology, pathology, pharmacology, physiology, and the behavioral sciences. Each subject contributes to the examination a large number of questions designed to test not only knowledge of the subject itself but also "the subtler qualities of discrimination, judgment, and reasoning." Questions in such fields as molecular biology, cell biology, and genetics are included, as are questions to test the "candidate's recognition of the similarity or dissimilarity of diseases, drugs, and physiologic, behavioral, or pathologic processes." Problems are presented in narrative, tabular, or graphic form, followed by questions designed to assess the candidate's knowledge and comprehension of the situation described.

Step 2 is also a two-day written test that includes questions in internal medicine, obstetrics and gynecology, pediatrics, preventive medicine and public health, psychiatry, and surgery. The questions, like those in Step 1, cover a broad spectrum of knowledge in each of the clinical fields. In addition to individual questions, clinical problems are presented in the form of case histories, charts, roentgenograms, photographs of gross and microscopic pathologic specimens, laboratory data, and the like, and the candidate must answer questions concerning the interpretation of the data presented and their relation to the clinical problems. The questions are "designed to explore the extent of the candidate's knowledge of clinical situations, and to test his [or her] ability to bring information from many different clinical and basic science areas to bear upon these situations."

The examinations of both Step 1 and Step 2 are scored as a whole, certification being given on the basis of performance on the entire part, without reference to disciplinary breakdown. The grade for the examination is derived from the total number of questions answered correctly, rather than from an average of the grades in the component basic science or clinical science subjects. A candidate who fails will be required to repeat the entire examination. Nevertheless, as noted above, in spite of the interdisciplinary character of the examinations, all of the traditional disciplines are represented in the test, and separate grades for each subject can be extracted and reported separately to students, to state examining boards, or to those medical schools that request them for their own educational and academic purposes.

This type of interdisciplinary examination and the method of scoring the entire test as a unit have definite advantages, especially in view of the changing curricula in medical schools. The former type of rigid, almost standardized, curriculum, with its emphasis on specific subjects and a specified number of hours in each, has been replaced by a more liberal, open-ended curriculum, permitting emphasis in one or more fields and corresponding deemphasis in others. The result has been rather wide variations in the totality of education in different medical schools. Thus, the scoring of these tests as a whole permits accommodation to this variability in the curricula of different schools. Within the total score, weakness in one subject that has received relatively little emphasis in a given school may be balanced by strength in other subjects.

The rationale for this type of comprehensive examination as replacement for the traditional department-oriented examination in the basic sciences and the clinical sciences is given in the National Board Examiner:

> The student, as he [or she] confronts these examinations, must abandon the idea of "thinking like a physiologist" in answering a question labeled "physiology" or "thinking like a surgeon" in answering a question labeled "surgery." The one question may have been written by a biochemist or a pharmacologist; the other question may have been written by an internist or a pediatrician. The pattern of these examinations will direct the student to thinking more broadly of the basic sciences in Step 1 and to thinking of patients and their problems in Step 2.

Until a few years ago, the Part I examination could not be taken until the work of the second year in medical school had been completed, and the Part II test was given only to students who had completed the major part of the fourth year. Now students, if they feel they are ready, may be admitted to any regularly scheduled Step 1 or Step 2 examination during any year of their medical course without prerequisite completion of specified courses or chronologic periods of study. Thus, emphasis is placed on the acquisition of knowledge and competence rather than the completion of predetermined periods.

Candidates are eligible for Step 3 after they have passed Steps 1 and 2, have received the M.D. degree from an approved medical school in the United States or Canada, and subsequent to the receipt of the M.D. degree, have served at least six months in an approved hospital internship or residency. Under certain circumstances, consideration may be given to other types of graduate training provided they meet with the approval of the National Board. After passing the Step 3 examination, candidates will receive their diplomas as of the date of the satisfactory completion of an internship or residency program. If candidates have completed the approved hospital training prior to completion of Step 3, they will receive certification as of the date of the successful completion of Step 3.

The Step 3 examination, as noted above, is an objective test of general clinical competence. It occupies one full day and is divided into two sections, the first of which is a multiple-choice examination that relates to the interpretation of clinical data presented primarily in pictorial form, such as pictures of patients, gross and microscopic lesions, electrocardiograms, charts, and graphs. The second section, entitled Patient Management Problems, utilizes a programmed-testing technique designed to measure the candidate's clinical judgment in the management of patients. This technique simulates clinical situations in which the physician is faced with the problems of patient management presented in a sequential programmed pattern. A set of four to six problems is related to each of a series of patients. In the scoring of this section, candidates are given credit for correct choices; they are penalized for errors of commission (selection of procedures that are unnecessary or are contraindicated) and for errors of omission (failure to select indicated procedures).

All parts of the USMLE are given in many centers, usually in medical schools, in nearly every large city in the United States as well as in a few cities in Canada, Puerto Rico, and the Canal Zone. In some cities, such as New York, Chicago, and Baltimore, the examination may be given in more than one center.

The examinations of the National Board have become recognized as the most comprehensive test of knowledge of the medical sciences and their clinical application produced in this country.

THE NATIONAL BOARD OF MEDICAL EXAMINERS

For years the National Board examinations have served as an index of the medical education of the period and have strongly influenced higher educational standards in each of the medical sciences. The Diploma of the National Board is accepted by 47 state licensing authorities, the District of Columbia, and the Commonwealth of Puerto Rico in lieu of the examination usually required for licensure and is recognized in the American Medical Directory by the letters DNB following the name of the physician holding National Board certification.

The National Board of Medical Examiners has been a leader in developing new and more reliable techniques of testing, not only for knowledge in all medical fields but also for clinical competence and fitness to practice. In recent years, too, a number of medical schools, several specialty certifying boards, professional medical societies organized to encourage their members to keep abreast of progress in medicine, and other professional qualifying agencies have called upon the National Board's professional staff for advice or for the actual preparation of tests to be employed in evaluating medical knowledge, effectiveness of teaching, and professional competence in certain medical fields. In all cases, advantage has been taken of the validity and effectiveness of the objective, multiple-choice type of examination, a technique the National Board has played an important role in bringing to its present state of perfection and discriminatory effectiveness.

Objective examinations permit a large number of questions to be asked, and approximately 150 to 180 questions can be answered in a 2½-hour period. Because the answer sheets are scored by machine, the grading can be accomplished rapidly, accurately, and impartially. It is completely unbiased and based on percentile ranking. Of long-range significance is the facility with which the total test and individual questions can be subjected to thorough and rapid statistical analyses, thus providing a sound basis for comparative studies of medical school teaching and for continuing improvement in the quality of the test itself.

QUESTIONS

Over the years, many different forms of objective questions have been devised to test not only medical knowledge but also those subtler qualities of discrimination, judgment, and reasoning. Certain types of questions may test an individual's recognition of the similarity or dissimilarity of diseases, drugs, and physiologic or pathologic processes. Other questions test judgment as to cause and effect or the lack of causal relationships. Case histories or patient problems are used to simulate the experience of a physician confronted with a diagnostic problem; a series of questions then tests the individual's understanding of related aspects of the case, such as signs and symptoms, associated laboratory findings, treatment, complications, and prognosis. Case-history questions are set up purposely to place emphasis on correct diagnosis within a context comparable with the experience of actual practice.

It is apparent from recent certification and board examinations that the examiners are devoting more attention in their construction of questions to more practical means of testing basic and clinical knowledge. This greater realism in testing relates to an increasingly interdisciplinary approach toward fundamental material and to the direct relevance accorded practical clinical problems. These more recent approaches to questions have been incorporated into this review series.

Of course, the new approaches to testing add to the difficulty experienced by the student or physician preparing for board or certification examinations. With this in mind, the author of this review is acutely aware not only of the interrelationships of fundamental information within the basic science disciplines and their clinical implications but also of the necessity to present this material clearly and concisely despite its complexity. For this reason, the questions are devised to test knowledge of specific material within the text and identify areas for more intensive study, if necessary. Also, those preparing for examinations must be aware of the interdisciplinary nature of fundamental clinical material, the common multifactorial characteristics of disease mechanisms, and the necessity to shift back and forth from one discipline to another in order to appreciate the less than clear-cut nature separating the pedagogic disciplines.

The different types of questions that may be used on examinations include the completion-type question, in which the individual must select one best answer among a number of possible choices, most often five, although there may be three or four; the completion-type question in the negative form, in which all but one of the choices is correct and words such as *except* or *least* appear in the question; the true-false type of question, which tests an understanding of cause and effect in relationship to medicine; the multiple true-false type, in which the question may have one, several, or all correct choices; one matching-type question, which tests association and relatedness and uses four choices, two of which use the word, *both* or *neither;* another matching-type question that uses anywhere from three to twenty-six choices and may have more than one correct answer; and, as noted above, the patient-oriented question, which is written around a case and may have several questions included as a group or set.

Many of these question types may be used in course or practice exams; however, at this time the most commonly used types of questions on the USMLE exams are the completion-type question (one best answer), the completion-type negative form, and the multiple matching-type question, designating specifically how many choices are correct. Often included within the questions are graphic elements such as diagrams, charts, graphs, electrocardiograms, roentgenograms, or photomicrographs to elicit knowledge of structure, function, the course of a clinical situation, or a statistical tabulation. Questions then may be asked in relation to designated elements of the same. As noted above, case histories or patient-oriented questions are more frequently used on these examinations, requiring the individual to use more analytic abilities and less memorization-type data.

For further detailed information concerning developments in the evolution of the examination process for medical licensure (for graduates of both U.S. and foreign medical schools), those interested should contact the National Board of Medical Examiners at 3750 Market Street, Philadelphia, PA 19104; telephone 215-590-9500; or http://www.usmle.org.

FIVE POINTS TO REMEMBER

In order for the candidate to maximize chances for passing these examinations, a few common sense strategies or guidelines should be kept in mind.

First, it is imperative to prepare thoroughly for the examination. Know well the types of questions to be presented and the pedagogic areas of particular weakness, and devote more preparatory study time to these areas of weakness. Do not use too much time restudying areas in which there is a feeling of great confidence and do not leave unexplored those areas in which there is less confidence. Finally, be well rested before the test and, if possible, avoid traveling to the city of testing that morning or late the evening before.

Second, know well the format of the examination and the instructions before becoming immersed in the challenge at hand. This information can be obtained from many published texts and brochures or directly from the testing service (National Board of Medical Examiners, 3750 Market Street, Philadelphia, PA 19104; telephone 215-590-9500). In addition, many available texts and self-assessment types of examination are valuable for practice.

Third, know well the overall time allotted for the examination and its components and the scope of the test to be faced. These may be learned by a rapid review of the examination itself. Then, proceed with the test at a careful, deliberate, and steady pace without spending an inordinate amount of time on any single question. For example, certain questions such as the "one best answer" probably should be allotted 1 to 1½ minutes each. The "matching" type of question should be allotted a similar amount of time.

Fourth, if a question is particularly disturbing, note appropriately the question (put a mark on the question sheet) and return to it later. Don't compromise yourself by so concentrating on a likely "loser" that several "winners" are eliminated because of inadequate time. One way to save this time on a particular "stickler" is to play your initial choice; your chances of a correct answer are always best with your first impression. If there is no initial choice, reread the question.

Fifth, allow adequate time to review answers, to return to the questions that were unanswered and "flagged" for later attention, and check every *n*th (e.g., 20th) question to make certain that the answers are appropriate and that you did not inadvertently skip a question in the booklet or answer on the sheet (this can happen easily under these stressful circumstances).

There is nothing magical about these five points. They are simple and just make common sense. If you have prepared well, have gotten a good night's sleep, have eaten a good breakfast, and follow the preceding five points, the chances are that you will not have to return for a second go-around.

Edward D. Frohlich, MD, MACP, FACC

Series Acknowledgments

In no other writing experience is one more dependent on others than in a textbook, especially a textbook that provides a broad review for the student and fellow practitioner. In this spirit, I am truly indebted to all who have contributed to our past and current understanding of the fundamental and clinical aspects related to the practice of medicine. No one individual ever provides the singular "breakthrough" so frequently attributed as such by the news media. Knowledge develops and grows as a result of continuing and exciting contributions of research from all disciplines, academic institutions, and nations. Clearly, outstanding investigators have been credited for major contributions, but those with true and understanding humility are quick to attribute the preceding input of knowledge by others to the growing body of knowledge. In this spirit, we acknowledge the long list of contributors to medicine over the generations. We also acknowledge that in no century has man so exceeded the sheer volume of these advances than in the twentieth century. Indeed, it has been said by many that the sum of new knowledge over the past 50 years has most likely exceeded all that had been contributed in the prior years.

With this spirit of more universal acknowledgment, I wish to recognize personally the interest, support, and suggestions made by my colleagues in my institution and elsewhere. I specifically refer to those people from my institution who were of particular help and are listed at the outset of the internal medicine volume. But, in addition to these colleagues, I want to express my deep appreciation to my institution and clinic for providing the opportunity and ambience to maintain and continue these academic pursuits. As I have often said, the primary mission of a school of medicine is that of education and research; the care of patients, a long secondary mission to ensure the conduct of the primary goal, has now also become a primary commitment in these more pragmatic times. In contrast, the primary mission of the major multidisciplinary clinics has been the care of patients, with education and research assuming secondary roles as these commitments become affordable. It is this distinction that sets the multispecialty clinic apart from other modes of medical practice.

Over and above a personal commitment and drive to assure publication of a textbook such as this is the tremendous support and loyalty of a hard-working and dedicated office staff. To this end, I am tremendously grateful and indebted to Mrs. Lillian Buffa and Mrs. Caramia Fairchild. Their long hours of unselfish work on my behalf and to satisfy their own interest in participating in this major educational effort is appreciated to no end. I am personally deeply honored and thankful for their important roles in the publication of the Rypins' series.

Words of appreciation must be extended to the staff of Lippincott Williams & Wilkins. It is more than 25 years since I have become associated with this publishing house, one of the first to be established in our nation. Over these years, I have worked closely with Mr. Richard Winters, not only with the Rypins' editions but also with other textbooks. His has been a labor of commitment, interest, and full support—not only because of his responsibility to his institution, but also because of the excitement of publishing new knowledge. In recent years, we discussed at length the merits of adding the intensive review supplements to the parent textbook and together we worked out the details that have become the substance of our present "joint venture." Moreover, together we are willing to make the necessary changes to assure the intellectual success of this series. To this end, we are delighted to include a new member of our team effort, Ms. Susan Kelly. She joined our cause to ensure that the format of questions, the reference process of answers to those questions within the text itself, and the editorial process involved be natural and clear to our readers. I am grateful for each of these facets of the overall publication process.

Not the least is my everlasting love and appreciation to my family. I am particularly indebted to my parents who inculcated in me at a very early age the love of education, the respect for study and hard work, and the honor for those who share these values. In this regard, it would have been impossible for me to accomplish any of my academic pursuits without the love, inspiration, and continued support of my wife, Sherry. Not only has she maintained the personal encouragement to initiate and continue with these labors of love, but she has sustained and supported our family and home life so that these activities could be encouraged. Hopefully, these pursuits have not detracted from the development and love of our children, Margie, Bruce, and Lara. I assume that this has not occurred; we are so very proud that each is personally committed to education and research. How satisfying it is to realize that these ideals remain a familial characteristic.

Edward D. Frohlich, MD, MACP, FACC
New Orleans, Louisiana

Contents

Chapter 1

The Big Picture: Blueprint of Metabolism

Biochemistry is the **study of life at the molecular level**. It includes a study of the molecular composition of **living systems** and the **chemical reactions** that living systems undergo. Biochemistry is also concerned with the production and use of fuel molecules, which provide living organisms with the chemical energy required to maintain their highly organized state and for biosynthesis. It considers the mechanisms responsible for the work of transport, intracellular movement, and muscle contraction. Biochemistry and molecular biology also entail a consideration of replication, differentiation, development, maintenance, healing or repair, and aging.

In addition to addressing **physiologic processes**, biochemistry plays an important role in understanding the **pathogenesis of diseases**. A complete understanding of pathologic processes occurs only after the biochemical mechanisms have been discovered. Current literature conveys and illustrates the importance of biochemistry through frequent references to enzymes and biochemical reactions and the elucidation of the gene structures associated with diseases.

The complexity of living systems and attempts to understand the biochemistry of humans and human pathogens are sometimes daunting and bewildering. The large number of components involved adds to the difficulty. The realization, however, that all forms of life are made up of approximately 50 fundamental building blocks and their derivatives constitutes a major step in simplifying the science of biochemistry.

THE BIG PICTURE

Biochemistry and molecular biology can be simplified not only by considering that there is only a finite number of building blocks, but that the broad chemical processes of life also can be organized based on a relatively simple blueprint. For the purpose of discussion, this blueprint is referred to as the **Big Picture** (Fig. 1-1). This figure will serve as a metabolic map for the remainder of this review and will be cited repeatedly to place subsequent discussions in the context of the overall chemistry of life.

For many years, discussion of biochemistry has centered on the mechanisms by which organisms derive energy. This process is called **metabolism** (see right side of Fig. 1-1). Foodstuffs (fats and sugars) are ingested and then subjected to a series of reactions that break them down while trapping their energy in the form of **adenosine triphosphate (ATP)**. ATP serves as the **energy currency** of all life forms. It is the fuel that drives the synthesis of all of the complex molecules that comprise cellular structures (e.g., membranes, structural proteins), the genetic machinery (e.g., DNA and RNA), and the engines and workhorses of the cell (e.g., enzymes). These **biosynthetic reactions** also are represented in the Big Picture (see left side of Fig. 1-1).

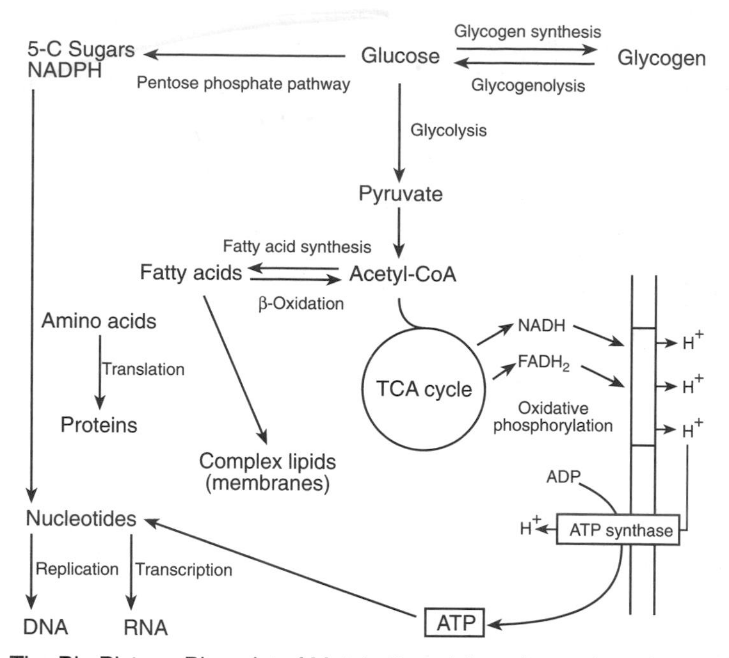

Figure 1-1. The Big Picture: Blueprint of Metabolism. A broad overview of the biochemistry of life. The biosynthetic reactions on the *left* (anabolic processes) represent the use of energy and simple building blocks to create the macromolecular structures, genetic information, and enzymes of the cell. The processes on the *right* (with the exception of the short-term storage of sugars as glycogen) are the central components of energy use (intermediary metabolism or catabolism) in which foods (primarily carbohydrates and lipids) are broken down to produce adenosine triphosphate (*ATP*), the unifying energy currency of the cell. Although not shown in this figure, there is a limited ability to convert amino acids into energy-yielding components [acetylcoenzyme A (*acetyl-CoA*), pyruvate, and intermediates of the tricarboxylic acid (*TCA*) cycle] and vice versa. ADP, adenosine diphosphate; $FADH_2$, flavin adenine dinucleotide (reduced form); NADH, nicotinamide adenine dinucleotide (reduced form).

Figure 1-1 also illustrates a fundamental **principle of nutrition**. When nutrients are plentiful, mechanisms exist for storing the materials as fatty acids and glycogen for when they are needed. Finally, it should be obvious that given a limited number of building blocks, there must be the potential for interconverting seemingly disparate molecules to generate an incredible complexity—the human body. Some of this chemical connectivity is also illustrated in the Big Picture.

As the discussions of specific aspects of biochemistry and molecular biology are undertaken, we will revisit Fig. 1-1. An important concept in the study and mastery of the field of biomedical science is that nature can be quite elegant in its integration of these different processes. Therefore, when studying one aspect of metabolism (e.g., glycolysis), knowledge of the Big Picture will immediately predict how other aspects of biochemistry will be regulated. These relationships follow the metabolic logic of the cell.

THE CELL

Humans, other animals, plants, and microorganisms are composed of fundamental units called **cells**. All cells are surrounded by a semipermeable plasma membrane. The **plasma membrane** is the boundary between the cell interior and the surrounding environment. The plasma membrane regulates and limits the influx and efflux of fuel molecules such as glucose and ions, which include sodium, potassium, and chloride.

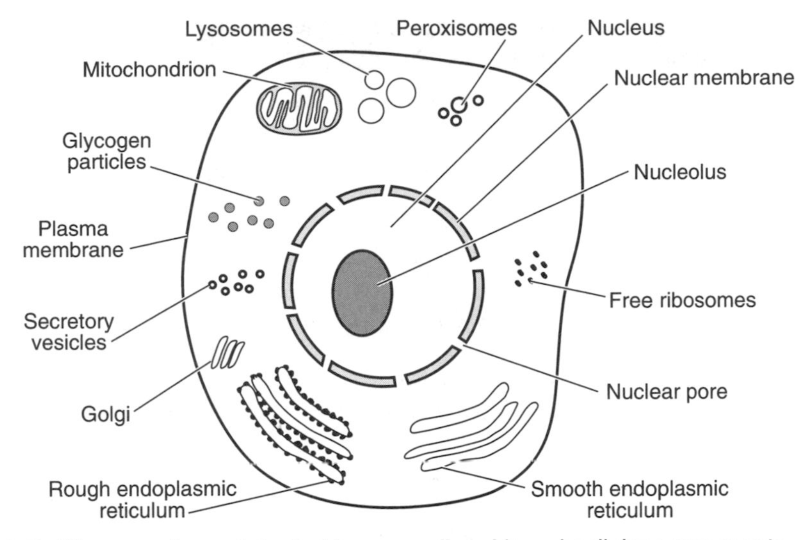

Figure 1-2. Diagram of a prototypical human cell and its subcellular components.

The importance of ion transport is indicated by the disorder **cystic fibrosis**, which is due to a defect in chloride transport. Intracellular metabolites, such as glucose 6-phosphate and citrate, bear electrical charges and pass through the plasma membrane with difficulty, if at all. Those charged molecules that cross the plasma membrane are transported by specific proteins called **translocases**.

Cells arise from other cells by the process of cell division. Organisms are divided into two major classes based on the presence or absence of a discrete cell nucleus. Humans and other organisms whose cells contain a nucleus are called **eukaryotes**. Cell division occurs by mitosis. The four stages of the cell cycle include G_1, S, G_2, and M. G_1 and G_2 are *g*rowth phases. S refers to the DNA *s*ynthesis phase, and M refers to *m*itosis. Microorganisms, such as *Escherichia coli* or *Streptococcus pneumoniae*, lack a well-defined nucleus and are called **prokaryotes**. Prokaryotes divide by binary fission.

A diagram of a prototypic animal cell is shown in Fig. 1-2. The constituent parts and the biochemical reactions associated with each part are listed in Table 1-1. The **nucleus** contains genetic information in the form of DNA. **Chromatin** refers to a combination of DNA, histone and nonhistone proteins, and nascent RNA. The nucleus is surrounded by a double membrane that contains nuclear pores and is essentially freely permeable (see Fig. 1-2).

The mitochondrion is the powerhouse of the cell and is responsible for most of the cellular ATP generation. The mitochondria contain an **outer membrane** that is freely permeable to small organic molecules and some larger proteins. The **inner mitochondrial membrane**, in contrast, exhibits restricted permeability. Except for a few uncharged substances, such as oxygen, carbon dioxide, and urea, the passage of metabolites through the mitochondrial inner membrane is mediated by specific transport proteins. For example, a specific protein carrier translocates ATP in exchange for adenosine diphosphate (ADP). This is an example of **antiport**: One substance moves in one direction, and the other moves in the opposite direction. In a **symport** process, both substances move in the same direction. The transport of glucose into the intestinal epithelium, for example, is accompanied and driven by the cotransport of sodium. The impermeability of the inner mitochondrial membrane to protons (H^+) is important in the mechanism of biosynthesis of ATP. The infolding of the inner mitochondrial membrane creates **cristae**, which increase membrane surface area. The compartment within the inner mitochondrial membrane is called the mitochondrial **matrix**.

In many ways, the mitochondrion is a small organism of its own. These organelles contain their own DNA and synthesize mRNA and proteins. In this sense, mitochon-

TABLE 1-1. Properties of Eukaryotic Cell Components

Component	General Properties	Associated Biochemical Processes	Percent Volume	Number Per Cell
Cytosol	Nonsedimentable	Glycolysis, glycogenesis; glycogenolysis, pentose phosphate pathway, gluconeogenesis, fatty acid synthesis, steroid synthesis, purine and pyrimidine formation, carbamoyl-phosphate synthetase II, protein synthesis (free ribosomes)	54	1
Nucleus	Repository and expression of genes	DNA replication, RNA synthesis and processing; contains chromatin, histones, and nonhistones	6	1
Mitochondrion	Powerhouse of the cell; major site of ATP formation	Citric acid cycle; β-oxidation of fatty acids; oxidative phosphorylation and ATP synthesis; pyruvate dehydrogenase, citrate synthase, carbamoyl-phosphate synthetase I (liver); some DNA, RNA, and protein synthesis; metabolic water formed here	22	1,700
Lysosome	Wastebasket of the cell	Acid phosphatase, cathepsins (degrade several classes of proteins), DNAse, RNAse, hexosaminidase, and many other hydrolytic activities	1	300
Peroxisome	Hydrogen peroxide metabolism	Catalase, peroxidase	1	400
Rough endoplasmic reticulum and Golgi bodies	Synthesis of membrane proteins and proteins for export	Membrane-bound ribosomes protein processing, and glycosylation	9	1
Smooth endoplasmic reticulum	Complex lipid biosynthesis	Cytochrome P450 electron transport, steroid hydroxylation, fatty acid desaturation, phospholipid biosynthesis	6	1
Plasma membrane	Boundary between cell exterior and interior	Na^+/K^+ ATPase, adenylyl cyclase, many receptors (e.g., insulin, β-adrenergic, HDL receptor), ion channels, glucose, and amino acid transport proteins	—	1

ATP, adenosine triphosphate; ATPase, adenosine triphosphatase; DNAse, deoxyribonuclease; HDL, high-density lipoprotein; RNAse, ribonuclease.

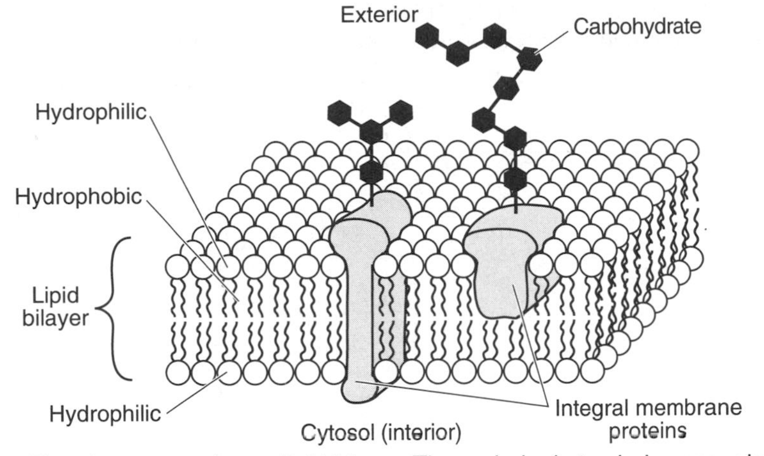

Figure 1-3. The plasma membrane lipid bilayer. The carbohydrate chains occur in the extracellular leaf.

dria are self-contained genetic entities; however, mitochondrial DNA is very small (approximately 16,500 base pairs) and encodes only a small percentage of the proteins of the organelle. Therefore, many constituents of the mitochondrion are synthesized within the cytoplasm of the cell and then transported into the organelle.

Membranes are composed of a phospholipid **bilayer** with a hydrophilic exterior and hydrophobic interior (Fig. 1-3). **Integral membrane proteins** are imbedded in or traverse the membrane. The lipids and proteins readily move laterally in a two-dimensional plane; lateral movement is termed **fluidity**. The existence of proteins within the lipid scaffold or sheet is termed a **mosaic**. Both properties (fluidity and mosaicism) are important and have given rise to the **fluid-mosaic model** of membranes.

Lysosomes are membrane-bound organelles that participate in the hydrolytic degradation of several types of compounds. The interior of **lysosomes** is acidic (pH 5) relative to the cytosol (pH 7). The degradative enzymes found in lysosomes exhibit an acid pH optimum. The hereditary absence of specific enzymes is associated with several **lysosomal storage diseases**, including **Niemann-Pick** (sphingomyelinase deficiency), **Tay-Sachs** (hexosaminidase A deficiency), and **Hurler's diseases**.

Ribosomes are large nucleoprotein complexes that serve as the protein synthesis factories. In the prokaryote *E. coli,* they are composed of 56 proteins and three distinct ribosomal RNA molecules. In a complex series of energy-requiring steps, the ribosomes "read" the messenger RNA blueprint and construct a protein from the information. Ribosomes function in two major states: free ribosomes within the cytoplasm, which are responsible for the synthesis of soluble proteins; and membrane-bound ribosomes, which are associated with the endoplasmic reticulum and synthesize membrane proteins and secreted proteins.

The **endoplasmic** (inside the cell) **reticulum** (network) is a membranous structure that participates in a wide variety of activities (see Table 1-1). When associated with ribosomes (**rough endoplasmic reticulum**), the endoplasmic reticulum plays a role in protein synthesis. The ribosomes attached to the endoplasmic reticulum exhibit a studded or roughened appearance, as observed by electron microscopy. Many reactions of lipid synthesis occur in the **smooth endoplasmic reticulum** because of the solubility of lipids in membranes.

The dimensions of cells vary considerably. A cuboidal liver cell is approximately 20 μm in diameter. The circular mature human erythrocyte is approximately 7 μm in diameter and 2 μm thick. Erythrocytes (7 μm in diameter) serve as an important relative standard when viewing tissues by microscopy. An *E. coli* cell and a liver mitochondrion are approximately 1 μm in diameter and 2 μm in length.

The mammalian erythrocyte lacks a nucleus, mitochondria, an endoplasmic reticulum, and other membranous organelles. They obtain their energy by anaerobic glycolysis. Mature erythrocytes cannot participate in DNA, RNA, or protein synthesis. Erythrocytes contain hemoglobin at a concentration of 5 mmol/L, and also contain 2,3-bisphosphoglycerate at a comparable concentration. Hemoglobin accounts for approximately 90% of the protein of erythrocytes. Erythrocytes function in both oxygen and carbon dioxide transport.

Chapter 2

Structure of Biomolecules

ELEMENTS

The elementary components of matter that constitute humans can be classified as elements of **organic matter and water**, **bulk minerals**, and **trace minerals**. The identity of these elements and their approximate mass in a 70-kg human are given in Table 2-1.

Oxygen is the most abundant element in the body (in terms of mass, not number of atoms). The average adult is composed of 55% water, 19% protein, 19% fat, 7% inorganic matter, and less than 1% carbohydrate. The preponderance of **water** accounts for the large mass of oxygen in humans. **Calcium** is predominantly an extracellular element, existing as solid hydroxyapatite in bones and teeth and as a 2-mmol/L solution in the extracellular space. **Hydroxyapatite** is a calcium phosphate–calcium hydroxide complex with the formula $Ca_{10}(PO_4)_6(OH)_2$; however, calcium also plays an important regulatory role within cells. Changes in intracellular concentration from 10^{-8} mol/L to 10^{-7} mol/L or more trigger muscle contraction, exocytosis, and other cellular processes. The other elements are mentioned as appropriate in the text and are discussed with nutrition.

CHEMICAL BONDS

There are four types of chemical bonds important in the formation of molecules in biological systems. These consist of (1) covalent bonds, (2) ionic bonds (salt bridges), (3) hydrogen bonds, and (4) hydrophobic interactions.

Covalent bonds are composed of a pair of electrons. Covalent bonds are strong (100 kcal/mol) and account for the stability of carbohydrates, fats, proteins, and nucleic acids.

In aqueous solution, **salt bridges** (ionic bonds) between positively and negatively charged species are weaker (5 kcal/mol) than covalent bonds.

Hydrogen bonds refer to the sharing of a hydrogen atom between electronegative oxygen atoms, nitrogen atoms, or a combination of the two. The hydrogen atom is covalently linked to one of the atoms of the pair and interacts electrostatically with the second. The strength of hydrogen bonds is very dependent on direction. Although the bonds are individually weak (2 to 5 kcal/mol), formation of a large number promotes stability. A prime example of such stability is illustrated by the DNA double helix. The millions of hydrogen bonds that comprise the base pairs between individual DNA strands combine to create a noncovalent associate of tremendous strength and stability.

Hydrophobic (water-fearing) bonds are apolar bonds between hydrocarbon-containing compounds. It is energetically favorable to sequester hydrocarbons in hydrophobic domains and minimize their contact with polar water molecules in solution.

TABLE 2-1. Elements of the Human Body

Element	Mass in 70-kg Human	Comments
Organic matter and water		
Carbon	12.6 kg	Organic chemicals
Hydrogen	7.0 kg	Organic chemicals and water
Oxygen	45.5 kg	Organic chemicals and water
Nitrogen	2.1 kg	Nucleic acids and amino acids
Phosphorous	0.7 kg	Nucleic acids and many metabolites; constituent of bones and teeth
Sulfur	0.175 kg	Connective tissue and proteins
Bulk minerals		
Sodium	105 g	Principal extracellular cation
Potassium	245 g	Principal extracellular cation; diffusion through cell membrane generates, in part, the negative intracellular electromotive force; obligatory loss of 40 mEq/day in urine
Magnesium	35 g	Cofactor for ATP and other nucleotide reactants; a calcium antagonist; $MgSO_4$ used in treatment of eclampsia to decrease nerve excitability
Calcium	1,050 g	Constituent of teeth and bones; intracellular second messenger; triggers muscle contraction and exocytosis
Chloride	105 g	Major extracellular anion; activates amylase
Fluoride	8 g	Increases hardness of bones and teeth; excess produces dental fluorosis
Trace minerals		
Manganese	20 mg	Mitochondrial superoxide dismutase
Iron	3,000 mg	Found in hemoglobin, myoglobin, cytochromes, iron-sulfur proteins; transported as transferrin and stored as ferritin; deficiency leads to a microcytic anemia
Cobalt	5 mg	Constituent of vitamin B_{12}
Copper	100 mg	Component of cytochrome aa_3 and tyrosinase (in melanin formation); transported in blood by ceruloplasmin; bound to erythrocuprein of the red blood cell; Wilson's disease (hepatolenticular degeneration) is a rare hereditary disorder involving brain and liver with abnormal copper metabolism; cytosolic superoxide dismutase
Zinc	2,300 mg	Cofactor of carbonic anhydrase, carboxypeptidase, cytosolic superoxide dismutase
Molybdenum	Trace	Xanthine dehydrogenase of purine metabolism and aldehyde oxidase in catecholamine metabolism
Iodine	Trace	Required for production of thyroid hormones T_4 and T_3 (formed from thyroglobulin); deficiency of thyroid hormone produces cretinism in children and myxedema in adults; hyperthyroidism with thyroid hyperplasia is treated with radioiodine
Selenium	Trace	Glutathione peroxidase

ATP, adenosine triphosphate; T_3, triiodothyronine; T_4, tetraiodothyronine.

Like hydrogen bonds, although individually weak, formation of a large number of hydrophobic bonds results in a stable structure. This can be seen in examining the nature of the lipid bilayer, which is responsible for formation of membranes (see Fig. 1-3).

BIOMOLECULES

In considering the reactions of metabolism, it is important to understand the chemistry of the participating functional groups. Examining the precise bonds that are made and broken during a chemical transformation aids in the understanding and analysis of a biochemical process. The main **functional groups** include hydrocarbons, alcohols, amines, amides, thiols, carbonyl groups of various types, multifunctional compounds, phosphates and their derivatives, and sulfates and their derivatives (Table 2-2). Most of these compounds are familiar from organic chemistry, which is often described as the chemistry of carbon. Bond making and breaking usually involve a carbon atom. In biochemistry, on the other hand, reactions often involve processes at phosphorus, oxygen, and nitrogen atoms as well as at carbon atoms.

Most biomolecules contain more than one functional group; for example, **carbohydrates** contain an aldehyde or ketone and two or more alcohol groups. **Amino acids**, as their name indicates, contain both amino and carboxylic acid groups. Although there is an incredible diversity in all forms of life, there are only approximately 50 fundamental compounds that constitute the major mass of living organisms. This unity in nature makes our task easier. In addition to the fundamental building blocks, there

TABLE 2-2. Functional Groups in Biochemicals

Group		Example with Group
Hydrocarbons		
Alkyl groups	$CH_3(CH_2)_n^-$	Leucine
Alkenes	>C=C<	Fumarate
Aromatic	(benzene ring)	Phenylalanine
	R—OH	Ethanol
	$R—NH_2$	Glycine
Alcohol amines sulfur derivatives		
Sulfhydryl group (mercaptan)	R—SH	Cysteine
Disulfide	R—S—S—R'	Cystine
Thioether	R—S—R'	Methionine
Sulfate	HO—S(=O)(=O)—O^-	

(continued)

TABLE 2-2. (*continued*)

Group		Example with Group
Sulfur derivatives		
Sulfate ester	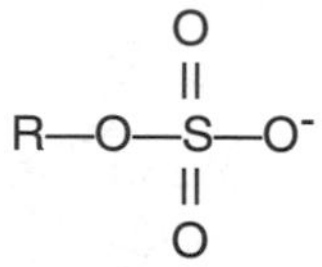 $R{-}O{-}S(=O)_2{-}O^-$	Chondroitin sulfate
Carbonyl groups	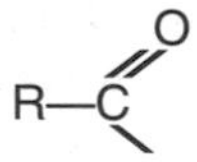 $R{-}C(=O){-}$	
Aldehyde	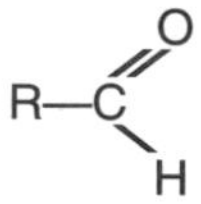 $R{-}C(=O){-}H$	Glyceraldehyde 3-phosphate
Ketone	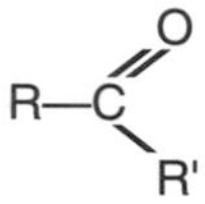 $R{-}C(=O){-}R'$	Dihydroxyacetonephosphate
Carboxylic acid	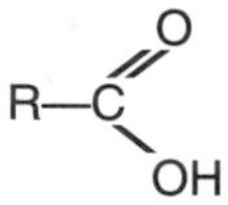 $R{-}C(=O){-}OH$	Palmitic acid
Ester	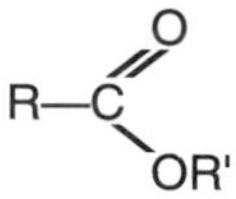 $R{-}C(=O){-}OR'$	Triglyceride
Amide	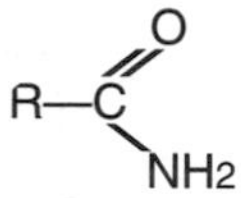 $R{-}C(=O){-}NH_2$	Glutamine
Thioester	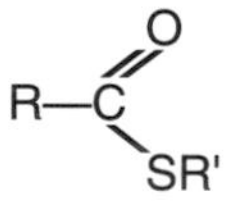 $R{-}C(=O){-}SR'$	Acetyl-CoA
Combinations		
Hemiacetal	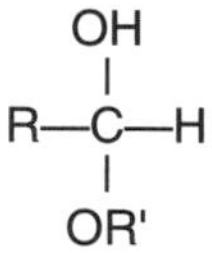 $R{-}C(OH)(OR'){-}H$	Glucopyranose
Acetal	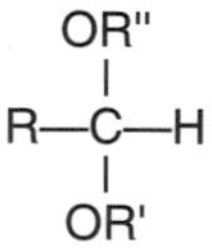 $R{-}C(OR'')(OR'){-}H$	Glycogen
Hydroxy acid	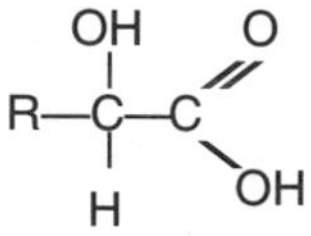 $R{-}C(OH)(H){-}C(=O){-}OH$	Lactate
Ketoacid	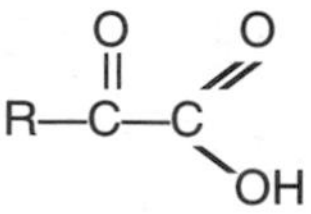 $R{-}C(=O){-}C(=O){-}OH$	Pyruvate
Dicarboxylate	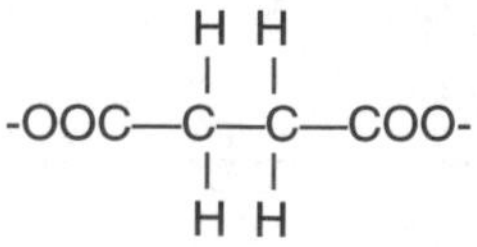 $^-OOC{-}CH_2{-}CH_2{-}COO^-$	Succinate

(*continued*)

TABLE 2-2. (*continued*)

Group	Structure	Example with Group
Phosphates		
Phosphoric acid	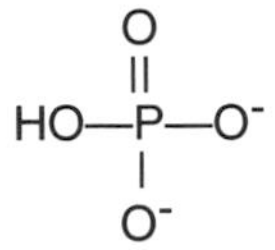 $HO{-}P(=O)(O^-){-}O^-$	
Pyrophosphate	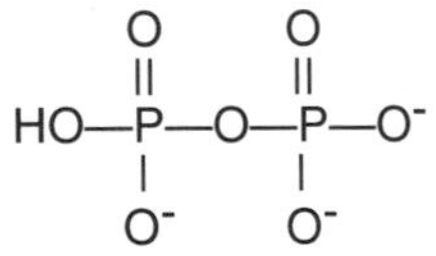 $HO{-}P(=O)(O^-){-}O{-}P(=O)(O^-){-}O^-$	
Phosphomonoester	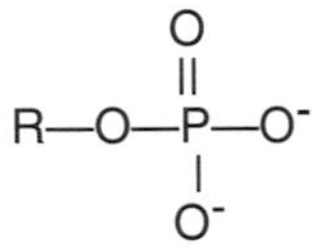 $R{-}O{-}P(=O)(O^-){-}O^-$	Glucose 6-phosphate
Phosphodiester	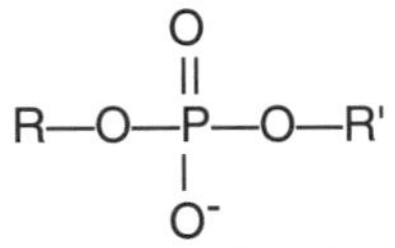 $R{-}O{-}P(=O)(O^-){-}O{-}R'$	Cyclic AMP, DNA, RNA
Bisphosphate	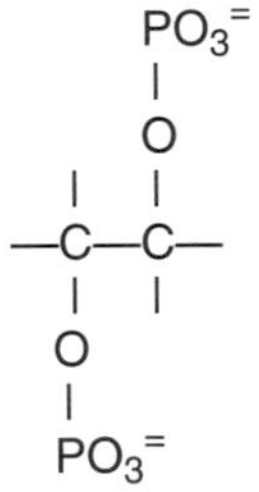 $-C(O{-}PO_3^{=})-C(O{-}PO_3^{=})-$	2,3-Bisphosphoglycerate
Trisphosphate	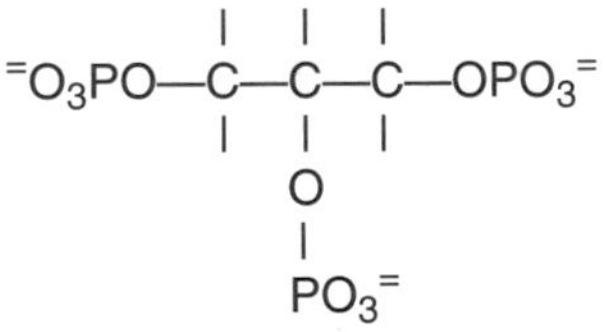 $^{=}O_3PO{-}C{-}C(O{-}PO_3^{=}){-}C{-}OPO_3^{=}$	Inositol trisphosphate
Diphosphate	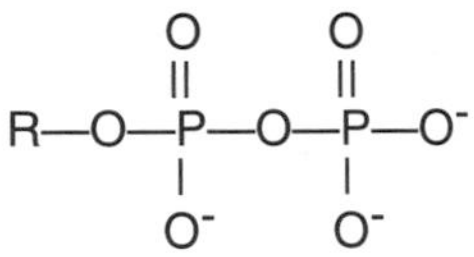 $R{-}O{-}P(=O)(O^-){-}O{-}P(=O)(O^-){-}O^-$	Adenosine diphosphate
Triphosphate	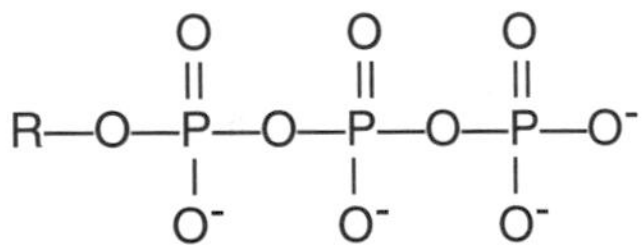 $R{-}O{-}P(=O)(O^-){-}O{-}P(=O)(O^-){-}O{-}P(=O)(O^-){-}O^-$	Adenosine triphosphate
Phosphoenol group	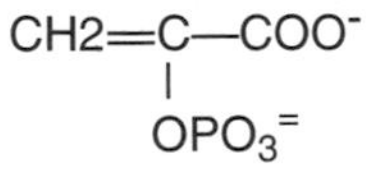 $CH2{=}C(OPO_3^{=}){-}COO^-$	Phosphoenolpyruvate
Phosphoramide	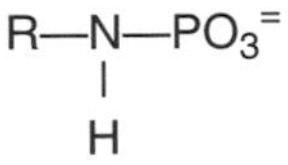 $R{-}N(H){-}PO_3^{=}$	Creatine phosphate
Acylphosphate	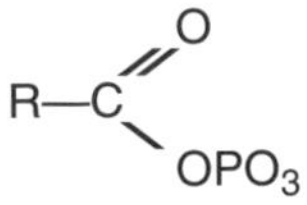 $R{-}C(=O){-}OPO_3$	1,3-Bisphosphoglycerate

Acetyl-CoA, acetylcoenzyme A; AMP, adenosine monophosphate.

are a few hundred other metabolites that constitute the vast majority of compounds with which biochemists are concerned; for example, the number and diversity of protein molecules in an organism greatly exceeds that of the low–molecular-weight compounds or metabolites. The next section covers individual classes of molecules beginning with proteins and enzymes; subsequent chapters consider carbohydrate, lipid, and nucleic acid structures and functions.

AMINO ACIDS, PEPTIDES, AND PROTEINS

Proteins perform a number of essential functions in all forms of life. They serve a structural role within the cell (cytoskeleton) and within the connective tissue and skeleton of the whole organism. They also function, *inter alia*, as catalysts, receptors, translocases, antibodies, and hormones. Approximately 19% of the human body is protein in nature. **Collagen**, a connective tissue protein, is the most abundant protein in humans. Proteins are polymers of α-amino acids. There are 20 **amino acids** that are genetically encoded and serve as precursors for protein biosynthesis on ribosomes. Some of these amino acid residues are modified or derivatized after biosynthesis (posttranslational modification); for example, specific protein serines, threonines, or tyrosines are phosphorylated to produce the corresponding phosphorylated residue. Other posttranslational modifications include hydroxylation, carboxylation, methylation, glycosylation, myristoylation, farnesylation, and acetylation.

COO^-
$H_3N^+ - C - H$
R

Figure 2-1. Structure of an α-amino acid.

Amino Acids and Peptides

Consider the identity and structures of the 20 genetically encoded **amino acids**. The amino acids that are found in proteins are α-amino acids (Fig. 2-1). With the exception of glycine, which lacks a chiral, or asymmetric, carbon atom (a carbon atom with four different substituents), the amino acids found in proteins possess the **L-configuration** (the absolute configuration corresponds to the standard L-glyceraldehyde).

The amino acids with hydrocarbon side chains are shown in Fig. 2-2. **Glycine** is the simplest of the amino acids and is so named because of its sweet taste (Gly,

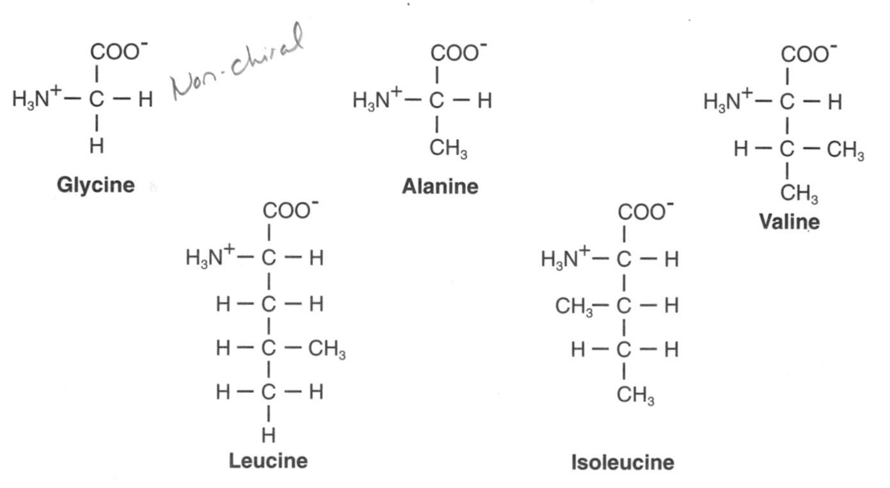

Figure 2-2. Aliphatic amino acids.

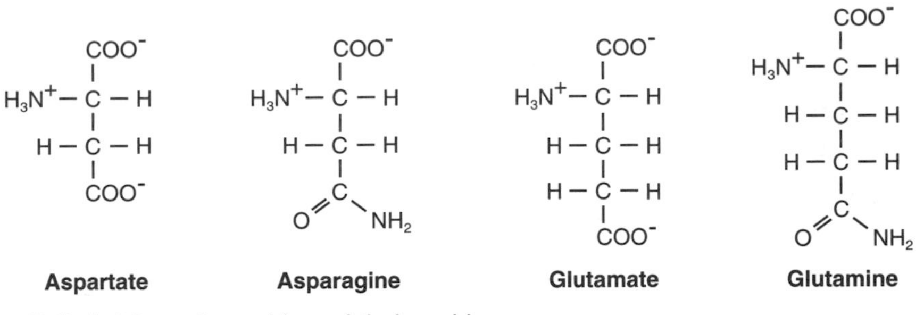

Figure 2-3. Acidic amino acids and their amides.

sugar). The side chains of **valine**, **leucine**, and **isoleucine**, which are branched, are hydrophobic. Four amino acids are dicarboxylic acids (**aspartate** and **glutamate**) or their derivatives (**asparagine** and **glutamine**) and are polar (Fig. 2-3). Asparagine was first isolated from asparagus. Three amino acids contain basic, nitrogen-containing, polar side chains: **lysine**, **arginine**, and **histidine** (Fig. 2-4). Three amino acids are aromatic and hydrophobic in nature: **phenylalanine**, **tyrosine**, and **tryptophan** (Fig. 2-5). Two amino acids contain sulfur: **cysteine** and **methionine** (Fig. 2-6). Two amino acids contain a polar alcohol side chain: **serine** and **threonine** (Fig. 2-7). The last of the genetically encoded amino acids is a cyclic amino acid (also termed an imino acid) named **proline** (Fig. 2-8). Note that the nitrogen atom of proline is linked to two carbon atoms.

There are a large number of amino acids in nature that are not found in proteins. Examples include **ornithine** and **citrulline**, which are important intermediates in urea biosynthesis.

Glycine lacks a chiral, or asymmetric, carbon atom. Two amino acids, isoleucine and threonine, contain two chiral carbon atoms. The β-carbon atom in each case constitutes the second chiral center. The amino acids are designated by a three- or single-letter abbreviation (Table 2-3).

The reaction of an α-amino group of one amino acid with a carboxyl group of a second amino acid along with the elimination of water results in the formation of a **peptide bond** (Fig. 2-9). The resulting compound is a **dipeptide**. A **tripeptide** contains three amino acid residues; an **oligopeptide** contains a few; and a **polypeptide** contains many amino acid residues. Note in Fig. 2-9 that all peptides contain an amino end (amino terminus) and a carboxyl terminus. As is noted in Chapter 13, protein syn-

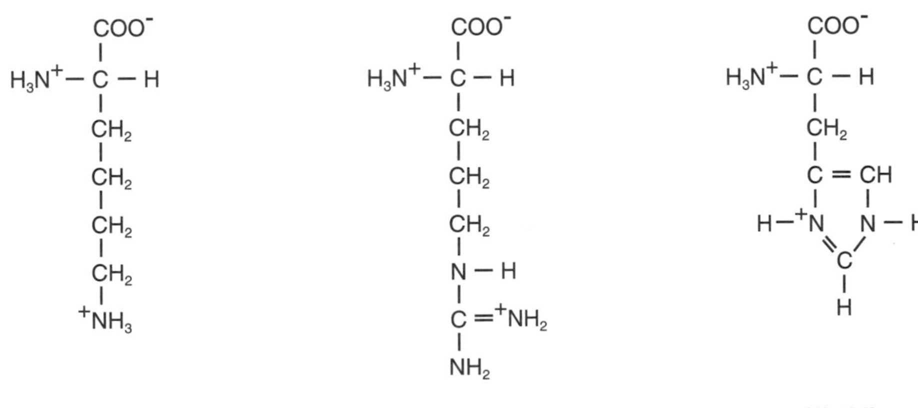

Figure 2-4. Basic amino acids.

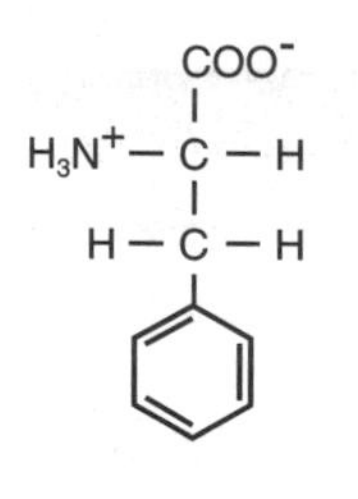

Phenylalanine

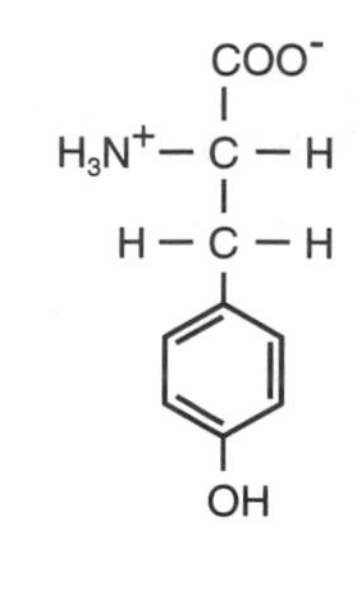

Tyrosine

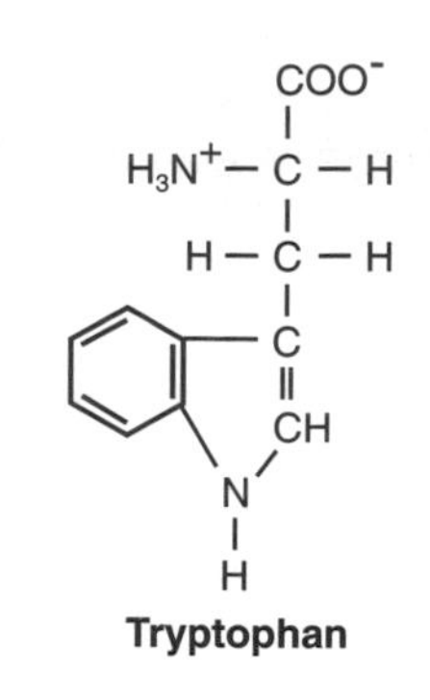

Tryptophan

Figure 2-5. Aromatic amino acids.

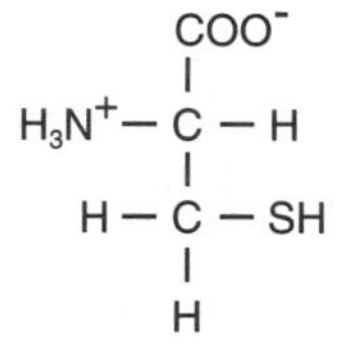

Cysteine

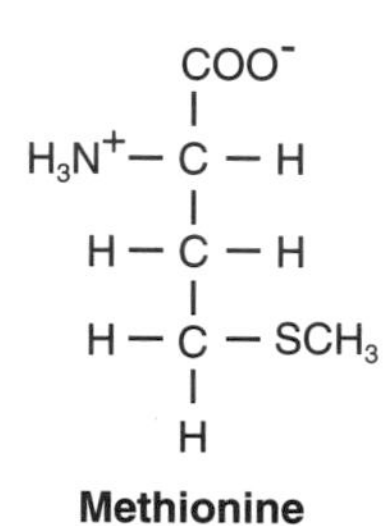

Methionine

Figure 2-6. Sulfur-containing amino acids.

Serine

Threonine

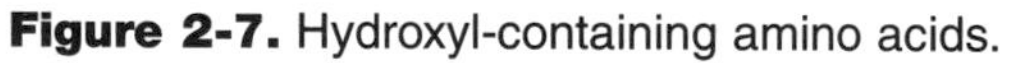

Figure 2-7. Hydroxyl-containing amino acids.

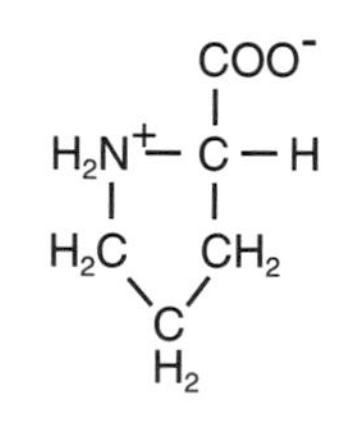

Figure 2-8. Proline, an imino acid.

TABLE 2-3. Genetically Encoded Amino Acids

Name	Abbreviations		Number of Codons	pKa of Side Chain	Comments
Aliphatic					
Glycine	Gly	G	4	—	Every third residue of collagen
Alanine	Ala	A	4	—	—
Valine	Val	V	4	—	Hydrophobic
Leucine	Leu	L	6	—	Hydrophobic
Isoleucine	Ile	I	3	—	Hydrophobic
Carboxylate-related					
Aspartate	Asp	D	2	4	Anionic
Glutamate	Glu	E	2	4	Anionic
Asparagine	Asn	N	2	—	—
Glutamine	Gln	Q	2	—	—
Basic					
Lysine	Lys	K	2	10.5	Cationic
Arginine	Arg	R	6	12.5	Cationic
Histidine	His	H	2	6	—
Aromatic					
Phenylalanine	Phe	F	2	—	Hydrophobic
Tyrosine	Tyr	Y	2	10.1	Rarely phosphorylated
Tryptophan	Trp	W	1	—	Hydrophobic; single codon
Sulfur-containing					
Methionine	Met	M	1	—	Initiator of protein synthesis
Cysteine	Cys	C	2	8.3	Oxidized to cystine
Hydroxyl-containing					
Serine	Ser	S	6	—	Chief phosphorylated residue of proteins
Threonine	Thr	T	4	—	Occasionally phosphorylated
Imino					
Proline	Pro	P	4	—	Occur at bends in protein chain

thesis *in vivo* occurs in an amino-to-carboxyl direction. The peptide bond is planar. The carbonyl group and substituted amine occur in a plane, and rotation about the C-N bond is prohibited. This limits the conformations that a polypeptide chain may assume (Fig. 2-10).

Proteins are polypeptides consisting of amino acid residues. The hormone insulin contains 51 amino acids (30 in the α-chain and 21 in the β-chain). Many biochemists consider this molecule a protein, and others regard it as a polypeptide. The distinction between polypeptide and protein is not absolute: All proteins are polypeptides, but the same is not true vice versa. An average polypeptide chain in a protein contains approximately 500 amino acid residues; a few contain more than 2,000 amino acid residues. The range of molecular weights of single polypeptide chains ranges from approximately

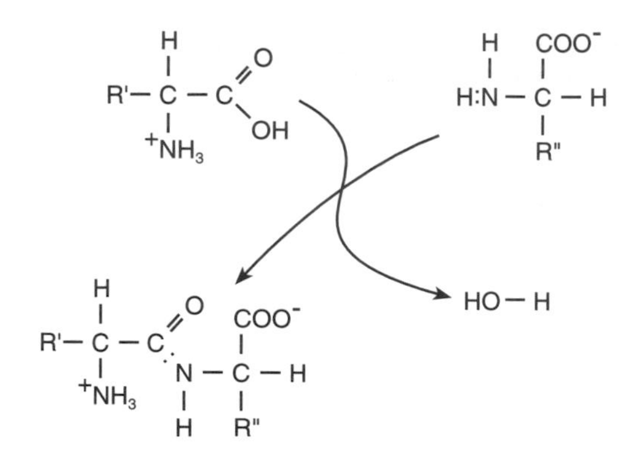

Figure 2-9. The peptide bond. The combination of two amino acids with the elimination of water yields a dipeptide.

5,000 to 300,000. To determine the approximate number of amino acids in a protein, divide the molecular weight by 110. This value (110) approximates the average molecular weight of an amino acid residue in an average protein.

Proteins consist of one or more polypeptide chains. **Myoglobin**, an intracellular oxygen storage protein containing heme, consists of a single polypeptide chain and is a monomer. **Hemoglobin**, the oxygen-transport protein found in erythrocytes, consists of two pairs of identical subunits, which form a tetramer. Hemoglobin contains two α-chains and two β-chains, and the tetramer is denoted as $\alpha_2\beta_2$.

Protein Structure

The structure of proteins is considered in a hierarchical fashion using four levels: primary, secondary, tertiary, and quaternary. The **primary structure** refers to the sequence of amino acids and the nature and position of any covalently attached derivatives. Peptides have a directionality with an amino group (not in peptide linkage) at one end and a carboxyl group (not in peptide linkage) at the other. The terminal amino or carboxyl groups may be free, or they may be derivatized. An example of the primary structure or sequence for a five–amino acid neuropeptide, methionine enkephalin, is illustrated in Fig. 2-11. By convention, structures are written with the amino terminus on the left and carboxyl terminus on the right; for instance, the dipeptide Gly-Ala differs from Ala-Gly.

The **secondary structure** of a protein refers to the pattern of hydrogen bonding. There are two major classes of hydrogen-bonded structures associated with secondary structure. The first to be described, the **α-helix**, refers to a helix stabilized by hydrogen bonding between a carbonyl group of one peptide bond and the nitrogen-hydrogen

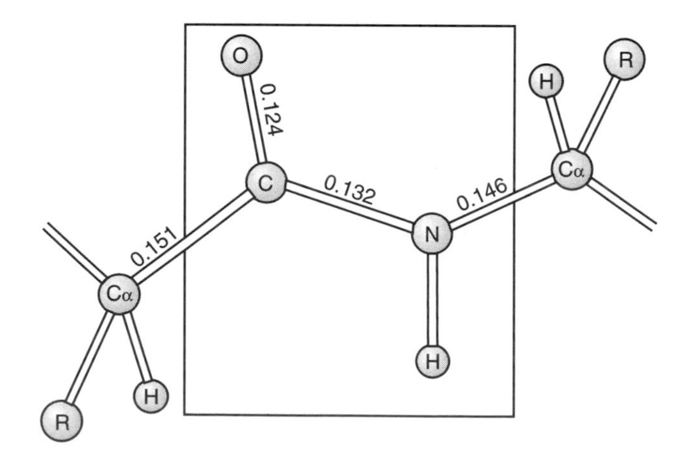

Figure 2-10. The planar peptide bond. The distances are given in nanometers (nm), where 1 nm is 1×10^9 m.

$H_2N-Tyr-Gly-Gly-Phe-Met-COO^-$

Figure 2-11. The primary structure of methionine enkephalin.

(N-H) group on the peptide bond on the chain four residues away (i.e., the residues are close together). The second form of secondary structure is called the **β-pleated sheet.** Here >N-H and >carbonyl (C=O) groups from residues very far apart on the polypeptide chain, or even residues on a different polypeptide chain, form hydrogen bonds. Two varieties of β-pleated sheets are recognized depending on the polarity of the participating polypeptide chains. When the chains are going in the same direction, from the amino to carboxyl end of the molecule, the structure is a **parallel β-pleated sheet.** When the participating chains are going in opposite directions with respect to the amino and carboxyl termini, the structure is an **antiparallel β-pleated sheet.** Reversal in the direction of the polypeptide chain occurs with the formation of a **β-bend.** This configuration involves the formation of a loop in which the carbonyl group of one amino acid residue forms a hydrogen bond with the amide N-H group of the residue three positions farther along the polypeptide chain.

The **tertiary** structure of a protein refers to the three-dimensional arrangement of the atoms of the molecule in space. For a monomeric protein such as myoglobin, the tertiary structure is the highest order of structure. The **quaternary structure** refers to the manner in which subunits of a multimeric protein interact. During the oxygenation of hemoglobin, a tetrameric protein, the subunits move relative to each other. This aspect of protein structure—the functional interaction of subunits as opposed to the merely physical interaction—is also a component of the quaternary structure.

The physiologically active conformation of a protein is called the **native** structure. The forces responsible for maintaining the active conformation include covalent bonds, salt bridges, hydrogen bonds, and hydrophobic bonds. The contributions of the latter three in maintaining the active conformation most likely varies among proteins. When these forces are disturbed as a result of exposure to extremes in pH (acid or alkali), high temperature (60°C or greater), 6-mol/L urea (a very high and nonphysiologic concentration), or treatment with charged detergents, such as sodium dodecylsulfate, the native conformation is destroyed and a **denatured** structure results. The native state corresponds to one or a few active conformations, but the denatured state may be associated with multiple but inactive conformations.

The **sequence of amino acids** in a polypeptide chain determines the structure and properties of the protein. As stated by Anfinsen's law, the primary structure of a protein governs its secondary, tertiary, and quaternary structures. The **substitution** of one amino acid by a similar one (e.g., the replacement of leucine by valine) is usually not of great consequence; however, substitution by unlike residues can result in a protein with greatly different properties. Substitution of valine, for example, for glutamate at position 6 (from the amino terminus) in the β-chain of human hemoglobin produces **hemoglobin S** (sickle cell hemoglobin). Deoxygenated **sickle cell hemoglobin** assumes an abnormal conformation and is poorly soluble in physiologic conditions. This leads to hemolysis and the circulatory abnormalities in the disease of **sickle cell anemia.** Single amino acid substitutions are associated with a remarkable number of additional diseases such as Alzheimer's disease.

As will become apparent in the discussion of macromolecular biosynthetic reactions (Chapters 11 to 13), the body devotes considerable attention and resources to ensuring that the primary structures of proteins are accurately created and maintained.

Chapter 3

pH and the Henderson-Hasselbalch Equation

The properties of **water** are very important in the maintenance of life's processes. Water constitutes approximately 70% of the lean body mass of humans and is an essential nutrient. Water spontaneously dissociates into a proton and a hydroxyl group:

$$H_2O \rightleftharpoons H^+ + OH^-$$

In **pure water**, $[H_2O]$ = 55.6 mol/L [(1,000 g/L)/(18 g/mol)], and the dissociation constant is such that

$$[H^+] = 10^{-7} \text{ mol/L}$$
$$[OH^-] = 10^{-7} \text{ mol/L}$$

The concentrations of $[OH^-]$ and $[H^+]$ are in a reciprocal relationship to each other. When $[H^+]$ increases, $[OH^-]$ decreases, and vice versa. Their product is 10^{-14} mol/L^2.

$$[H^+] \times [OH^-] = 10^{-14}$$

When $[H^+] = 10^{-4}$ mol/L, for example, then $[OH^-] = 10^{-10}$ mol/L.

The **pH** is defined by the following equation:

$$pH = -\log[H^+]$$

At neutrality (when $[H^+] = [OH^-]$)

$$pH = -\log[10^{-7}]$$
$$= -(-7) = +7$$

The **pH of blood** and most physiologic fluids is **7.4**.

$$7.4 = -\log[H^+]$$
$$3.98 \times 10^{-8} \text{ mol/L} = [H^+]$$

The pH of blood is maintained at 7.4 ± 0.05 through the actions of buffers. **Buffers** are substances that diminish the change in pH when acid or alkali is added to a solution. Buffers are composed of weak acids and their conjugate bases or weak bases and their conjugate acids. Recall that a base is a chemical species that accepts H^+, and an acid is one that donates H^+. The physiologically important buffer pairs in **blood and saliva** are (1) $H_2CO_3 \rightleftharpoons HCO_3^-$, (2) $H_2PO_4^- \rightleftharpoons HPO_4^{2-}$, and (3) protein $\rightleftharpoons$ protein$^-$.

The **Henderson-Hasselbalch equation** provides a convenient way to describe and think about buffers and pH, where

$$pH = pK_a + \log\frac{[\text{unprotonated species}]}{[\text{protonated species}]}$$

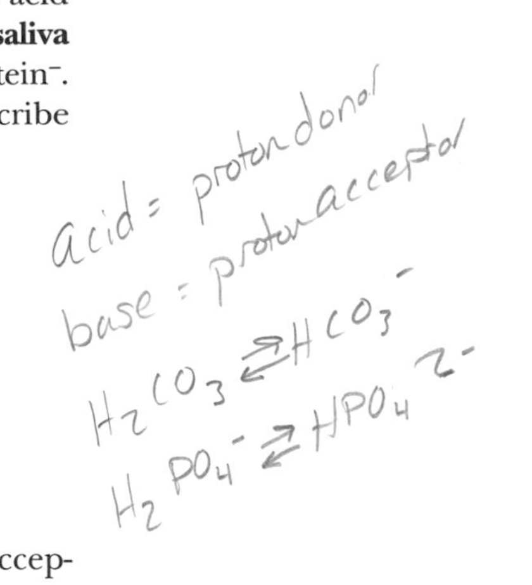

or equivalently

$$pH = pK_a + \log\frac{[\text{salt}]}{[\text{acid}]}$$

Many texts describe the ionization ratio as "[base]/[acid]" or "[proton acceptor]/[proton donor]." In any case, the meaning remains the same.

For the **phosphate** system,

$$H_2PO_4^- \rightleftharpoons H^+ + HPO_4^{2-}$$

$$pH = pK_a + \log\frac{[HPO_4^{2-}]}{[H_2PO_4^-]}$$

Note several potentially confusing aspects of the **phosphate buffering system**. First, at approximately pH 6.8, $H_2PO_4^-$ serves as the acid, and HPO_4^{2-} serves as the base. At much lower pH values (approximately 2), $H_2PO_4^-$ can serve as a base (accepting H^+ to become H_3PO_4). Similarly, at higher pH values (approximately 10), HPO_4^{2-} can serve as an acid, donating an H^+ ion to become PO_4^{3-}. When $[HPO_4^{2-}] = [H_2PO_4^-]$, the concentration of the protonated form equals that of the unprotonated form, and their ratio is 1. Some of these relationships can be seen in the titration of a phosphoric acid solution (Fig. 3-1).

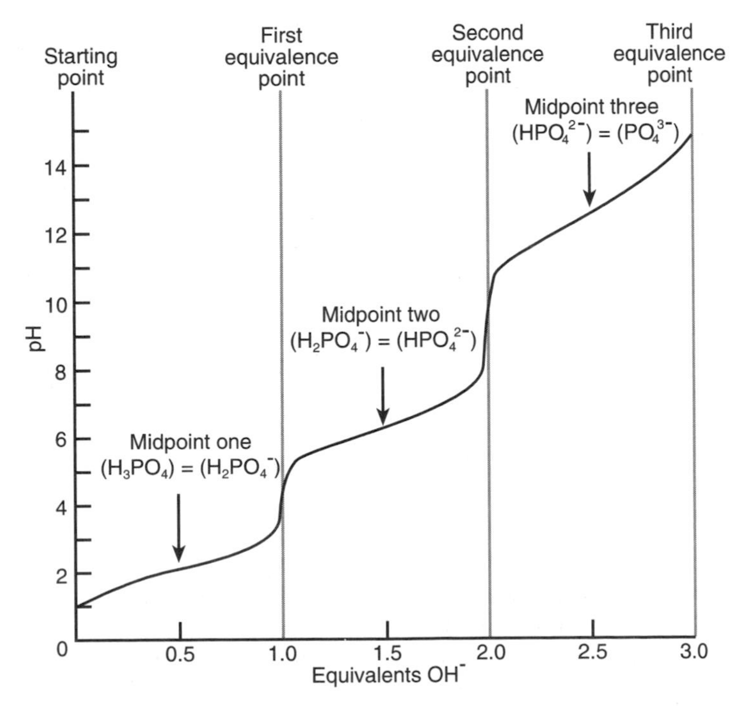

Figure 3-1. Titration curve of phosphoric acid (H_3PO_4).

$$pH = pK_a + \log 1$$
$$pH = pK_a$$

For **phosphate** at physiologic ionic strength, the **pK_a = 6.8**. The pK_a is the pH at which the concentration of salt and acid is identical. With this value, we can calculate the pH when the concentrations of $[HPO_4^{2-}]$ and $[H_2PO_4^-]$ are known. At a given pH, we can calculate the ratios of the two forms. For example, when $[HPO_4^{2-}]/[H_2PO_4^-] = 10$, then

$$\begin{aligned} pH &= 6.8 + \log 10 \\ &= 6.8 + 1.0 \\ &= 7.8 \end{aligned}$$

When $[HPO_4^{2-}]/[H_2PO_4^-] = 0.1$, then

$$\begin{aligned} pH &= 6.8 + \log 0.1 \\ &= 6.8 - 1.0 \\ &= 5.8 \end{aligned}$$

The **CO_2 system** can be expressed as follows

$$CO_2 + H_2O \rightleftharpoons H_2CO_3$$
$$H_2CO_3 \rightleftharpoons H^+ + HCO_3^-$$
$$pH = 6.1 + \log \frac{[HCO_3^-]}{[CO_2 + H_2CO_3]}$$

The **pK_a** of the CO_2 system is **6.1**. At pH 7.4, the ratio of $[HCO_3^-]/[CO_2 + H_2CO_3]$ is approximately 20. In this fairly unique case, the denominator contains values for both the carbon dioxide and carbonic acid because they can be considered the same.

When the pK_a of an **aspartyl group** in a protein-containing solution is 4.1, this means that this specific carboxylic acid residue is protonated in half of the molecules and unprotonated in the other half of the molecules at pH 4.1. When the pH is increased to 7.4, protons are liberated and the aspartyl group bears a net negative charge.

When a **lysyl residue** in a protein has a pK_a of 8.3, this means that at pH 8.3, the concentrations of $R–NH_2$ and $R–NH_3^+$ of the lysyl group are the same. When the pH is decreased to 7.4, more lysyl residues become protonated (because the solution is more acidic) and bear positive charges.

The **imidazole of histidine** is the only side chain with a pK_a in the physiologic range near 7 (see Table 2-3). If a histidine residue has a pK_a of 6.0, then at this pH, half are protonated (bearing a positive charge) and half are unprotonated (uncharged). At pH 7.4, approximately 4% of the imidazoles of this histidine residue are positively charged.

The **pK_a of the side chain of amino acids** varies with the protein and the specific residue in the protein. That is, not all carboxylic acids (whether the alpha carboxyl or the side group residue of aspartic or glutamic acid) display the same ionization constant. Moreover, an amino acid present in a protein generally has a different pK_a than the free amino acid residue, because the microenvironment (the presence of other residues in the vicinity of an ionizable moiety) alters the tendency of that moiety to accept or donate H^+. For this reason, the pK_a values given in Table 2-3 are representative or approximate.

In Fig. 3-2, titration curves for the various classes of amino acids are presented. Every amino acid has an α-amino and an α-carboxyl group; therefore, each amino acid shares those two ionization characteristics. Together, these curves are the basis for determining the net charge on a given peptide or protein at any pH.

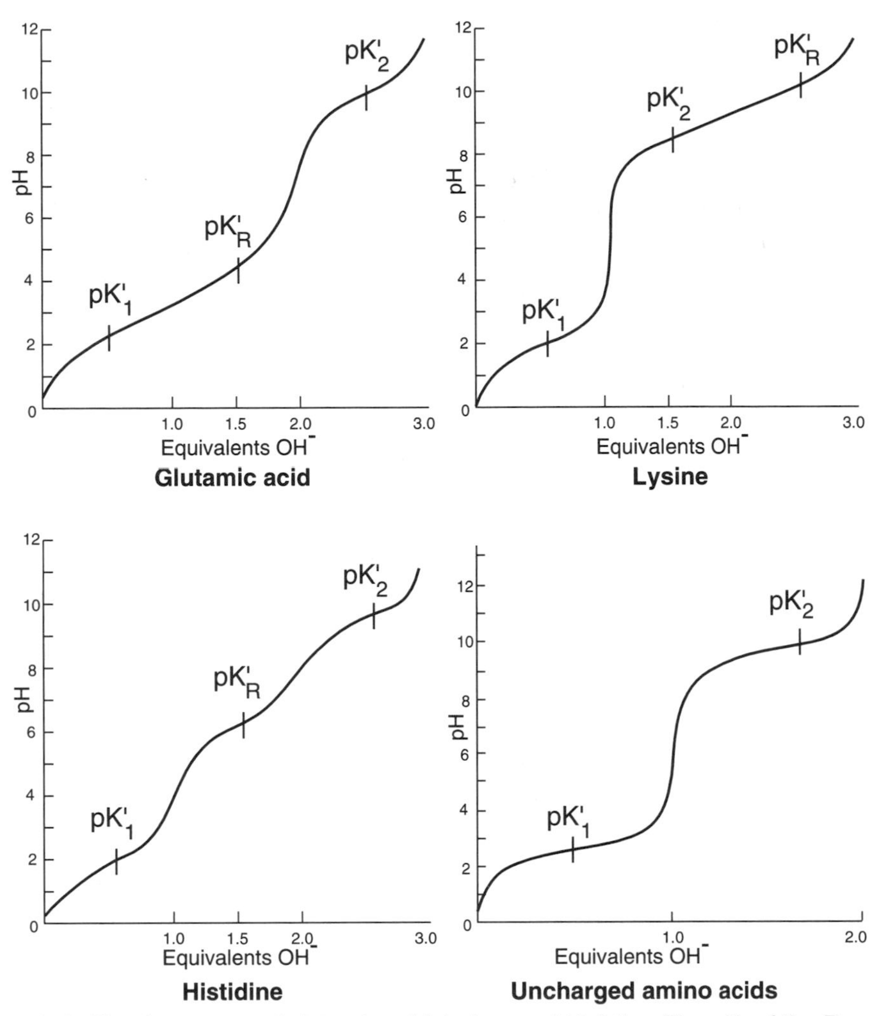

Figure 3-2. Titration curves of glutamic acid, lysine, and histidine. The pK_a of the R group is given by pK'_R.

Chapter 4

Bioenergetics

FREE ENERGY CHANGES

Bioenergetics is the study of energy changes that accompany biochemical reactions. A **chemical reaction** is the process whereby one or more substances are converted into other substances. For example, dihydroxyacetone phosphate (one biochemical compound) is converted into glyceraldehyde 3-phosphate (another compound) in the cell. The reaction is catalyzed by triose phosphate isomerase.

On a more fundamental level, consider the following reaction, $A + B \rightleftharpoons C + D$. In this case, two substrates (or reactants), A and B, combine to form two products, C and D. The double arrow indicates that the reaction can move in either direction; however, *in which direction will it go?* The short answer to this question depends on **concentration** and **energy**. If a vessel contains all A and B, then in time, we would expect the system to reach some equilibrium mixture of A, B, C, and D. If you think about the concentration component of the reaction, when you add solid salt to a glass of water, the crystals settle to the bottom of the glass. At this point, the concentration of salt is very high in the bottom of the vessel. With time, however, the salt dissolves and uniformly distributes itself throughout the liquid. This kind of process is thermodynamic and called **entropy**.

This model does not, however, take into account the influence of the **intrinsic energy** of the system. Imagine a hill and a collection of bowling balls in one of two energy states: bottom of the hill or top of the hill. Based on the concentration (or entropy) model, the balls would be uniformly distributed throughout the length of the hill; however, they do not uniformly distribute because of the **intrinsic energy differences** between the two energy states of the balls. The balls at the top of the hill have more energy potential than those at the bottom of the hill due to gravity. No matter how many balls are placed at the bottom, they will not lift themselves to the top; therefore, energy must be imparted to the balls at the bottom to raise them to the top.

The most important thermodynamic parameter in bioenergetics is the **free energy change** denoted by **ΔG**. This term is described by the equation **$\Delta G = \Delta H - T\Delta S$**, in which ΔH is the change in **enthalpy or intrinsic energy**, and ΔS is the change in **entropy**. ΔG, therefore, is defined in terms of energy and concentration. ΔG is an energy change occurring at constant temperature and pressure (the usual condition for biochemical reactions). The practical (experimental) expression corresponding to this reaction is

$$\Delta G = \Delta G^{\circ} + RT \ 1n \frac{[G\ 3\text{-}P]}{[DHAP]}$$

ΔG = free energy change

ΔG^{-} = standard free energy change

R = gas constant $(1.98 \text{ cal } K^{-1} \text{ mol}^{-1})$, where K is the absolute temperature $(273 + {}^{\circ}C;\ {}^{\circ}C$ corresponds to degrees Celsius$)$

T = absolute temperature

[G 3-P] = concentration of glyceraldehyde 3-phosphate (the product)

[DHAP] = concentration of dihydroxyacetone phosphate (the reactant)

For a reaction $A + B \rightleftharpoons C + D$.

$$\Delta G = \Delta G^{\circ} + RT \ln \frac{[C][D]}{[A][B]}$$

ΔG° corresponds to the free energy change when 1 mol of each reactant is converted to 1 mol of product, and all are present at 1 mol/L concentration. Because biochemical reactions occur in aqueous solutions and the concentration of water is constant, the effective **concentration of water** is given a **value of 1**; moreover, it would be impractical to perform biochemical reactions in 1 mol/L water because pure water has a molarity of approximately 55.5 mol/L. When $[H^+]$ is a reactant or product, it is also given a value of 1 because its concentration under physiologic conditions (10^{-7} mol/L) is also constant. The constant pH is usually designated by including a prime (') with ΔG and ΔG° as $\Delta G'$ and $\Delta G^{\circ\prime}$.

Thermodynamically favorable reactions proceed with the liberation of free energy and are exergonic. The **free energy** of the products is less than that of the reactants; that is, ΔG is negative. This means that in going from reactants to products, energy is released. Conversely (and more intuitively), energy must be added to the system to make products go back to reactants.

$$\Delta G = \text{free energy of products} - \text{free energy of reactants}$$

At equilibrium, $\Delta G = 0$ and the **reaction is isoergonic**. When the free energy of the products is greater than that of the reactants, the reaction is thermodynamically unfavorable (**endergonic**), and ΔG is positive (Table 4-1).

At **equilibrium**, no free energy is obtainable, and $\Delta G = 0$. The following important result can then be derived:

$$\Delta G = \Delta G^{\circ} + RT \ln \frac{[C][D]}{[A][B]}$$

$$0 = \Delta G^{\circ} + RT \ln \frac{[C][D]}{[A][B]}$$

$$\Delta G^{\circ} = -RT \ln \frac{[C][D]}{[A][B]}$$

TABLE 4-1. Free Energy Changes and Reaction Directionality

ΔG	Direction of Reaction Favored	Category
Negative	Toward products	Exergonic
Zero	Equilibrium	Isogonic
Positive	Toward reactants	Endergonic

Because the reaction is at equilibrium, the values of A, B, C, and D are the concentrations at equilibrium that reflect this situation.

$$\Delta G° = -RT \ln K_{eq}$$

With the **equilibrium constant (K_{eq})**, you can calculate the standard free energy change (ΔG°), and vice versa. The free energy change (ΔG) can be larger or smaller than the standard free energy change (ΔG°). Increasing the concentrations of reactants or decreasing the concentration of products produces a greater decrease in free energy or ΔG (the reaction is more exergonic). Decreasing the concentration of reactants or increasing the concentration of products produces a smaller decrease in free energy (the reaction is less exergonic).

The following examples may clarify these issues:

1. If the **concentrations of reactants and products are equal**, such that $K_{eq} = 1$, then $\ln K_{eq} = 0$ and ΔG = ΔG°. Entropy does not play a role in these conditions, and the direction of the reaction is determined strictly by the intrinsic energy term. If ΔG° is negative, the reaction moves to the right. If it is positive, the reaction moves to the left (see Table 4-1).

2. In a case in which the **ΔG° is near zero**, the free energy change is determined by concentrations of reactants and products; for example, assume that ΔG° = 0. If there are 100 times more products than reactants, $\Delta G = 0 + (1.98) \times 310 \times \ln 100$. ΔG = 2,827 cal/mol (this number can be rounded off to 2,800 cal/mol). The reaction, as written, requires energy to go to the right, or spontaneously go to the left (liberating energy).

A note of caution is appropriate here. The **free energy changes** indicate whether a reaction under specified conditions is thermodynamically feasible. It fails to provide any information about the rate or kinetics of a reaction. Many thermodynamically feasible reactions are unimportant or irrelevant in biochemistry because of the absence of an enzyme to mediate the reaction in a reasonable amount of time.

A nonbiological example may help to clarify the distinction between thermodynamics and kinetics. The reaction of **oxygen** with **gasoline** to form CO_2 and H_2O is exergonic and proceeds with the liberation of considerable free energy. The gasoline, however, is stable in the presence of oxygen for a very long time. The reaction occurs only under appropriate conditions, such as encountered in an internal combustion engine with an electrical spark initiating the process. The system (gasoline and oxygen) is said to be **kinetically stable** (unreactive) but **thermodynamically unstable** (exhibiting the potential for reacting).

ENERGY-RICH COMPOUNDS

The complete **oxidation of glucose to carbon dioxide and water** is associated with the liberation of 686 kcal of free energy ($\Delta G° = -686$ kcal). These changes in living systems occur in a graded and inexplosive fashion. Energy is released in a stepwise fashion and is coupled to the biosynthesis of **adenosine triphosphate** (ATP) from **adenosine diphosphate** (ADP) and **inorganic phosphate** (P_i). The ATP-ADP couple receives and distributes chemical energy in all living systems, a statement of **Lipmann's law** and a cornerstone of biochemistry. ATP serves as the common currency of energy exchange in living systems.

ATP is an energy-rich compound and serves as a donor of chemical energy for muscle contraction, ion transport, and biosynthetic reactions. The structure of ATP is shown in Fig. 4-1. ATP is composed of a nitrogen-containing base (**adenine**), a five-carbon sugar (**ribose**), and three **phosphates**. The three phosphates are designated α, β, and γ from ribose to the terminus. The β- and γ-linkages are **acid anhydrides** (two phosphoric acids linked in a dehydration reaction). These bonds are **energy rich** in nature and are associated with the following two reactions:

$$ATP + H_2O \rightarrow ADP + P_i$$
$$ADP + H_2O \rightarrow AMP + P_i$$

The $\Delta G°'$ for these reactions is approximately –7 kcal/mol. Compounds with a standard free energy of hydrolysis of –7 kcal/mol or a higher negative are classified as energy-rich compounds. The reaction ATP + H_2O → AMP (adenosine monophosphate) + PP_i (**inorganic pyrophosphate**) is also associated with the liberation of considerable free energy (approximately –7 kcal/mol). The **α-phosphate bond** (AMP + H_2O → adenosine + P_i) is a low-energy (–3 kcal/mol) bond because it is a simple phosphate ester (phosphoric acid + alcohol) and not an acid anhydride. The pyrophosphate (1), ADP (1), ATP (2), phosphoenolpyruvate (1), acyl phosphate as found in 1,3-bisphosphoglycerate (1), and phosphoamidate as found in creatine phosphate (1) are all high-energy compounds, where the *number in parentheses denotes the quantity of high-energy bonds per molecule.* Thioesters, such as acetylcoenzyme A (acetyl-CoA), also contain a high-energy bond.

The following examples illustrate the usefulness of energy-rich and energy-poor compounds in understanding whether a reaction is favorable (exergonic) or unfavor-

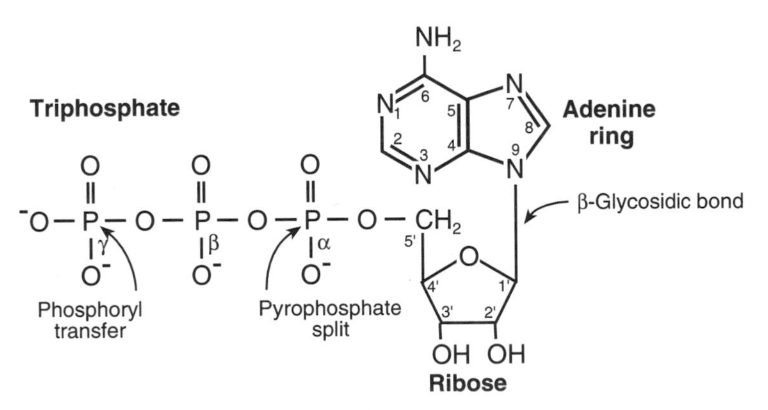

Figure 4-1. Structure of adenosine triphosphate.

able (endergonic). **Isoergonic reactions** are equipoise and may proceed in either direction, depending on the circumstances. The following is a prominent reaction in biochemistry:

$$\text{ATP}\ (2) + \text{glucose}\ (0) \rightarrow \text{ADP}\ (1) + \text{glucose 6-phosphate}\ (0)$$

There are two **high-energy bonds** on the left side; on the right side there is one, as indicated. The reaction proceeds with the loss of a high-energy bond. The **reaction is exergonic and proceeds to the right**. The **reaction from right to left is endergonic** and does not proceed to a physiologically significant extent.

Another example involves the interconversion of two **low-energy compounds**:

$$\text{Glucose 6-phosphate}\ (0) \rightleftharpoons \text{fructose 6-phosphate}\ (0)$$

The structure and energy richness of the two compounds are similar and the reaction is **isoergonic**.

A similar analysis was done for a case in which the number of high-energy bonds is the same:

$$\text{ADP}\ (1) + \text{1,3-biphosphoglycerate}\ (1) \rightleftharpoons \text{ATP}\ (2) + \text{3-phosphoglycerate}\ (0)$$

This reaction is **approximately isoergonic**; the reaction proceeds without providing or using much free energy.

In general, major types of biological reactions can be predicted to be favored or not, depending on the nature of the chemical transformation. The hydrolysis of both energy-rich and energy-poor compounds is exergonic, and the equilibrium constant is much greater than 1. That is, the concentration of product is much greater than the concentration of reactant at equilibrium. The reverse reaction is endergonic and is not thermodynamically favored. **Decarboxylation** reactions are exergonic; **carboxylation** reactions usually require the input of energy in the form of ATP. **Simple dehydrogenation** reactions (not associated with decarboxylation) are generally reversible in nature. Reactions with molecular oxygen are exergonic, and the reverse reaction fails to occur to a meaningful extent.

CLASSES OF MEMBRANE-ASSOCIATED ATP UTILIZING ENZYMES

ATP is the common currency of energy exchange, with several types of membrane-associated ATPase playing a pivotal role in the generation and use of ATP. Class F ATPase (also known as ATP synthase or F_0/F_1 complex) occurs in the inner mitochondrial membrane and is the main source of ATP production in the body (see Chapter 6). A proton gradient across the membrane provides the energy that drives the synthesis of ATP from ADP and P_i that is catalyzed by this class of enzyme.

Class V ATPase (V, vesicle) transports protons into intracellular organelles such as lysosomes and secretory vesicles. **Class P ATPase** (P, phosphorylated aspartyl-enzyme intermediate) is made up of integral membrane enzymes that mediate the translocation of ions across membranes. These enzymes include the **sodium/potassium ATPase**, or sodium pump, that maintains a high intracellular potassium ion concentration and a low intracellular sodium ion concentration.

The **gastric proton/potassium ATPase**, also a class P ATPase, is responsible for pumping protons from gastric cells to provide acidic gastric juice. The proton/potassium ATPase, which is a target for drugs used in the treatment of **peptic ulcer disease**, also has a phosphoaspartyl enzyme intermediate. ATP donates its terminal phosphoryl group to form an energized enzyme that is responsible for translocating the substrate ions. The class F and class V enzymes lack a phosphorylated-enzyme intermediate.

Chapter 5

Enzymes

GENERAL PROPERTIES

One important function of proteins is that of a catalyst. Almost all reactions of a biochemical nature occur under physiologic conditions of temperature and pH because of the existence of protein catalysts. **Enzyme** is the term ascribed to a **protein catalyst.** A protein is a polypeptide made up of α-amino acids (see Chapter 2). A **catalyst** is a substance that alters the rate of a chemical reaction without itself being permanently changed into another compound or consumed in the process. A catalyst increases the rate at which a thermodynamically feasible reaction attains its equilibrium without altering the position of the equilibrium. The rate of an enzyme-catalyzed reaction ranges from 10^3- to 10^{11}-fold greater than that of a reaction that has not been catalyzed. A catalyst accelerates a reaction by decreasing the free energy of activation $(\Delta G)^{\ddagger}$ (Fig. 5-1). An enzyme provides an alternative and speedier reaction route. Note that the ΔG for the illustrated reaction is unchanged by the enzyme catalysis; the intrinsic energies and equilibrium distributions of reactants and products are not altered by the enzyme. The speed with which equilibrium is obtained can be dramatically increased, however. The development of the science of biochemistry has proceeded concurrently with the development of the science of enzymology.

Enzymes fall into two general classes: simple proteins and complex proteins. **Simple proteins** contain only amino acids (e.g., the digestive enzymes ribonuclease, trypsin, and chymotrypsin). **Complex proteins** contain amino acids and a non–amino acid cofactor. The complete enzyme is called a **holoenzyme**, which is made of a protein portion (apoenzyme) and cofactor.

$$\text{Holoenzyme} \rightleftharpoons \text{apoenzyme} + \text{cofactor(s)}$$

A metal ion may serve as a cofactor for an enzyme; for example, zinc is a cofactor for the enzymes carbonic anhydrase and carboxypeptidase. An organic molecule such as pyridoxal phosphate or biotin may serve as a cofactor. Cofactors such as biotin, which are covalently linked to the enzyme, are called **prosthetic groups**.

In addition to their enormous **catalytic power**, which accelerates reaction rates, enzymes exhibit exquisite **specificity** in the types of reactions that they catalyze as well as specificity for the substrates on which they act. Phosphofructokinase catalyzes a reaction between adenosine triphosphate (ATP) and fructose 6-phosphate. The enzyme does not catalyze a reaction between other nucleoside triphosphates and other sugars to a physiologically meaningful extent. Hexokinase catalyzes a reaction between ATP (and not other nucleoside triphosphates) and glucose, fructose, or mannose (but not galactose).

Note that **trypsin** catalyzes the hydrolysis of peptides and proteins only on the carboxyl side of polypeptidic lysines or arginines (positively charged basic residues). **Chy-**

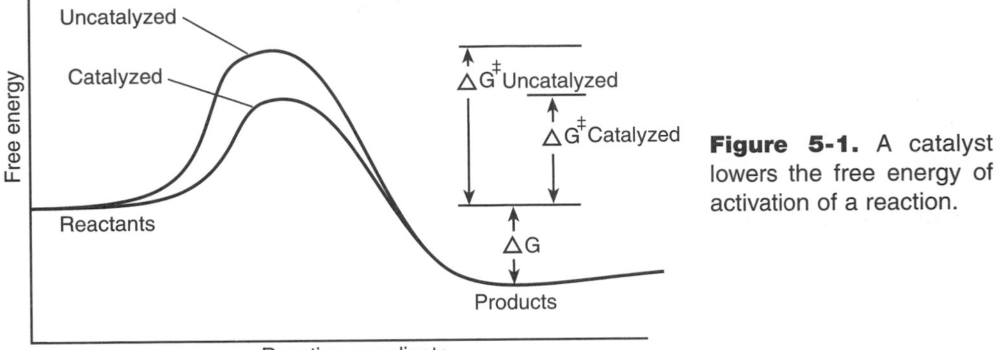

Figure 5-1. A catalyst lowers the free energy of activation of a reaction.

motrypsin catalyzes the hydrolysis of peptides and proteins on the carboxyl side of polypeptidic phenylalanine, tyrosine, and tryptophan (aromatic residues) and other hydrophobic amino acids. Many enzymes exhibit trypsinlike specificity. These include blood-clotting factors and enzymes that are important in processing hormonal peptides.

Enzymes are divided into six classes based on the type of reaction that they catalyze (Table 5-1). **Oxidation-reduction reactions** are important in energy metabolism. **Kinases** are a class of **transferase.** Kinases catalyze the transfer of the terminal phosphoryl group of ATP to acceptor substrates. Transfer of acyl groups and amino groups is prevalent in biochemistry. **Hydrolases** catalyze the hydrolysis of proteins, nucleic acids, and a variety of other compounds. Pancreatic digestive enzymes, such as trypsin,

TABLE 5-1. Enzyme Classification

Class	Reaction
Major classes	
Oxidoreductases	Transfer hydrogen atom or hydride ion (H:⁻); act on H_2O_2; act on O_2
Transferases	Transfer carbon, phosphoryl, glycosyl, acyl, and amino groups
Hydrolases	Cleave wide variety of substrates by adding water across bond
Lyases	Cleave carbon bound to carbon, nitrogen, or oxygen
Isomerases	Racemases, epimerases, intramolecular oxidoreductases, intramolecular transferases
Ligases	ATP- or nucleoside triphosphate–dependent condensation reaction
Selected subclasses	
Kinases	Transfer phosphoryl group from ATP and other nucleotides; transferases
Mutases	Move phosphoryl or other group intramolecularly; isomerases
Phosphorylases	Cleave by adding phosphate across bond; transferases
Decarboxylases	Carboxylate liberated as CO_2; lyases
Hydratases	Add water to double bond and the reverse; lyases
Synthetases	ATP (or equivalent nucleotide)-dependent synthesis; ligases
Synthases	ATP-independent synthesis (e.g., UDPG + glycogen → $glycogen_{n+1}$ + UDP); transferases

ATP, adenosine triphosphate; UDP, uridine diphosphate; UDPG, uridinediphosphoglucose.

lipase, and amylase, are examples of hydrolases. **Lyases** catalyze the nonhydrolytic cleavage of molecules (and the reverse reunification reaction). Cleavage by lyases results in the formation of an additional double bond in the products (the number of double bonds in the products is greater than the number of double bonds in the reactants). Aldolase of the glycolytic pathway is an example of a lyase. **Isomerases** catalyze the conversion of aldehydes to ketones and of L-compounds to D-compounds. **Ligases** (ligate, to bind or to tie) catalyze the ATP-dependent condensation of one molecule with another. The combination of an amino acid with its corresponding transfer RNA is one example of a ligase reaction.

ENZYME ACTIVE SITES AND REACTION MECHANISMS

A fundamental question is how enzymes act to increase the rate of a biological reaction. A simple way of envisioning enzyme mechanisms is presented in Fig. 5-2. This **lock-and-key model** illustrates how an enzyme might accelerate a reaction described as $A + B \rightleftharpoons C + D$. In this example, the enzyme possesses sites for the binding of the two reactant (A and B) substrates. After **reactant binding**, conformational changes occur, resulting in the transfer of part of B to A (creating the products C and D). Although the exact amino acid side chains involved in this process are not shown, imagine that the enzyme is acting on several levels to accelerate the reaction.

First, the enzyme, based on its affinity for the individual reactants, increases the local concentration of A and B. This serves to increase the chances of their interaction to greater than that which would be experienced in solution. Not only are the reactants brought into proximity, but they are properly oriented to permit the **transferase reaction** to occur. In addition, the local environment within the active site of the enzyme (charge-transfer relationships, polarizing electrophiles and nucleophiles) may activate one or both of the reactants, placing them in a more conducive conformation.

Similarly, for a **lyase reaction**, conformational changes in the enzyme after reactant binding may strain the specific bond that is destined for cleavage. Taken together, these considerations establish the conditions under which the enzyme can promote the acceleration of a given reaction.

Temperature contributes, in a positive manner, to enzyme-catalyzed reactions. In general, the rate of a reaction doubles for every 10°C increase in temperature. This is known as the law of Q_{10} and is illustrated in Fig. 5-3. Although the reaction rate increases with temperature, at some point the enzyme protein will denature, resulting in loss of enzyme activity (the sharp decline illustrated in Fig. 5-3). A notable exception to this latter observation consists of enzymes derived from thermostable organisms. Specifically, the thermostable DNA polymerases are the basis for the polymerase chain reaction (PCR). These enzymes are stable even at temperatures approaching 100°C.

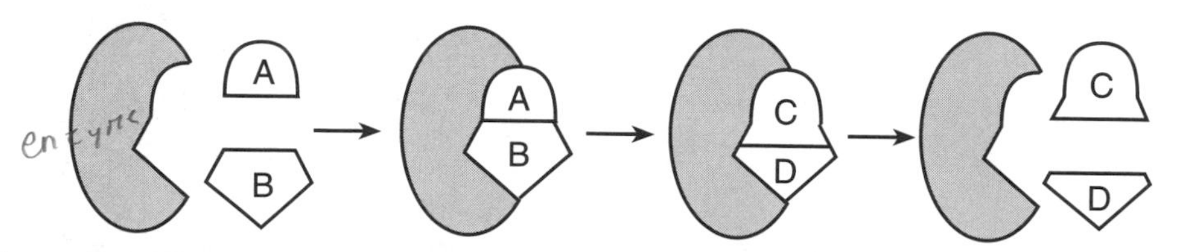

Figure 5-2. The Fischer lock-and-key hypothesis of enzyme action. In this example, the enzyme possesses sites for the binding of the two reactant (*A* and *B*) substrates. After reactant binding, conformational changes occur, resulting in the transfer of part of *B* to *A* (creating the products *C* and *D*).

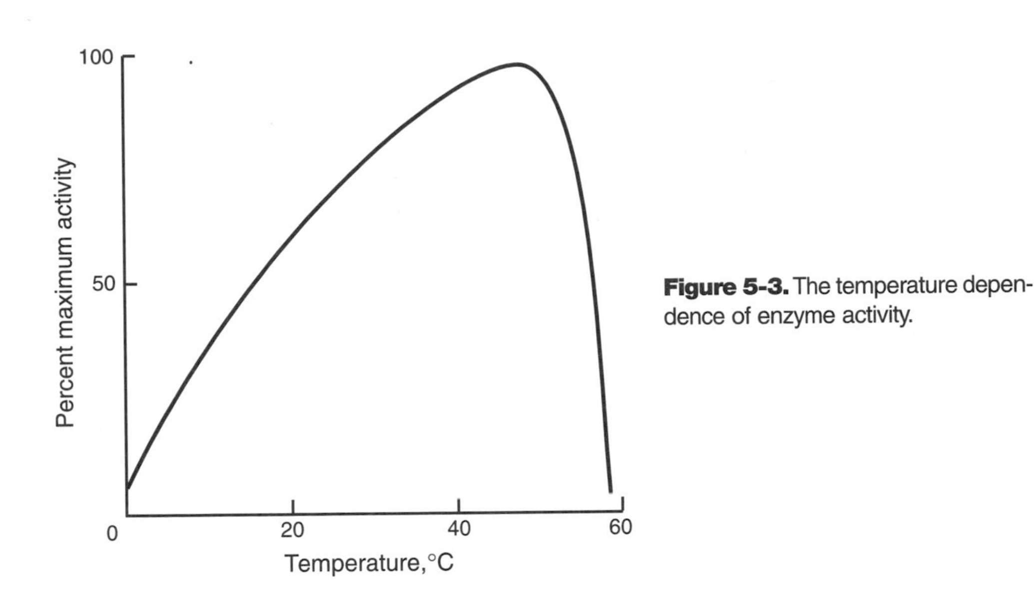

Figure 5-3. The temperature dependence of enzyme activity.

ENZYME KINETICS AND INHIBITORS

Most experimental characterization of enzyme activity is based on quantification of the enzyme-catalyzed reaction. This generally involves the **measurement of product accumulation** as a function of time, as shown in Fig. 5-4. There are a number of different biophysical modalities that can be used to monitor reactant conversion to product. These can include special spectrophotometric characteristics of the product or the use of radioactive tags to follow the conversion. Note that in most cases, as the reaction is followed, there is a **linear phase** at the beginning of the reaction (initial velocity) followed, in time, by a **plateau** representing substrate depletion or enzyme inactivation. The slope of the initial phase of the product accumulation curve is termed the **initial velocity** and has units of moles of product formed per unit time.

In an **enzyme-catalyzed reaction**, there is an increase in the initial reaction velocity with an increase in substrate concentration. In practical terms, this relationship indicates that as substrate concentrations increase, the enzyme in the solution can more efficiently create product. Keep in mind that the **interaction of reactant** (also known as

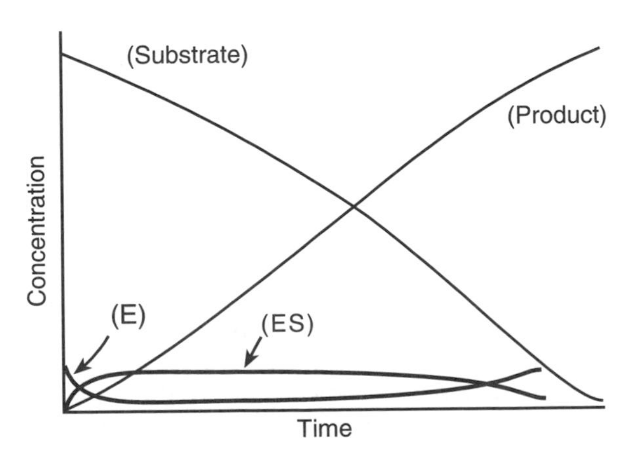

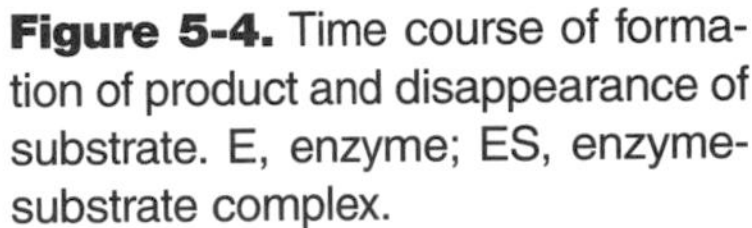
Figure 5-4. Time course of formation of product and disappearance of substrate. E, enzyme; ES, enzyme-substrate complex.

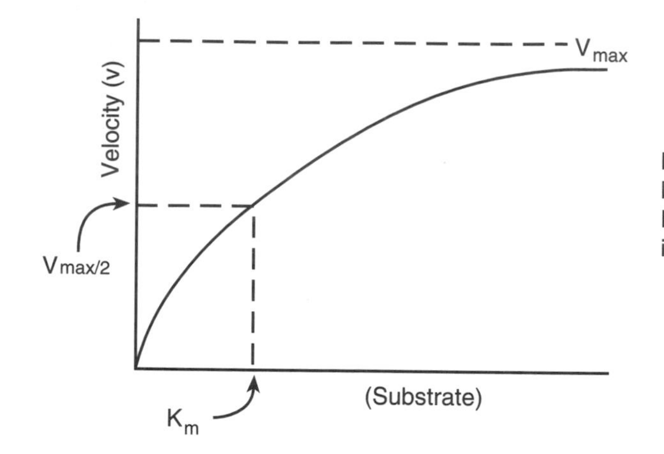

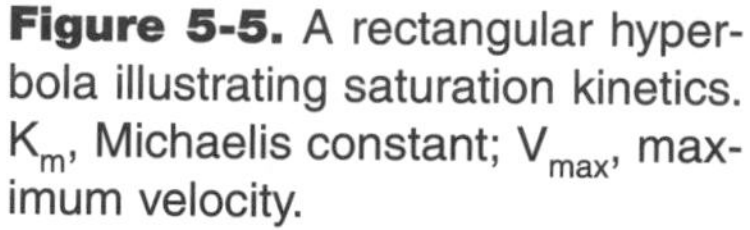
Figure 5-5. A rectangular hyperbola illustrating saturation kinetics. K_m, Michaelis constant; V_{max}, maximum velocity.

the **substrate**) and enzyme is a second-order process; thus, the higher the substrate concentration, the more likely it will find its way into the active site of the enzyme.

In most cases, a plot of **velocity** as a function of substrate concentration yields a **rectangular hyperbola** (Fig. 5-5). At progressively higher substrate concentrations, the increase in activity is progressively smaller. Such data demonstrate that enzymes exhibit saturation. This is interpreted to mean that the substrates interact with a finite number of catalytic molecules and are converted into products. In an uncatalyzed process, the reaction would increase indefinitely as reactant concentration increases.

In defined conditions with specific amounts of protein, an enzyme exhibits a **maximum velocity** (V_{max}), which approaches a limit (or mathematical asymptote) as the substrate concentration approaches infinity. At this point, every enzyme molecule is engaged in catalysis, such that increasing the amount of substrate has no further effect on product formation. Obviously, the magnitude of this plateau (on the *Y*-axis) is dependent on the amount of enzyme. By doubling the amount of enzyme, the V_{max} doubles.

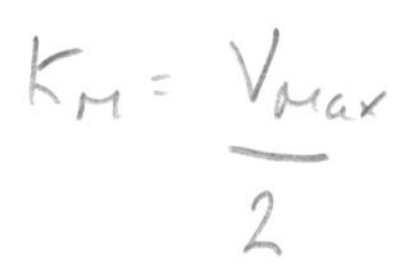

The **Michaelis constant (K_m)** is the substrate concentration at half the **maximal velocity ($V_{max}/2$)**, as shown in Fig. 5-5. This constant is characteristic of the enzyme being studied and is independent of the amount of enzyme. The V_{max} per enzyme molecule is designated as k_{cat}, a first-order rate constant. The **specificity constant** of an enzyme is given by k_{cat}/K_m, a second-order rate constant.

An enzyme such as **hexokinase** can catalyze the phosphorylation of several sugars, including glucose, fructose, and mannose. The best substrate for hexokinase is glucose. The specificity constant of hexokinase for glucose is higher than that for the other two sugars. The K_m incorporates a number of different factors, including substrate affinity, the efficiency with which the enzyme can convert that particular substrate to product as well as kinetic relationships between the varied substrate and other participants in the reaction (cosubstrates and cofactors). The K_m is frequently, and incorrectly, referred to as a **binding constant**; however, it is more than that.

The **Michaelis-Menten equation** is an expression for the reaction velocity (v) as a function of substrate concentration ([S]) and the kinetic constants (V_{max} and K_m):

$$v = \frac{V_{max} \times [S]}{K_m + [S]}$$

When $v = V_{max}/2$, one can verify the result that $K_m = [S]$. It is difficult to determine the K_m and V_{max} from a rectangular hyperbola. Several methods are available for obtaining accurate values for these parameters, including the use of computer programs. A traditional way of determining the kinetic constants is through the use of a double reciprocal equation. When the reciprocal of the **substrate concentration (1/S)**

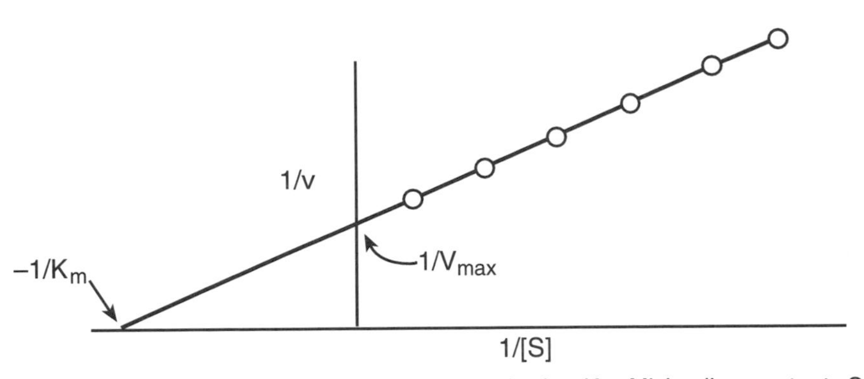

Figure 5-6. A double-reciprocal or Lineweaver-Burk plot. K_m, Michaelis constant; S, substrate; v, velocity; V_{max}, maximum velocity.

is plotted versus the **reciprocal of the velocity (1/v)**, results similar to those in Fig. 5-6 are obtained. This is called a **Lineweaver-Burk plot**. The value of $1/V_{max}$ is obtained by extrapolation, and it corresponds to the velocity at an infinite substrate concentration. The plot also yields $-1/K_m$. Because this is a reciprocal plot, note that a larger V_{max} corresponds to a smaller value of the ordinate (*Y*-axis). Similarly, a larger K_m corresponds to a less negative value of the abscissa (*X*-axis).

Double reciprocal plots are helpful in studying enzyme inhibition. Enzyme inhibitors are classified as reversible and irreversible. **Irreversible inhibitors** usually react covalently with an enzymic amino acid residue and render the enzyme inactive. The rate constant for the reaction of the inhibitor with the enzyme can be measured. **Reversible inhibitors** generally interact noncovalently and virtually instantaneously with an enzyme; the interactions of this class of inhibitor with enzymes can be studied by steady-state enzyme kinetics.

There are two major classes of reversible inhibitor: competitive and noncompetitive. **Competitive inhibitors** are structural analogues of the substrate whose concentration is being varied. Consider the following hypothetical enzyme-catalyzed reaction:

$$A + B \rightleftharpoons C + D$$

Assume that **B'** is a **substrate analogue** of B and interacts with the enzyme but fails to undergo a reaction (Fig. 5-7). If we measure the velocity as a function of the concentration of B at a few fixed concentrations of B', we will find that at a higher concentration of B (fixed B') the magnitude of inhibition is decreased. In fact, at infinite concentrations of B (determined by extrapolation), inhibition is overcome.

The location on an enzyme where catalysis occurs is termed the **active site**. In this example, a portion of the active site corresponds to A, and another portion corresponds to B. B' inhibits the enzyme by interacting with the enzyme at the site corresponding to B, and it thereby prevents catalysis. By increasing the concentration of B, its effect overrides that of B', and inhibition is overcome. This is illustrated by the unchanged V_{max} for competitive inhibition shown in Fig. 5-8.

Next, consider the effects of **varying the concentration** of A while keeping B constant. Increasing B' increases the degree of inhibition. Increasing concentrations of A, however, cannot completely override the effects of B', and the V_{max} at infinite A is decreased (noncompetitive, see Fig. 5-8). This is because A and B' do not bind to the same site. No matter how large the concentration of A, B' may still bind to and inhibit the enzyme. This type of inhibition cannot be overcome by increased concentrations of substrate not homologous to the reversible inhibitor. Moreover, B' does not alter the K_m for A. In the case of competitive inhibition, B' increases the apparent K_m of B (see Fig. 5-8).

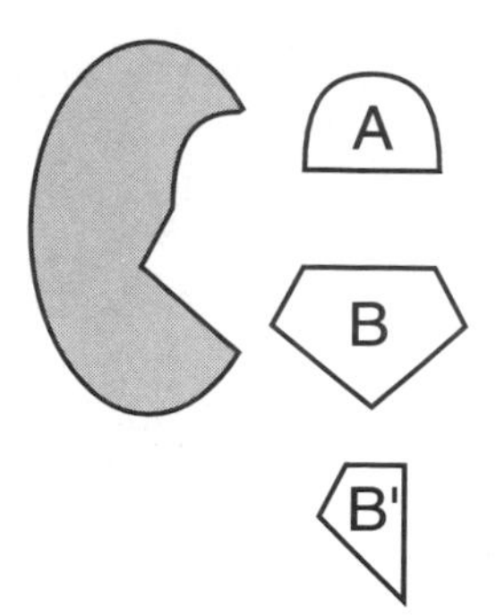

Figure 5-7. The lock-and-key concept of enzyme inhibition. *A* and *B* represent the two substrates; *B'* is an inhibitory analog of *B* that does not participate in the reaction.

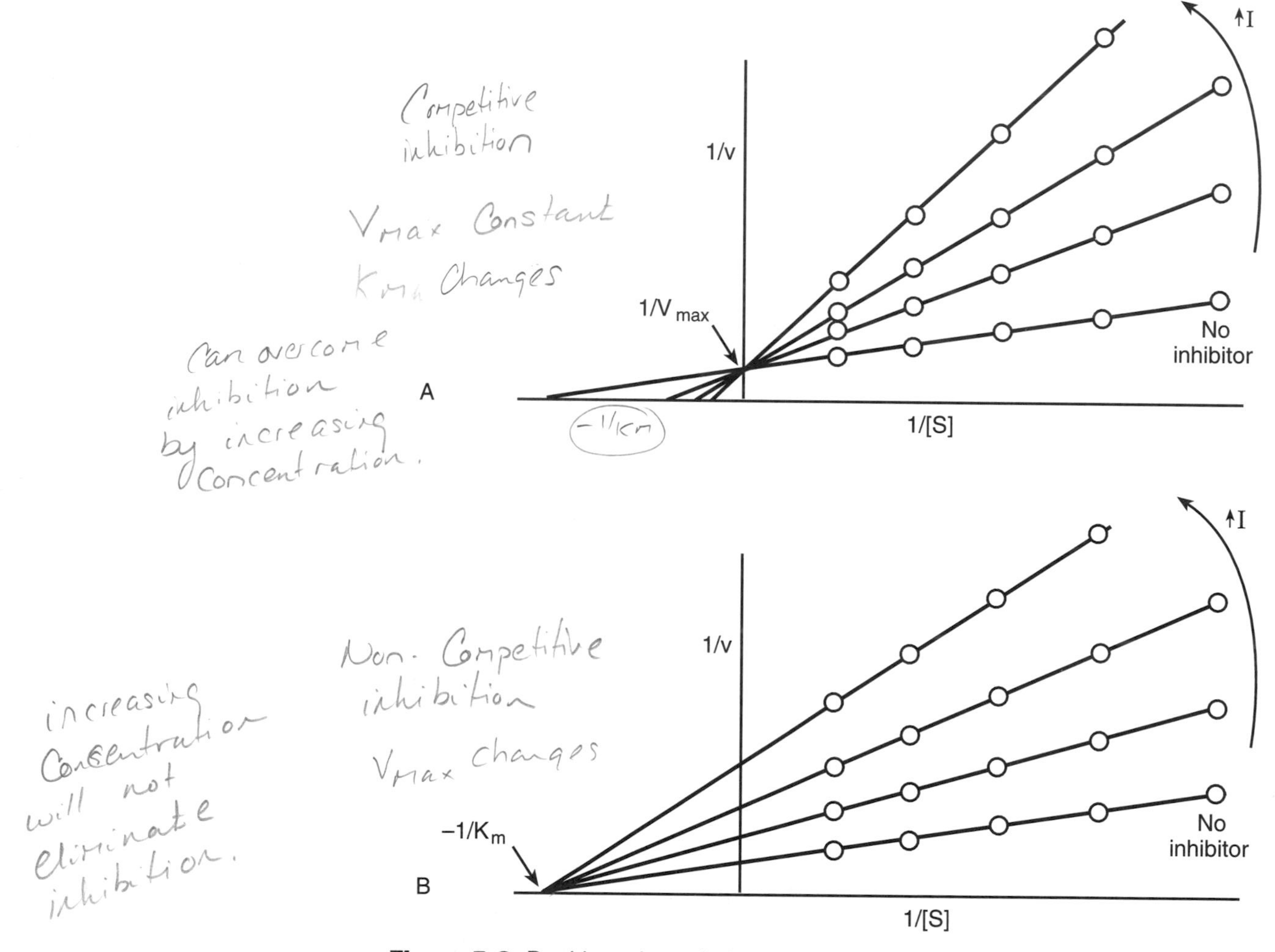

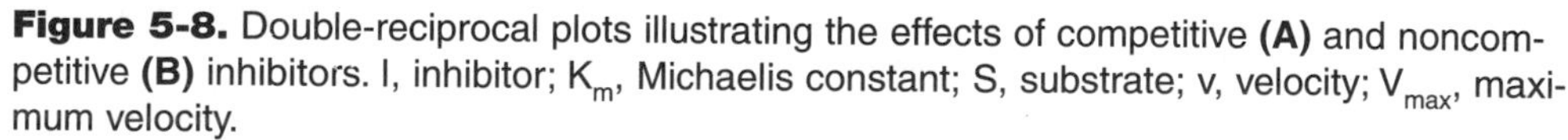

Figure 5-8. Double-reciprocal plots illustrating the effects of competitive **(A)** and noncompetitive **(B)** inhibitors. I, inhibitor; K_m, Michaelis constant; S, substrate; v, velocity; V_{max}, maximum velocity.

In examining **double reciprocal plots** to determine the type of inhibition (see Fig. 5-8), examine the V_{max}. If the V_{max} is unchanged, the inhibition is competitive. If the V_{max} is decreased and the K_m is unchanged, inhibition is noncompetitive. If similar experiments are carried out with an irreversible inhibitor, a pattern similar to that of noncompetitive inhibition is seen. The irreversible inhibitor inactivates some of the enzyme. The underivatized enzyme is normal (no change in K_m), but there is less active enzyme, which is reflected by a decrease in V_{max}. As noted above, the chief use of steady-state enzyme kinetics is in the study of instantaneous, reversible inhibitors and not in the study of irreversible inhibitors.

One of the concepts to emerge from a theoretical consideration of steady-state enzyme kinetics is that the enzyme binds with a substrate to form an enzyme-substrate complex. The enzyme is said to contain an **active** or **catalytic site**. After binding of the substrate(s), the enzyme promotes a reaction and the products dissociate. The active site contains residues that participate in the reaction. Amino acids that have been shown to participate in enzymic catalysis include the serine hydroxyl, the cysteine sulfhydryl, the γ-carboxyl of aspartate, the imidazole of histidine, and the ε-amino group of lysine, as well as others. Only one to three residues usually participate in reactions in a particular enzyme-active site; additional residues may participate in binding the substrate to the enzyme.

REGULATION OF ENZYME ACTIVITY

The activity of enzymes is subject to a variety of regulatory mechanisms. The amount of enzyme can be altered by increasing or decreasing its synthesis or degradation. Enzyme **induction** refers to an enhancement of enzyme biosynthesis. Enzyme **repression** refers to a decrease in enzyme biosynthesis. Enzyme activity can also be altered by **covalent modification**. Phosphorylation of specific serine residues by protein kinases increases or decreases catalytic activity, depending on the enzyme. Proteolytic cleavage of proenzymes (chymotrypsinogen, trypsinogen, proelastase, clotting factors) converts an inactive form to an active form. **Enterokinase** is an intestinal enzyme that initiates the conversion of trypsinogen to trypsin; trypsin can catalyze the conversion of chymotrypsinogen, proelastase, and additional trypsinogen to the active enzyme forms.

Enzyme activity can also be regulated by **noncovalent or allosteric mechanisms**. Isocitrate dehydrogenase is an enzyme in the Krebs citric acid cycle that is activated by adenosine diphosphate (ADP). ADP is not a substrate or a substrate analogue, and it is postulated to bind to a site distinct from the active site called the **allosteric site**. Allosteric regulation is common, and the changes in activity in response to allosteric effectors make physiologic sense and therefore are called the **molecular logic of the cell**.

When it is realized that one of the primary functions of the Krebs cycle is to provide reducing equivalents for ATP biosynthesis, Krebs cycle regulation can be rationalized by ADP. When the concentration of ATP is decreased, the concentration of ADP increases and serves as a signal to activate ATP formation. ADP regulates one of the early reactions of the Krebs cycle (**isocitrate dehydrogenase**) and promotes greater activity.

In addition to activation, some enzymes are subject to **allosteric inhibition**. In *Escherichia coli*, for example, threonine deaminase catalyzes the first step in the reaction pathway for isoleucine biosynthesis. When isoleucine is plentiful, it produces **feedback inhibition** of the first enzymatic reaction and the committed step of the pathway. This inhibition decreases the synthesis of an already abundant compound.

Some enzymes fail to conform to simple saturation kinetics and do not exhibit a rectangular hyperbola when velocity is measured as a function of substrate concentration. In the most common case, a **sigmoidal curve** is observed (Fig. 5-9). A sigmoidal curve is the *sine qua non* for **positive cooperativity**.

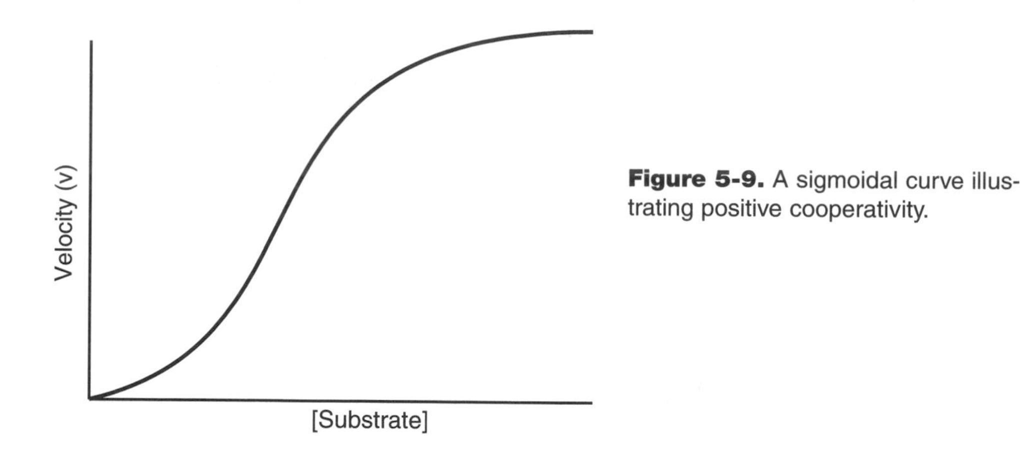

Figure 5-9. A sigmoidal curve illustrating positive cooperativity.

Positive cooperativity is a condition in which the binding of one substrate (or ligand) makes it easier for the second to bind. In the case of four binding sites per protein, binding of the second molecule facilitates binding of a third molecule, and this facilitates the binding of the fourth molecule. This is reflected by the increasing slope on the initial portion of the sigmoidal curve. The sigmoidal curve indicates only positive cooperativity. Oxygen binding to hemoglobin is cooperative. Allosteric enzymes often exhibit positive cooperativity, but not invariably. A sigmoidal binding curve does not indicate that an enzyme has an allosteric site distinct from an active site. The phenomena of cooperativity and allosterism are distinct.

CLINICAL SIGNIFICANCE OF ENZYMES

Besides the central role of enzymes in biology, the activity of enzymes in the serum provides information that is valuable in the diagnosis of various diseases. **Aspartate aminotransferase**, **alanine aminotransferase**, and **lactate dehydrogenase** are elevated in liver disease and after myocardial infarction. **Alkaline phosphatase** is elevated in some forms of liver disease and some bone diseases, including metastasis of cancer cells to bone. **Acid phosphatase** is elevated in metastatic prostate cancer, and such a finding in an elderly male is pathognomonic of this disease. **γ-Glutamyltranspeptidase** activity is a sensitive indicator of liver function, and this enzyme is often the first enzyme that is elevated in individuals who have an excessive ethanol intake.

Creatine phosphokinase is elevated within 12 hours after myocardial infarction. Creatine phosphokinase is a dimer, and there are two distinct monomers (B, brain; M, muscle) that constitute this protein. There are three possible **isozymes**, or enzyme forms, for creatine kinase: MM, MB, and BB. The MB isozyme is the chief form derived from heart cells.

Chapter 6

Carbohydrate Metabolism

OVERVIEW

Metabolism refers to all of the chemical reactions performed by an organism. Nearly all reactions in living systems are catalyzed by enzymes. The chemical reactions, or metabolisms, of living organisms are not random; rather, they are directed along specific sequences called **metabolic pathways**. A metabolic pathway may be composed of from 2 to 20 enzyme-catalyzed steps necessary for the conversion of one molecule into a product. Each of the participant compounds is a **metabolite**.

The Big Picture of cellular metabolism (see Fig. 1-1) can be approximately divided into two components: anabolism and catabolism. The process of **catabolism** refers to the conversion of large, complex molecules to simpler, small molecules. Some of these reactions release chemical energy; a portion of this chemical energy is captured as adenosine triphosphate (ATP), which is formed from adenosine diphosphate (ADP). This then is the source for the energy needs of the cell.

Examples of **catabolic processes** include glycogenolysis (breakdown of glycogen to glucose), glycolysis (conversion of glucose to pyruvate, with a small yield of ATP), Krebs citric acid cycle [conversion of acetylcoenzyme A (acetyl-CoA) to CO_2 and reducing equivalents], oxidative phosphorylation (conversion of reducing equivalents and O_2 to H_2O, with a large yield of ATP, β-oxidation of fatty acids (yielding acetyl-CoA for the Krebs cycle and ultimately for oxidative phosphorylation), and amino acid and purine metabolism (generally a last resort for energy production).

Anabolism refers to the conversion of small molecules to larger ones in the process of biosynthesis. These reactions require chemical energy, which is ultimately derived from ATP. Obviously, anabolic processes are designed as a reverse of catabolic processes and include glycogen synthesis, pentose phosphate shunt [for creation of five-carbon sugars for nucleic acids and a specialized form of reducing equivalent, reduced nicotinamide adenine dinucleotide phosphate (NADPH), which is used in a variety of anabolic processes], fatty acid synthesis, amino acid synthesis and translation (protein synthesis), purine and pyrimidine biosynthesis, transcription (RNA synthesis), and replication (DNA synthesis).

In general, the **pathway for biosynthesis** of a compound is not the simple reversal of its pathway for catabolism. The pathways may be completely independent or may share common intermediates; for instance, the conversion of glucose to pyruvate (glycolysis) involves ten specific enzyme-catalyzed reactions. The conversion of pyruvate to glucose requires 11 reactions. Of these reactions, seven are common to both **synthesis** and **degradation**, and the others are unique. The different or unique steps occur in such a fashion that the reactions of the entire pathway are bioenergetically favorable and exergonic.

A consequence of this biochemical strategy of independent pathways for synthesis and degradation is that it permits **independent regulation** of the flux of metabolites through the pathway; for example, the biosynthetic rates can be increased and degradative rates can be decreased because of the occurrence of reactions unique to biosynthesis and degradation. If all the steps were common, alteration of an enzyme activity would increase or decrease both pathways simultaneously. Regulatory enzymes are generally unique to the synthetic or degradative pathway. Moreover, **metabolic regulation** generally occurs at the first step or an early step in a pathway. The regulated reaction is often physiologically irreversible and usually constitutes a committed step in metabolism.

The topic of **intermediary metabolism** traditionally includes the catabolic reactions described in Fig. 1-1: glycolysis and the conversion of pyruvate and fatty acids to acetyl-CoA, the oxidation of acetyl-CoA in the citric acid cycle, and the reactions of electron-transport phosphorylation-yielding ATP. Amino acids are also degraded to pyruvate, acetyl-CoA, or intermediates of the tricarboxylic acid (TCA) cycle and then oxidized.

In **electron transport** or oxidative phosphorylation, reducing equivalents are transported sequentially and stepwise in an electron transport chain to oxygen. Electron transport is exergonic. Some of the energy is conserved when a proton gradient is established across a membrane in a process energized by electron transport. Protons then move down the established electrochemical gradient to drive ATP formation from ADP and inorganic phosphate (P_i). ATP formation is an endergonic process energized by the proton motive force.

CARBOHYDRATE CHEMISTRY

Carbohydrates are polyhydroxy **aldehydes** or **ketones**. Formulas representing D-glucose, the most common sugar in nature, are shown in Fig. 6-1. The middle figure is the open-chain form, and it shows the positions of the hydroxyl group on each of the asymmetric carbon atoms in the Fischer projection formula. The asymmetric hydroxyl group on C^5 forms a hemiacetal adduct with the carbonyl group on C^1, and a stable six-membered ring results. This generates another asymmetric carbon atom at position one. The hydroxyl group on C^1 occurs with the α- or β-configuration, as shown. It is noteworthy that glucose with the β-configuration is one of the most stable sugar structures in nature. First, the six-membered ring is stable. Second, the hydroxyl groups occupy equatorial positions and are as far apart from one another as possible.

The **ring structures of sugars** are shown in the Haworth projection format. If you recognize and can draw the structure of the β-enantiomer of D-glucose, then the other common hexoses can be deduced from it. If you draw the six-membered ring (five car-

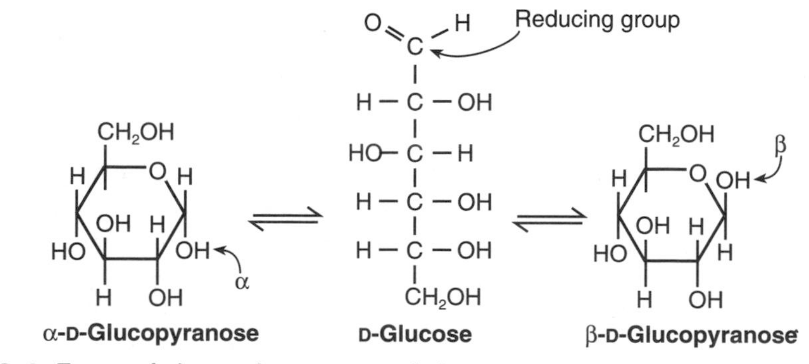

Figure 6-1. Forms of glucose in aqueous solution.

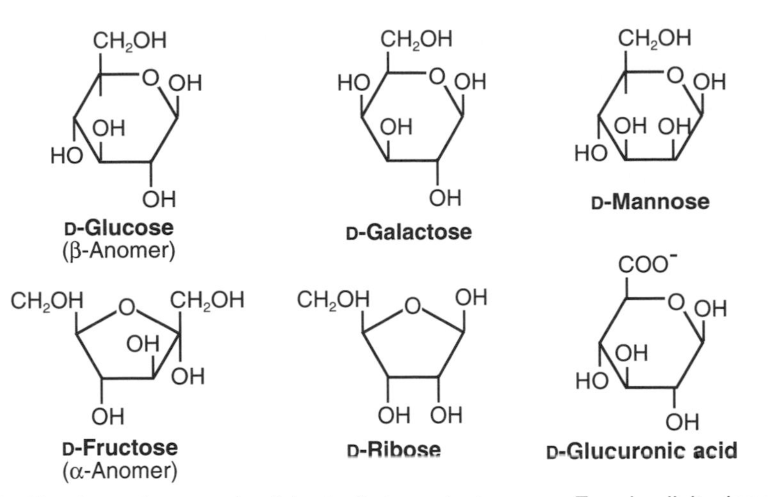

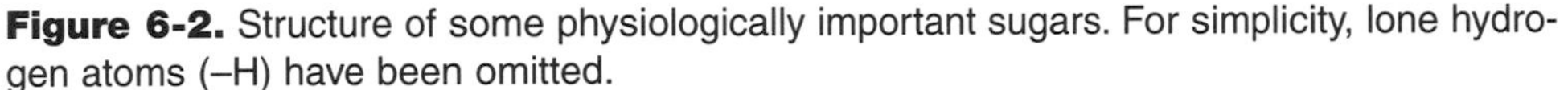
Figure 6-2. Structure of some physiologically important sugars. For simplicity, lone hydrogen atoms (–H) have been omitted.

bons and one oxygen) and place C^6 with its hydroxyl group above the ring, then the other hydroxyl groups alternate from top to bottom in a regular fashion on C^4 (*bottom*), C^3 (*top*), C^2 (*bottom*), and C^1 (*top*, as the β-anomer) as shown. If the configuration of the hydroxyl on C^1 is below, then the α-anomer results. Mannose differs from glucose by the hydroxyl configuration about C^2; galactose differs at C^4 (Fig. 6-2). These three compounds (glucose, mannose, and galactose) are epimers.

Anomers refer to differences of configuration at the hemiacetal (e.g., α-D-glucose and β-D-glucose) or hemiketal (e.g., α-D-fructose and β-D-fructose) carbon; **epimers** refer to differences of configuration at the other carbons (exclusive of the hemiacetal or hemiketal linkage).

The aldehyde of **glucose**, **mannose**, and **galactose** or the ketone of **fructose** constitutes a reducing component in alkaline copper solutions. These substances are **reducing sugars**. The disaccharide **lactose** contains a hemiacetal group, which also makes it a reducing sugar (Fig. 6-3). In the disaccharide **sucrose**, the hemiacetal linkage of glucose and hemiketal linkage of fructose are further derivatized to form a gly-

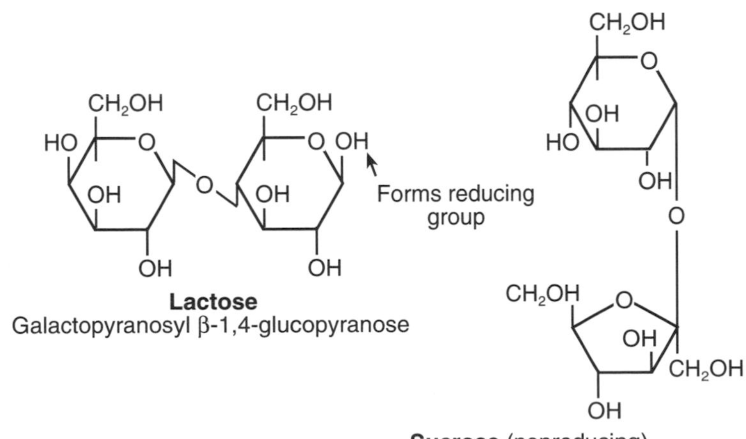

Figure 6-3. Structures of lactose and sucrose.

coside bond. The absence of a simple hemiacetal or hemiketal bond makes sucrose a nonreducing sugar in alkaline copper solution, a noteworthy property.

The **ring form of glucose** is in equilibrium with the open-chain form, which possesses a free aldehyde group. The aldehyde group can react nonenzymatically with the amino group proteins to form glycoproteins. The amino group may be contributed by the free amino-terminal residue or by protein-lysine side chains.

The nonenzymatic glycosylation of proteins may account for the **retinopathy**, **nephropathy**, and **neuropathy** of diabetes mellitus. Furthermore, the percentage of hemoglobin that is glycosylated can be used as an approximate index of average serum glucose concentrations. In normoglycemic people, the percentage of hemoglobin that is glycosylated is less than 7%. In individuals with **diabetes mellitus**, the therapeutic goal is to keep the percentage of glycosylated hemoglobin at less than 10%. The significance of diabetes mellitus in contributing to morbidity and mortality is considerable. Diabetes mellitus is the most common cause of blindness in the United States.

GLYCOLYSIS

The initial steps in the catabolism of glucose constitute the **Embden-Meyerhof glycolytic pathway**. This pathway and its enzymes are present in all human cells and are located in the cytosol (see Table 1-1). The overall reaction is abbreviated:

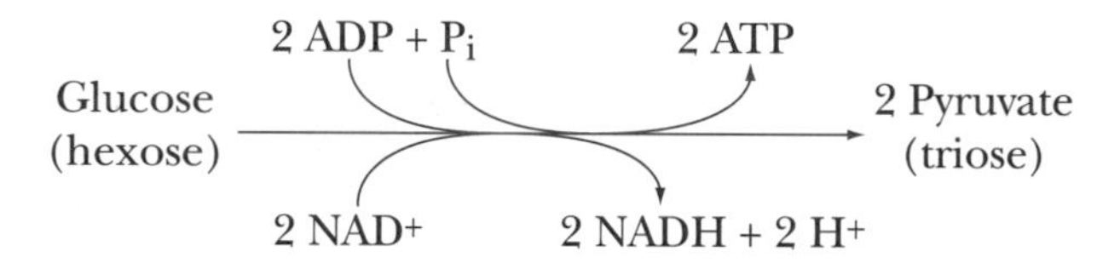

During the stepwise conversion of **glucose to pyruvate**, two molecules of reduced nicotinamide adenine dinucleotide (NADH) are formed, with a net production of two ATP molecules. In the first stage of glycolysis, two ATP molecules are consumed in priming reactions to produce fructose 1,6-bisphosphate. This substance is cleaved into two triose phosphates. Two mol of glyceraldehyde 3-phosphate undergo oxidation to yield 2 mol of NADH and 2 mol of 1,3-bisphosphoglycerate. Two mol of 1,3-bisphosphoglycerate contain an energy-rich acyl phosphate linkage, which will be donated to 2 mol of ADP to yield 2 mol of ATP. Subsequently, 2 mol of phosphoenolpyruvate are formed, which will donate their energy-rich phosphoryl group to 2 mol of ADP to yield an additional 2 mol of ATP. Two net ATP molecules result (–2 + 4 = 2).

The **first step in glycolysis** involves the phosphorylation of glucose by ATP (Fig. 6-4). The enzyme that catalyzes this exergonic and irreversible reaction is **hexokinase**, which is found in all cells. Hexokinase will also catalyze the phosphorylation of mannose and fructose to yield the respective hexose 6-phosphate. Liver and β-cells of the pancreas contain a second enzyme that catalyzes glucose phosphorylation named **glucokinase**.

Glucokinase will not catalyze the phosphorylation of mannose or fructose. **Glucokinase** exhibits a higher Michaelis constant (K_m) for glucose than hexokinase (10 mmol/L vs. 50 μmol/L; note the three orders of magnitude difference in units) and, thus, is not saturated by the high glucose concentrations delivered to the liver by the portal vein postprandially. Glucokinase is also not inhibited by glucose 6-phosphate as is hexokinase. In pancreatic β-cells, glucokinase functions as a glucose sensor and participates in the regulation of insulin secretion. In terms of the metabolic Big Picture (see Fig. 1-1), in times of nutrient excess the high capacity (low efficiency) glucokinase in the liver will trap glucose in this organ in the form of glycogen. At low nutrient levels, glucokinase will be ineffective (substrate will be below the K_m for glu-

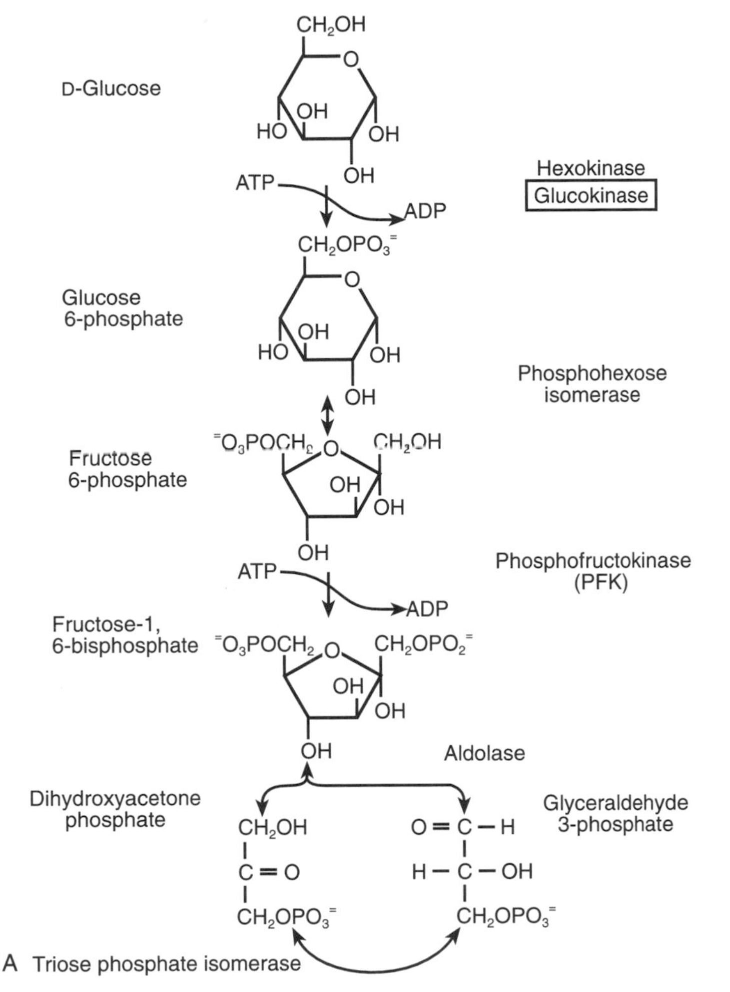

Figure 6-4. Embden-Meyerhof glycolytic pathway. Note that all the components of part **B** of the pathway are doubled.

cokinase) and all tissues will compete for the glucose using their hexokinase (high efficiency; low capacity).

Phosphohexose isomerase catalyzes the conversion of glucose 6-phosphate to fructose 6-phosphate. The interconversion of these two energy-poor compounds is nearly isoergonic. **Phosphofructokinase** (PFK) catalyzes the second phosphorylation. This reaction is exergonic and physiologically irreversible. PFK catalyzes the rate-limiting or pacemaker reaction of glycolysis. Aldolase then catalyzes the cleavage of fructose 1,6-bisphosphate to glyceraldehyde 3-phosphate and dihydroxyacetone phosphate. **Triose phosphate isomerase** catalyzes the interconversion of these two compounds. **Glyceraldehyde 3-phosphate dehydrogenase** mediates a reaction between the designated compound, the oxidized form of nicotinamide adenine dinucleotide (NAD^+), and inorganic phosphate (P_i) to yield 1,3-bisphosphoglycerate.

Next, **phosphoglycerate kinase** catalyzes the reaction of the latter, an energy-rich compound, with ADP to yield ATP and phosphoglycerate. **Phosphoglycerate mutase** catalyzes the transfer of the phosphoryl group from C^3 to C^2 to yield 2-phosphoglycerate. This compound contains an energy-poor phosphate-ester linkage. **Enolase** catalyzes an isoergonic dehydration to yield phosphoenolpyruvate (PEP). This compound contains a very energy-rich phosphate bond (standard free energy of hydrolysis of −14.8 kcal/mol) and then donates its phosphoryl group to ADP to yield ATP and pyruvate in a reaction catalyzed by **pyruvate kinase**. Although the number of

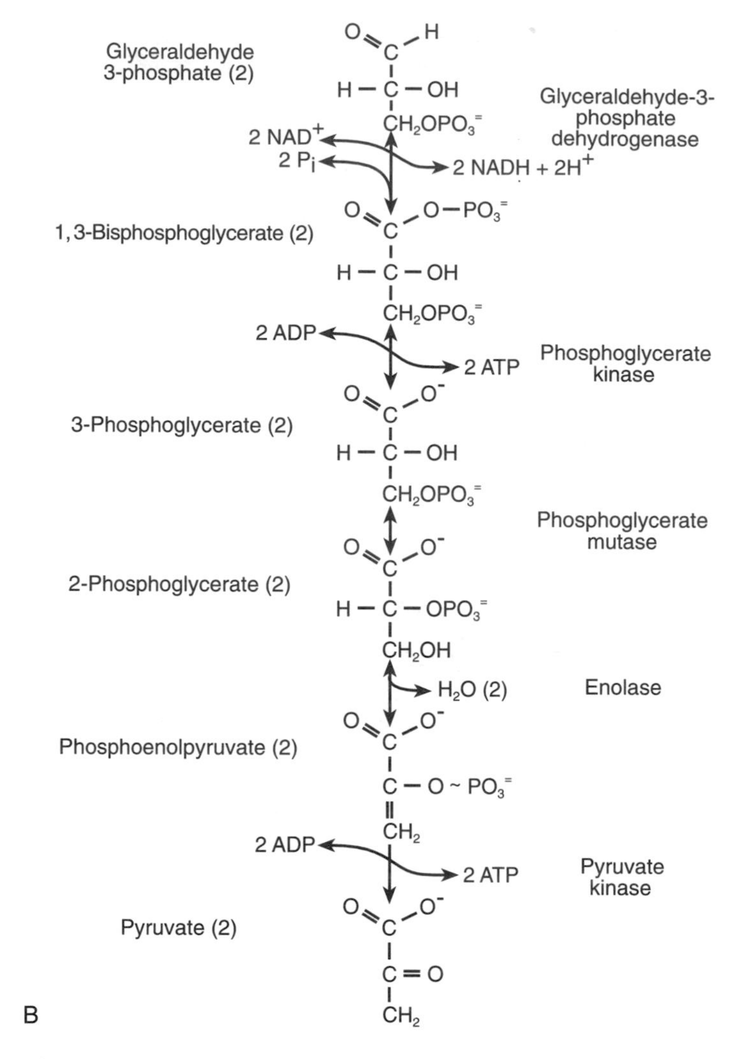

Figure 6-4. *Continued*

high-energy bonds is the same in the reactants and products (two), the reaction is highly exergonic and is physiologically irreversible.

To recapitulate, the three irreversible steps of glycolysis include the reactions catalyzed by hexokinase, phosphofructokinase (PFK), and pyruvate kinase. PFK is the rate-limiting enzyme of the pathway and the main regulatory enzyme. PFK is activated by **adenosine monophosphate** (AMP) and **fructose 2,6-bisphosphate** and inhibited by ATP and citrate. Note that ATP is both a substrate and allosteric modulator. One function of glycolysis is to generate chemical energy as ATP. When the cellular concentrations of ATP are high, glycolysis is inhibited. When ATP levels fall, ADP and AMP are formed and AMP activates PFK. In conditions in which fatty acids serve as a fuel, citrate levels increase and glycolysis decreases. (This regulatory process will be revisited in the discussion of gluconeogenesis, and the regulatory role of fructose 2,6-bisphosphate will be considered.)

Fluoride, at high and nonphysiologic concentrations, inhibits the enolase reaction. The ATP formed in glycolysis results from **substrate-level phosphorylation**. An energy-rich metabolite is produced, leading to ATP formation. This is in contrast to **oxidative phosphorylation**, in which ATP is produced by reactions involving electron transport, a proton motive force, and an ATP synthase.

A note regarding nomenclature is appropriate here. When phosphates are linked together, as in ADP and ATP, the appropriate prefixes are **di-** or **tri-**, respectively. When the phosphates are not attached to each other, as in fructose 1,6-bisphosphate or inositol trisphosphate, the prefixes **bis-** or **tris-** are appropriate (see Table 2-2).

HEXOSE METABOLISM

Consider the metabolism of fructose (from fruit) and galactose (from milk). In addition to hexokinase, the liver contains a specific **fructokinase** that catalyzes the phosphorylation of fructose by ATP to yield **fructose 1-phosphate** in an exergonic and physiologically irreversible process. The product is cleaved in a reaction catalyzed by aldolase to form glyceraldehyde and dihydroxyacetone phosphate. A specific kinase mediates the phosphorylation of glyceraldehyde to yield glyceraldehyde 3-phosphate. In this way, the metabolites of fructose processing enter the glycolytic pathway in midstream.

The catabolism of galactose is more complex than that of the sugars covered to this point. Galactose (see Fig. 6-2) is derived from the disaccharide lactose (see Fig. 6-3) found in milk. Lactose is hydrolyzed to galactose and glucose by a digestive enzyme (lactase), absorbed by the gut, and transported to the liver. In the liver, galactose is phosphorylated by ATP in a reaction catalyzed by **galactokinase** to yield ADP and galactose 1-phosphate. Galactose is the only common aldohexose that is not a substrate for hexokinase. The reaction is also unusual in that the hydroxyl group of a hemiacetal linkage is derivatized (the hydroxyl of fructose on carbon one is not in hemiacetal or hemiketal linkage). Galactose 1-phosphate reacts with uridine diphosphoglucose (UDPG) to yield uridine diphosphogalactose (UDP-gal) and glucose 1-phosphate. The reaction is catalyzed by **galactose-1-phosphate uridyltransferase** (Fig. 6-5).

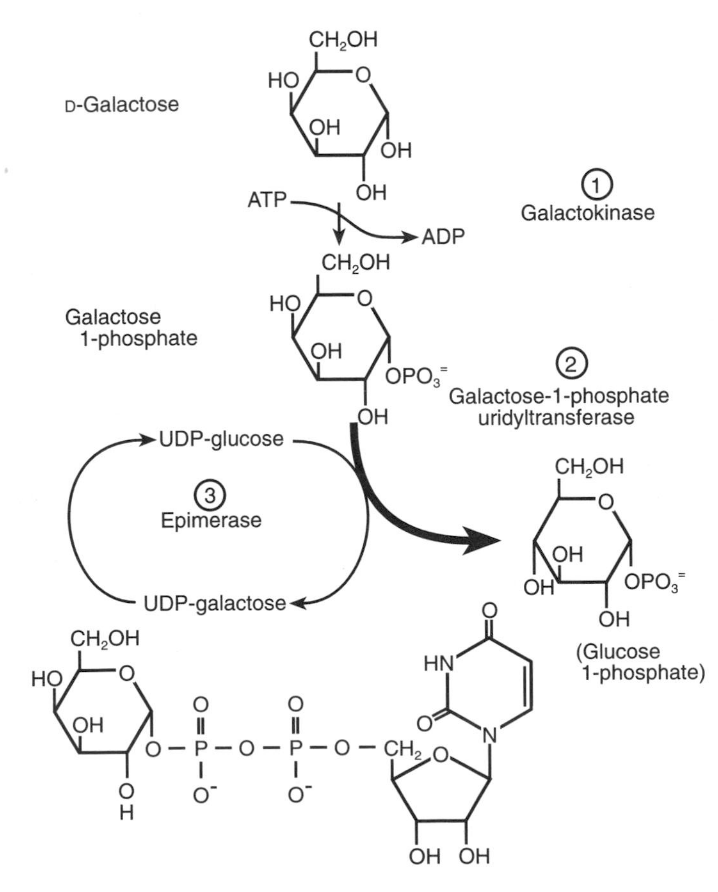

Figure 6-5. Galactose catabolism. The numbers represent the three key steps in galactose utilization. Enzyme-catalyzed steps *1* and *2* create uridine diphosphate (*UDP*) galactose that can then be converted to UDP-glucose (in step *3*) for entry into the glycolytic pathway.

Glucose 1-phosphate is converted to glucose 6-phosphate by **phosphoglucomutase** in an isoergonic reaction. Glucose 6-phosphate is metabolized by the reactions of the Embden-Meyerhof pathway, as previously considered.

UDP-gal must be converted to UDPG to regenerate the initial reactant. This reaction is catalyzed by an epimerase that converts the hydroxyl on C^4 to a ketone, and then reduces it to give the hydroxyl of the alternative configuration (UDPG) shown in Fig. 6-5. The epimerase contains tightly bound NAD^+ as a cofactor. The UDPG can now react with a second molecule of galactose 1-phosphate to yield glucose 1-phosphate and UDP-gal. The UDP moiety is thus used repeatedly in a cyclic fashion to mediate the conversion of appreciable galactose 1-phosphate to glucose 1-phosphate. The UDPG is said to function in a catalytic fashion; UDPG is regenerated after every reaction.

The disease called **galactosemia** is due to a deficiency of **galactose-1-phosphate uridyltransferase**. Excessive galactose 1-phosphate accumulates in cells with consequent deleterious effects. A diet lacking milk and milk products (specifically lactose) constitutes treatment. Galactose forms an essential component of many carbohydrate-containing glycoproteins. Withholding galactose is not harmful because the epimerase can catalyze the formation of UDP-gal from UDPG as necessary. A milder form of galactosemia is due to a hereditary deficiency of galactokinase. The second type of galactosemia is less severe than classical galactosemia because there is no accumulation of charged intracellular metabolites.

ANAEROBIC GLYCOLYSIS

For continued glycolysis, it is necessary to **regenerate ADP and NAD^+**. ATP is used in many cellular reactions as the common currency of energy exchange, resulting in the formation of ADP to be regenerated in the normal course of cellular metabolism; however, such is not necessarily the case for NAD^+ regeneration. Normally, the NADH-reducing equivalents are used in oxidative phosphorylation (electron transport), and NAD^+ is regenerated in the process. In erythrocytes, which lack mitochondria, and in other cells in which NADH production exceeds the capacity for electron transport (e.g., exercising muscle), NAD^+ is regenerated by the **lactate dehydrogenase** reaction. Pyruvate and NADH + H^+ react to yield lactate and NAD^+. The regenerated NAD^+ can participate again in glycolysis.

Streptococcus mutans is a bacterium found in the oral cavity and is a causative agent of dental caries. Bacteria are able to generate a variety of acids anaerobically (e.g., acetic, propionic, and butyric acid), but *S. mutans* uses the classical glycolytic pathway to generate lactic acid. Lactic acid dissolves dental enamel, which initiates the caries-forming process.

Lactate is released from the erythrocyte or exercising skeletal muscle and is carried to the liver by the circulation. In opportune conditions, lactate reacts with NAD^+ to yield NADH and pyruvate. Pyruvate may be reconverted to glucose by the process of gluconeogenesis. The glucose can be released from the liver and return to other tissues. The conversion of glucose to lactate in extrahepatic tissues, the resynthesis of glucose from lactate in liver and subsequent transport to extrahepatic tissues is called the **Cori cycle** (Fig. 6-6). The reoxidation of NADH to NAD^+ under aerobic conditions involving the **malate-aspartate shuttle** system or the **glycerol phosphate shuttle** is considered in Metabolite Transport Across the Inner Mitochondrial Membrane (see Fig. 6-13).

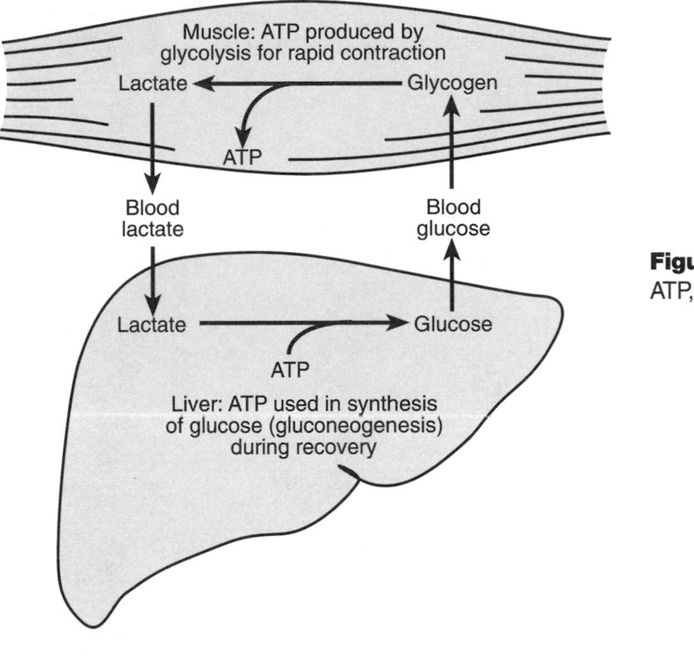

Figure 6-6. The Cori cycle. ATP, adenosine triphosphate.

PYRUVATE DEHYDROGENASE

The complete oxidation of pyruvate occurs within mitochondria. Pyruvate is transported through the inner mitochondrial membrane by a specific carrier protein or translocase. Pyruvate is oxidized to acetyl-CoA and CO_2 by the **pyruvate dehydrogenase** multienzyme complex found in the mitochondrial matrix (see Table 1-1). A **multienzyme complex** is an aggregate of enzymes that catalyzes a series of reactions. The intermediates in the sequence may be covalently bound to the complex (as in the case of pyruvate dehydrogenase) and are not free to diffuse throughout the surrounding solution. The pyruvate dehydrogenase complex contains three enzyme activities and five cofactors (Fig. 6-7 and Table 6-1).

Pyruvate is converted to CO_2 and a two-carbon hydroxyethyl moiety covalently linked to the thiamine pyrophosphate of E1. The hydroxyethyl group is transferred to lipoate on E2. In the process, it becomes the more oxidized acetyl group, and it is covalently linked to sulfur (as an energy-rich thioester). E2 transfers the acetyl group to CoA, yielding acetyl-CoA. E3 catalyzes the oxidation of reduced lipoate, and **flavin adenine dinucleotide** (FAD) is converted to reduced FAD ($FADH_2$). $FADH_2$ is then oxidized to FAD, resulting in NADH + H^+. After this reaction, the enzyme complex is now in its original form.

Pyruvate dehydrogenase catalyzes the following net reaction:

$$\text{Pyruvate}^{-1} + \text{CoA} + \text{NAD}^+ \rightarrow CO_2 + \text{acetyl-CoA} + \text{NADH}$$

The reaction is highly exergonic and physiologically irreversible.

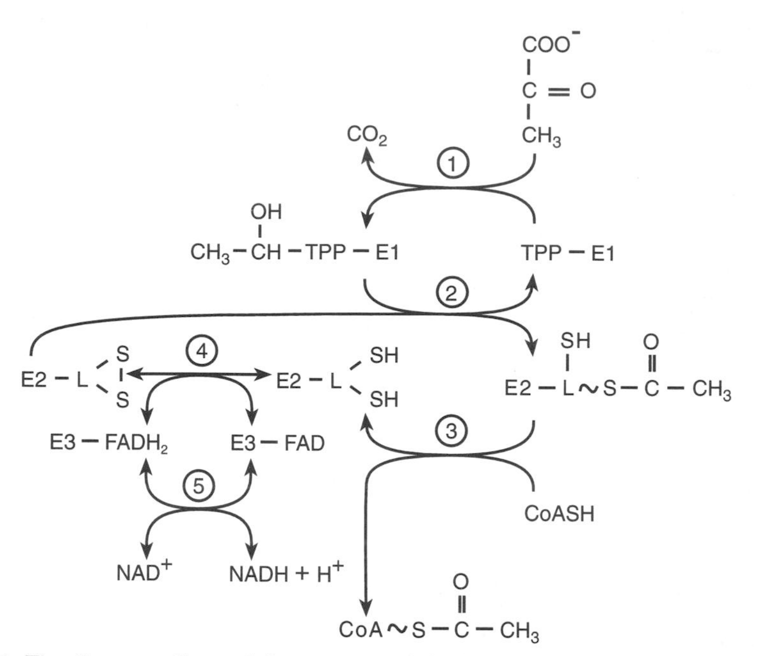

Figure 6-7. The three portions of the pyruvate dehydrogenase reaction. The numbers refer to the five sequential steps in acetylcoenzyme A production described within the text. CoA, coenzyme A; CoASH, uncombined coenzyme A; FAD, oxidized flavin adenine dinucleotide; $FADH_2$, reduced flavin adenine dinucleotide; NAD^+, oxidized form of nicotinamide adenine dinucleotide; NADH, reduced nicotinamide adenine dinucleotide; TPP, thiamine pyrophosphate.

Pyruvate dehydrogenase is activated by NAD^+ and coenzyme A and inhibited by NADH and acetyl-CoA. The mechanism, however, is indirect. **Pyruvate dehydrogenase kinase** is a specific protein kinase associated with the dehydrogenase in mitochondria. After phosphorylation by ATP, pyruvate dehydrogenase exhibits less activity. Acetyl-CoA and NADH activate pyruvate dehydrogenase kinase; CoA and NAD^+ inhibit it. A phosphoprotein phosphatase catalyzes the hydrolytic removal of phosphate from pyruvate dehydrogenase to generate the initial enzyme form.

TABLE 6-1. Pyruvate Dehydrogenase Multienzyme Complex

Enzyme	Cofactor	Vitamin
E1 Pyruvate decarboxylase	Thiamine pyrophosphate	Thiamine
E2 Dihydrolipoyltransacetylase	Lipoate	
	CoA	Pantothenate
E3 Dihydrolipoyldehydrogenase	FAD	Riboflavin
	NAD^+	Niacin

CoA, coenzyme A; FAD, flavin adenine dinucleotide; NAD^+, nicotinamide-adenine dinucleotide.

THE KREBS CYCLE

The Krebs citric acid or **tricarboxylic acid (TCA) cycle** is often called the final common pathway of metabolism. As illustrated in Fig. 1-1, for aerobic organisms, all nutrient sources ultimately lead to acetyl-CoA and the Krebs TCA cycle. The catabolism of glucose and fatty acids yields acetyl-CoA; metabolism of amino acids yields acetyl-CoA or actual intermediates of the cycle. The TCA cycle provides a pathway for the oxidation of acetyl-CoA. The reactions occur in the mitochondria of eukaryotes (see Table 1-1). The pathway includes eight discrete steps. Seven of the enzyme activities are found in the mitochondrial matrix; the eighth (succinate dehydrogenase) is associated with the electron-transport chain within the inner mitochondrial membrane.

The net reaction catalyzed during each revolution of the **TCA cycle** can be depicted as follows:

$$\begin{array}{ccc} \text{Acetyl-CoA} & & 2\ CO_2 \\ + & & + \\ 2\ H_2O & & \text{CoA} \\ + & & + \\ 3\ NAD^+ & \rightarrow & 3\ NADH + 3\ H^+ \\ + & & + \\ FAD & & FADH_2 \\ + & & + \\ GDP + P_i & & GTP \end{array}$$

CO_2 is an end product of metabolism. Stop for a moment to consider what has been accomplished in this metabolic process. The carbon atoms of glucose and fatty acids have been converted to CO_2 molecules. A small amount of high-energy currency [ATP from glycolysis and guanosine triphosphate (GTP) from substrate-level phosphorylation within the TCA cycle] has been obtained; however, a large amount of energy potential has resulted in the form of reducing equivalents [two NADH from glycolysis and two $FADH_2$ and six NADH from two turns of the TCA cycle (two acetyl-CoA are formed from each glucose molecule)]. CoA can be reused for a variety of reactions. The reduced cofactors will donate their reducing equivalents to the electron-transport chain and oxygen to yield water and the oxidized cofactors.

GTP is formed by substrate-level phosphorylation. It is bioenergetically equivalent to ATP and serves a variety of functions. An overview of the cyclic pathway depicts the location of CO_2 production and NADH and $FADH_2$ formation (Fig. 6-8). The production of 11 ATP equivalents by oxidative phosphorylation and one GTP by substrate-level phosphorylation is also noted. The source of the ATP (oxidative phosphorylation) is presented in more detail in the section Oxidative Phosphorylation (see Figs. 6-10 through 6-12).

The reactions of the cycle are shown in Fig. 6-9. **Citrate synthase** catalyzes a reaction between acetyl-CoA, oxaloacetate, and water to yield citrate (a tricarboxylic acid) and CoA. This reaction occurs with the hydrolytic removal of CoA, which is a highly exergonic process and renders the reaction physiologically irreversible.

A note about nomenclature is appropriate here. The distinction between **synthases** and **synthetases** is that synthetases require ATP or an equivalent nucleoside triphosphate as an energy source. Citrate synthase, in contrast, is a biosynthetic

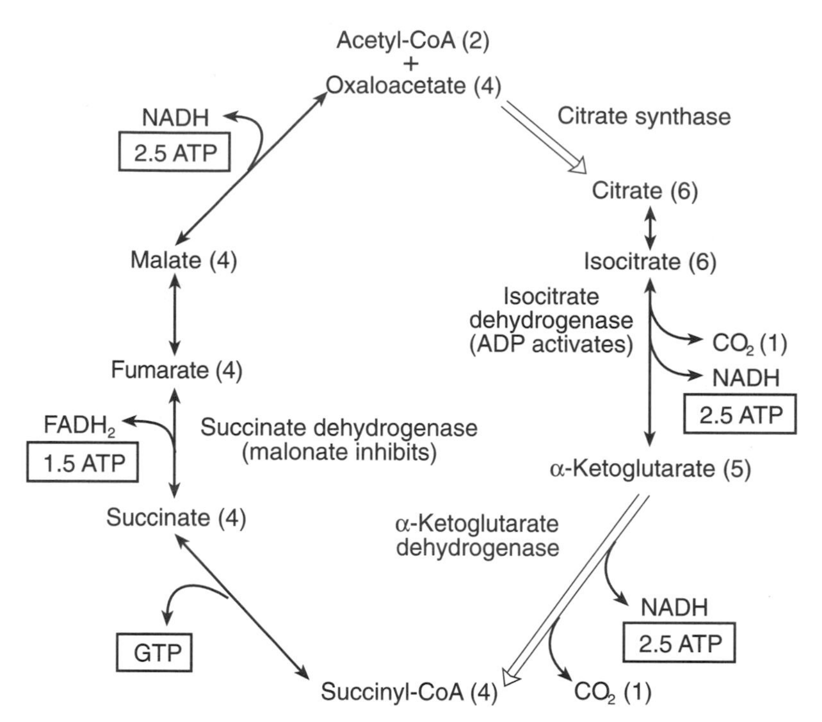

Figure 6-8. Overview of the Krebs citric acid cycle. The numbers in parentheses represent the number of component carbon atoms. Acetyl-CoA, acetylcoenzyme A; ADP, adenosine diphosphate; ATP, adenosine triphosphate; $FADH_2$, reduced flavin adenine dinucleotide; GTP, guanosine triphosphate; NADH, reduced nicotinamide adenine dinucleotide.

enzyme that does not use a nucleoside triphosphate such as ATP as a reactant. **Aconitase** catalyzes an isoergonic isomerization to yield isocitrate. Although citrate lacks an asymmetric carbon, it is prochiral, and the hydroxyl is moved away from the two-carbon end just derived from the acetyl group. Isocitrate undergoes an oxidative decarboxylation reaction involving NAD^+, which is catalyzed by **isocitrate dehydrogenase**; the products include α-ketoglutarate, CO_2, NADH, and H^+. The reaction is modestly exergonic. The high ratio of NAD^+ to NADH in mitochondria pulls the reaction in the forward direction. Isocitrate dehydrogenase is the rate-limiting enzyme in the Krebs cycle. ADP serves as a positive allosteric effector. High ADP levels serve as a signal to enhance the flux of substrates through the cycle.

α-Ketoglutarate dehydrogenase catalyzes the reaction between substrate, NAD^+, and CoA to yield succinyl-CoA, CO_2, NADH, and H^+. The enzyme consists of a multienzyme complex that is completely analogous to the pyruvate dehydrogenase complex considered in the previous section. The reaction is highly exergonic and physiologically irreversible. The α-ketoglutarate dehydrogenase reaction ensures that the cycle is unidirectional. The next reaction is catalyzed by **succinate thiokinase**. This reaction is an example of substrate-level phosphorylation conserving the energy-rich thioester bond of succinyl-CoA as the terminal phosphoanhydride bond of GTP. Succinyl-CoA, P_i, and guanosine diphosphate (GDP) form succinate and GTP. (Because GDP + $P_i \rightarrow$ GTP + H_2O, this reaction serves indirectly as a source of water. This property is used in calculating the stoichiometry of the cycle.)

Succinate dehydrogenase has three properties worth noting. First, it is the only enzyme of the Krebs cycle found within the inner mitochondrial membrane. Succinate dehydrogenase is localized contiguous to the electron-transport chain where it passes its reducing equivalents. Second, only 1.5 mol of ATP are produced

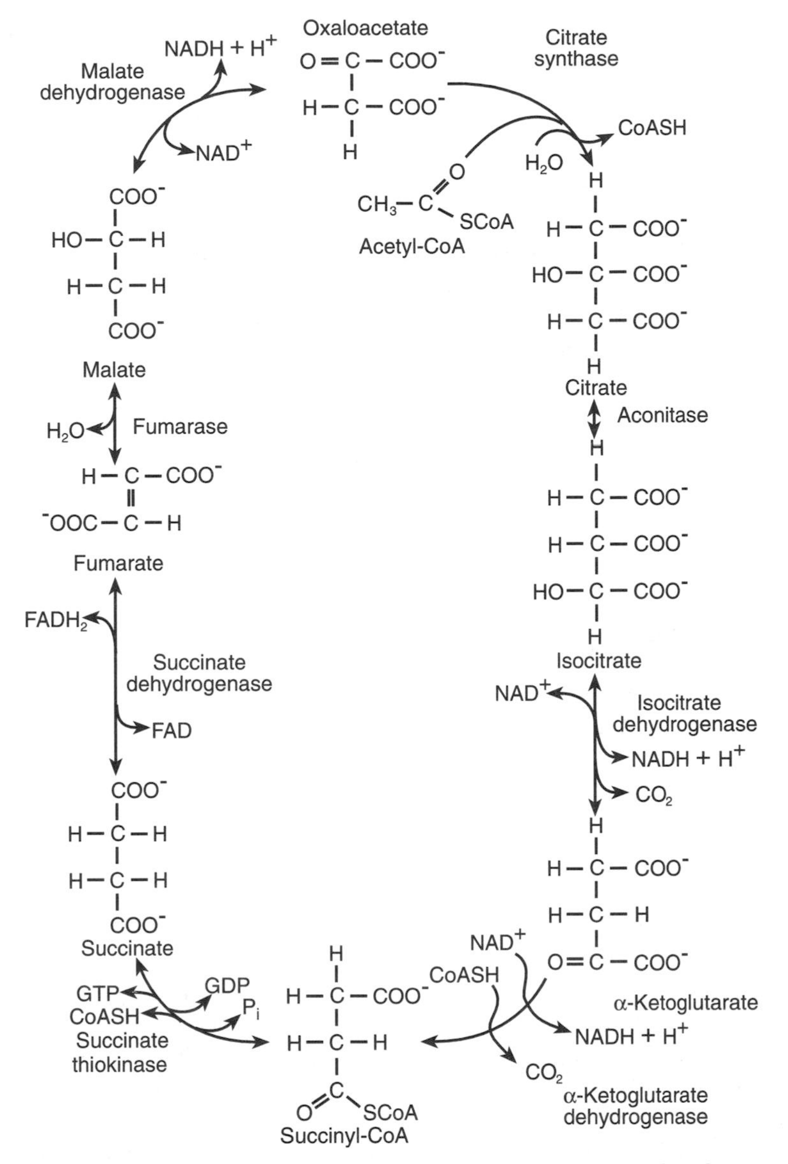

Figure 6-9. Krebs citric acid cycle. Acetyl-CoA, acetylcoenzyme A; CoASH, uncombined coenzyme A; FAD, flavin adenine dinucleotide; $FADH_2$, reduced flavin adenine dinucleotide; GDP, guanosine diphosphate; GTP, guanosine triphosphate; NAD^+, oxidized form of nicotinamide adenine dinucleotide; NADH, reduced nicotinamide adenine dinucleotide; SCoA, derivated coenzyme A.

by electron transport per mol of the $FADH_2$ (in contrast to 2.5 mol of ATP per mol of NADH). Third, succinate dehydrogenase is inhibited competitively by malonate ($^-OOCCH_2COO^-$). Succinate dehydrogenase mediates the conversion of substrate to fumarate (note that fumarate has the *trans* configuration at the double bond). **Fumarase** catalyzes the addition of water to fumarate (a hydration and lyase reaction) yielding malate. The reaction is isoergonic. **Malate dehydrogenase** catalyzes the regeneration of oxaloacetate by the reduction of NAD^+ to NADH and H^+. Although the malate dehydrogenase reaction is readily reversible, it is endergonic.

To recapitulate, the citrate synthase and α-ketoglutarate dehydrogenase reactions are physiologically irreversible. Isocitrate dehydrogenase is the pacemaker reaction of the pathway and is allosterically activated by ADP. Succinate thiokinase catalyzes a substrate-level phosphorylation. The two molecules of CO_2 given off during a single turn of the cycle are not those immediately derived from the acetyl group.

In addition to their role in catabolism, metabolites of the Krebs cycle serve as precursors for the biosynthesis of amino acids and heme. A process playing a role in both catabolism and anabolism is called an **amphibolic** process. The following reaction, catalyzed by **pyruvate carboxylase**, plays the important role of replenishing intermediates that are used for biosynthesis. The reaction involves ATP, HCO_3^-, and pyruvate; oxaloacetate, ADP, and P_i are products. The cofactor for the enzyme is **biotin**. Biotin is covalently linked to the enzyme and serves as a prosthetic group. Pyruvate carboxylase is allosterically activated by acetyl-CoA and requires acetyl-CoA for the expression of its activity. The overall reaction is depicted by the following chemical equation:

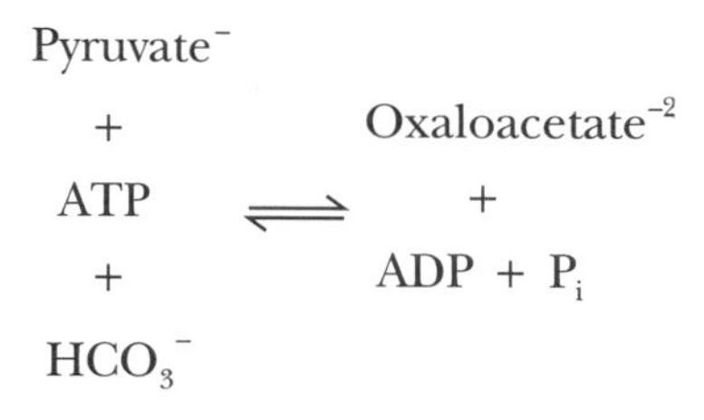

OXIDATIVE PHOSPHORYLATION

The term **oxidative phosphorylation** refers to reactions associated with oxygen consumption and the phosphorylation of ADP to yield ATP. Oxidative phosphorylation is associated with an **electron-transport** or **respiratory chain**, which is found in the inner mitochondrial membrane of eukaryotes (Fig. 6-10; see Table 1-1). A similar process occurs within the plasma membrane of prokaryotes such as *Escherichia coli.*

At this juncture, the importance of oxidative phosphorylation is that it accounts for the **reoxidation** of reducing equivalents generated in the reactions of the Krebs cycle as well as in glycolysis. Oxidative phosphorylation accounts for the preponderance (90% or more) of ATP production in humans. Although some short-term energy needs can occasionally be accommodated by anaerobic glycolysis, as noted, oxygen and oxidative phosphorylation are required for life. The electron-transport chain transfers electrons from reductants to oxygen in a series of exergonic reactions.

According to **Mitchell's chemiosmotic theory**, a portion of the liberated free energy energizes the transport of protons from inside to outside of the inner mitochondrial membrane (or the plasma membrane of prokaryotes). Such reactions result in energy conservation or storage of chemical energy in the form of a proton gradient. The **proton motive force** exhibits a voltage component (inside negative) and a concentration component (external pH < internal pH). The protons then move down their electrochemical gradient (from outside to inside of the mitochondrion or the bacterial cell) in an exergonic process and drive the conversion of ADP + $P_i \rightarrow$ ATP + H_2O in a reaction catalyzed by **ATP synthase** (see Figs. 6-10 and 1-1). The key to this process is that the mitochondrial membrane is impermeable to protons, except through the portal provided by the ATP synthase. In traversing the synthase, the energy of the gradient is somehow converted to stored chemical energy in the form of ATP. The enzyme is located on the inner face of the inner mitochondrial membrane; ATP synthase catalyzes a reversible reaction and can synthesize or hydrolyze substrate (adenosine triphosphatase) depending on the experimental conditions. Its function in humans *in vivo* is ATP synthesis.

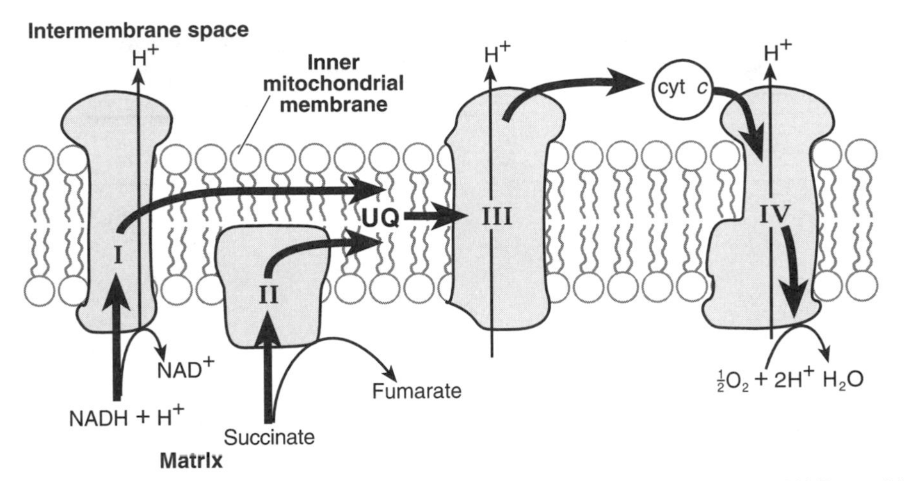

Figure 6-10. Electron transport and proton translocation. cyt, cytochrome; NAD^+, oxidized nicotinamide adenine dinucleotide; NADH, reduced nicotinamide adenine dinucleotide; UQ, ubiquinone.

THE ELECTRON TRANSPORT CHAIN

The components of the electron transport chain include iron-sulfur proteins, cytochromes c_1, c, b, aa_3, and coenzyme Q or ubiquinone. The pathway of electrons along the chain is shown in Fig. 6-11. Reducing equivalents can enter the chain at two locations. Electrons from NADH are transferred to NADH dehydrogenase.

In reactions involving **flavin mononucleotide** (FMN) and **iron-sulfur proteins**, electrons are transferred to **coenzyme Q**. During the process, protons are translocated from the interior to exterior of the mitochondrion. Electrons entering from succinate dehydrogenase ($FADH_2$) and other flavoproteins via an electron transfer protein are donated to coenzyme Q (bypassing the NADH dehydrogenase reaction). This transfer is not associated with proton translocation. Electrons are transported from reduced coenzyme Q to cytochrome *b* and then cytochrome c_1. This step is associated with active **proton translocation**. Electrons are then carried by cytochrome *c* to cytochrome aa_3. Cytochrome aa_3 is also known as **cytochrome oxidase**, and catalyzes a reaction of electrons and protons with molecular oxygen to produce water. Cytochrome oxidase also actively translocates protons across the inner mitochondrial membrane.

The three positions for active **proton translocation** by the electron transport chain are called sites I, II, and III. The precise localization of these sites is not as well

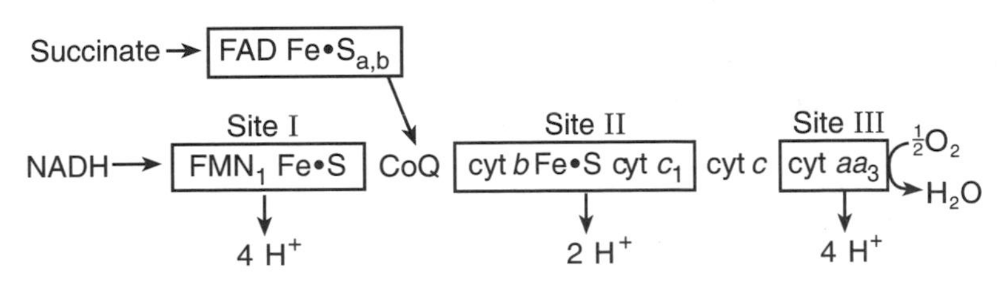

Figure 6-11. Mitochondrial electron transport chain. CoQ, coenzyme Q; cyt, cytochrome; FAD, flavin adenine dinucleotide; Fe•S, iron sulfur; FMN, flavin mononucleotide; NADH, reduced nicotinamide adenine dinucleotide.

TABLE 6-2. Inhibitors of the Electron Transport or Respiratory Chain

Site	Inhibitor
I	Rotenone
II	Antimycin A
III	Cyanide
	Azide (N_3^-)
	Carbon monoxide

established as their designations might signify. **Site I** occurs between NADH dehydrogenase and coenzyme Q, **site II** occurs between coenzyme Q and cytochrome *c*, and **site III** occurs between cytochrome *c* and molecular oxygen. The sites were named on the basis of the effects of inhibitors of electron transport and before the notion of discrete proton translocating locations. The site-specific inhibitors are given in Table 6-2. Rotenone is commonly used as a rat poison. Cyanide is a powerful inhibitor of cytochrome oxidase in humans and other animals, accounting for cyanide toxicity and its lethal effects. Carbon monoxide also binds tightly to cytochrome oxidase. Its major toxicity, however, is related to the formation of a complex with hemoglobin, which abolishes its oxygen-binding and transport capacity.

Amyl nitrate is used in the treatment of **cyanide poisoning**. Amyl nitrate oxidizes the ferrous heme of hemoglobin to ferric heme-yielding methemoglobin. Methemoglobin binds cyanide and decreases its effective concentration. Cyanide is metabolized in a reaction catalyzed by rhodanese to form nontoxic thiocyanate (SCN^-).

Cytochrome *c* is a small (a molecular weight of approximately 10,000 daltons), water-soluble heme protein. All the other proteins of the electron transport chain are water insoluble and are found embedded in the inner mitochondrial membrane as integral membrane proteins. Coenzyme Q is a lipid-soluble organic compound. NAD^+, FAD, and coenzyme Q are two-electron carriers; the cytochromes (with their ferrous heme) and iron-sulfur proteins are one-electron carriers. Cytochrome oxidase contains two iron atoms and two copper atoms, each of which is thought to function in a series of one-electron transfers. The reduction of oxygen to water, which is catalyzed by cytochrome oxidase, is a four-electron reaction.

Note that more than 90% of oxygen consumed by humans involves a reaction catalyzed by **cytochrome oxidase**. When oxygen transport to tissues is blocked as the result of an arterial occlusion, serious pathology or death ensues. The occlusion produced by **coronary artery disease** resulting in a myocardial infarction or by **cerebral vascular disease** resulting in a stroke cogently illustrates the importance of the cytochrome oxidase reaction. The product of oxygen reduction is water. In humans, reduction of oxygen accounts for the production of approximately 300 mL of metabolic water per day.

The precise mechanism of **proton translocation** at each of the three sites is unknown. Uncertainty exists concerning the number of protons transported per electron per site. It appears to be between two (site II) and four (sites I and III). Moreover, there is some question concerning the number of pumped protons that are required to generate an ATP and the exact energy yield from oxidative phosphorylation. An important aspect of the chemiosmotic theory is that a membrane is required, and the membrane must be relatively impermeable to protons. Protons are transported by specific transport proteins and do not simply diffuse through the membrane.

ATP SYNTHESIS

An intricate enzyme called **ATP synthase** is associated with the inner aspect of the inner mitochondrial membrane. As protons move down their electrochemical gradient in an exergonic fashion, they provide the energy for the reaction of ADP and P_i to give ATP and H_2O. The synthase is made up of two domains. The **F_o domain** (o, oligomycin—an inhibitor of the overall synthase reaction—and not zero) is embedded in the inner membrane.

The **F_1 complex** forms a knoblike structure in association with F_o. In the intact structure, the membrane is proton impermeable. When F_1 is removed from F_o, the membrane transmits protons, evidence that protons course directly through the ATP synthase. The F_1 complex contains binding sites for ATP, ADP, and P_i. Movement of protons down their thermodynamic gradient provides the energy to drive the endergonic portion of the reaction ($ADP + P_i \rightarrow ATP + H_2O$). Three protons move down their gradient to drive the synthesis of each ATP. An additional proton is required to translocate P_i from the cytosol to the interior of the mitochondrion; thus, a total of four protons is required to sustain the synthesis of each molecule of ATP.

One of the triumphs of the **chemiosmotic theory** is the explanation of the necessity of a membrane in oxidative phosphorylation. The theory also explains the effects of uncouplers of oxidative phosphorylation. **2,4-Dinitrophenol** is the prototype of an uncoupler, which enhances the rates of oxygen transport. 2,4-Dinitrophenol abolishes phosphorylation or ATP formation. This explains the term "uncoupler," because oxidation occurs but phosphorylation does not. 2,4-Dinitrophenol and other uncouplers dissipate the proton gradient. Uncouplers ferry protons across the membrane and abolish the proton-motive force. In the presence of an uncoupler, proton translocation by the electron transport system is not opposed by an existing proton gradient, and electron transport to oxygen is increased.

Experiments show that the P to O ratio (number of ATPs formed from ADP + P_i per gram atom of oxygen consumed) using **NADH** as substrate is 2.5, and the P to O ratio with **succinate** as substrate is 1.5. Sites I and III are associated with the generation of a proton gradient sufficient for the formation of one ATP, and site II is associated with 0.5 ATP. Succinate circumvents site I and results in the production of only 1.5 ATP molecules (see Fig. 6-11). Studies show that 2.5 ATP molecules are derived from oxidation of NADH (instead of the traditional 3.0) and that 1.5 are derived from $FADH_2$ (instead of 2 ATP). In the Big Picture, this has had no effect on our understanding of intermediary metabolism.

To recapitulate, note the four properties of **oxidative phosphorylation**. First, the transport of electrons from reductant to oxygen along the respiratory chain is a very

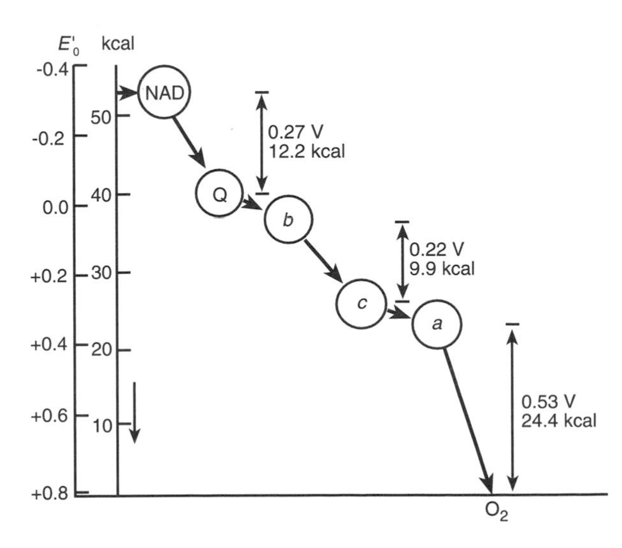

Figure 6-12. Thermodynamics of the electron transport chain. Electrons are passed from a donor with a low electromotive potential to one with higher electromotive potential. NAD, nicotinamide adenine dinucleotide.

exergonic process (Fig. 6-12). Second, part of the chemical energy is conserved as protons are translocated from the inside to the outside of the inner mitochondrial membrane to establish a proton gradient. Third, the membrane is not freely permeable to protons. Fourth, protons then move down their electrochemical gradient (an exergonic process) and drive ATP formation in a process involving the ATP synthase of the inner mitochondrial membrane (see Fig. 6-10). A total of four protons are required for the synthesis of each ATP: one for the translocation of phosphate and three for the ATP synthase reaction.

METABOLITE TRANSPORT ACROSS THE INNER MITOCHONDRIAL MEMBRANE

The lipid portion of the inner mitochondrial membrane is relatively impermeable to ionic metabolites, phosphate, hydroxide, and especially protons. This is a central component of the chemiosmotic theory of oxidative phosphorylation; however, it also creates a problem in that it represents an effective barrier for translocation of metabolites from the inside of the mitochondrion to the cytoplasm and vice versa. It is important in metabolism to transport compounds generated in one cellular compartment into another. Specific proteins, called **translocases**, mediate transport across membranes. The identification of a number of such translocases and their physiologic role are considered in this section.

ATP is generated within the mitochondrion and functions predominantly in the cytosol. In the cytosol, **ADP** and $\mathbf{P_i}$ are formed. Two different translocases are necessary for transporting these substances. ATP is transported out of the mitochondrion in exchange for ADP (an antiport system). Both ATP and ADP are transported down a concentration gradient. In physiologic conditions, ADP is transported down its gradient into the mitochondrial matrix. These processes are inhibited by a plant-derived toxin called **atractyloside**. Phosphate is transported into the mitochondrion along with a proton. The extramitochondrial proton concentration exceeds that of the mitochondrion (hence the proton gradient used to generate ATP), and this proton gradient also provides osmotic energy for phosphate transport. Carriers also exist for the exchange of α-ketoglutarate for malate, citrate for malate, phosphate for malate, and aspartate for glutamate.

Transport proteins do not exist for the following substances (the biochemical ramifications are considered in appropriate sections): NAD^+, NADH, oxidized form of nicotinamide adenine dinucleotide phosphate ($NADP^+$), NADPH, CoA, acyl-CoA, and oxaloacetate.

Now, consider the transport of reducing equivalents into the mitochondrion. One important source of cytosolic reductant includes the **NADH** (two per molecule of glucose metabolized) generated during glycolysis. Only a small proportion of $\mathbf{NAD^+}$ is regenerated by the lactate dehydrogenase reaction in most cells (except mature erythrocytes) in aerobic conditions. Because a mitochondrial translocase for NADH is nonexistent, the cell uses indirect methods called the **malate-aspartate shuttle** or the **glycerol phosphate shuttle** for moving reducing equivalents from the cytosol into the mitochondrion.

Two membrane translocases are required for the operation of the malate-aspartate shuttle: One is specific for malate and α-ketoglutarate, and the second exchanges the amino acids aspartate and glutamate. Two sets of two enzymes also are required: mitochondrial and cytosolic **malate dehydrogenase** and mitochondrial and cytosolic **aspartate aminotransferase**.

Beginning with NADH in the cytosol, shown in Fig. 6-13 (*upper left*), malate dehydrogenase catalyzes a reaction between oxaloacetate and NADH + H^+ to yield NAD^+

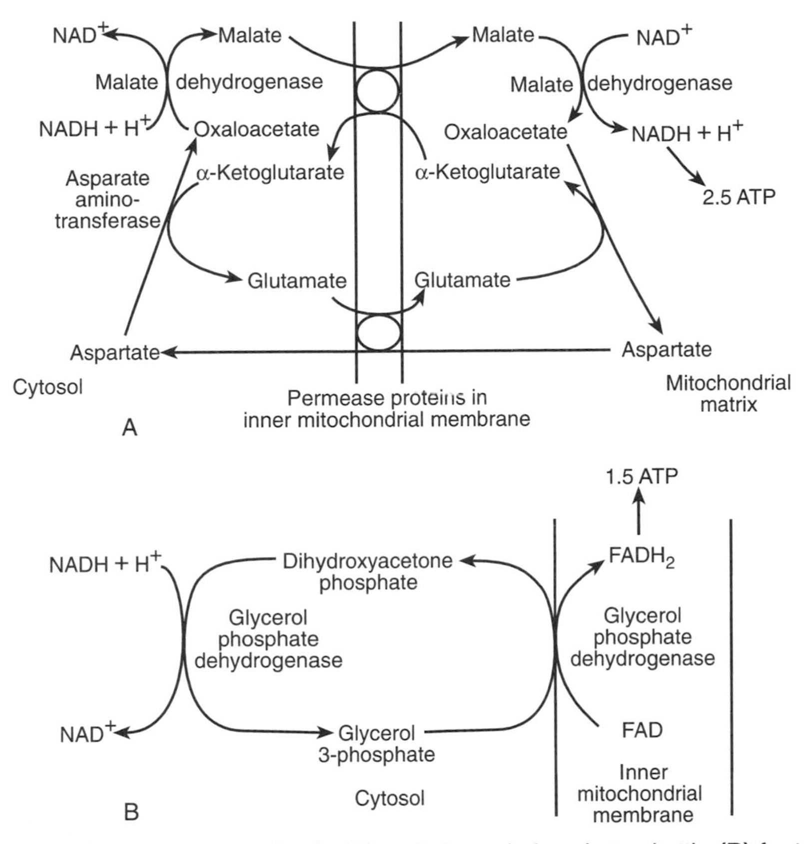

Figure 6-13. Malate-aspartate shuttle **(A)** and glycerol phosphate shuttle **(B)** for transporting reducing equivalents into the mitochondrion. ATP, adenosine triphosphate; FAD, flavin adenine dinucleotide; $FADH_2$, reduced flavin adenine dinucleotide; NAD^+, oxidized form of nicotinamide adenine dinucleotide; NADH, reduced nicotinamide adenine dinucleotide.

and malate. The NAD^+ can now participate in the **glyceraldehyde-3-phosphate dehydrogenase** reaction of glycolysis. Malate is translocated into the mitochondrial matrix, and α-ketoglutarate is transported out. Inside the mitochondrion, malate dehydrogenase catalyzes a reaction of substrate with NAD^+ to yield oxaloacetate and NADH + H^+. The latter serves as reductant for the respiratory chain and leads to the formation of 2.5 mol of ATP per mol of NADH. Two NADH equivalents in the cytosol generated from 1 mol of glucose (2 mol of triose phosphate) will yield 5 mol of ATP.

If a malate-oxaloacetate exchange protein occurred, the shuttle would be much less complex. Such a system, however, does not exist, and additional processes are necessary to reestablish the initial conditions. **Oxaloacetate** (a C_4 compound derived from malate) reacts with glutamate to yield aspartate (C_4) and α-ketoglutarate. **Aspartate** is transported in exchange for glutamate. To resume the initial conditions, external aspartate (C_4) reacts with α-ketoglutarate to yield oxaloacetate (C_4) and glutamate. The biochemical machinery for the malate-aspartate shuttle exists in the liver and heart.

Brain and muscle contain a different shuttle, the **glycerol phosphate shuttle**, for transporting reducing equivalents into the mitochondrion. The components include a cytosolic glycerol phosphate dehydrogenase and an inner mitochondrial membrane glycerol phosphate dehydrogenase (see Fig. 6-13). NADH converts dihydroxyacetone phosphate to glycerol phosphate in the cytosol. Glycerol phosphate flows through the outer mitochondrial membrane and donates electrons to the glycerol phosphate dehydrogenase imbedded within the inner mitochondrial membrane with the attendant formation of dihydroxyacetone phosphate. The electrons are passed from the dehydrogenase to coenzyme Q, and 1.5 mol of ATP are formed via oxidative phosphorylation from each mol of glycerol phosphate. Dihydroxyacetone phosphate is able to recycle, or it is metabolized by its usual reactions.

ATP YIELD FROM GLUCOSE METABOLISM

We have seen that the conversion of 1 mol of glucose to 2 mol of pyruvate during glycolysis results in the net formation of two ATP equivalents. Consider the energy yield after the complete oxidation of glucose by the Krebs cycle and oxidative phosphorylation. We can also determine the yield of ATP from selected intermediates. We noted that 1.0 mol of intramitochondrial NADH yields 2.5 mol of ATP. FAD-containing enzymes such as succinate dehydrogenase, which donate reducing equivalents into the respiratory chain at the level of coenzyme Q, yield 1.5 mol of ATP per mol of coenzyme QH_2. Table 6-3 summarizes the various reactions in the catabolism of glucose. A net yield of 32 ATP equivalents is associated with the use of the malate-aspartate shuttle. A net yield of 30 ATP equivalents is associated with the use of the glycerol phosphate shuttle. We can also deduce that 1.0 mol of pyruvate yields 12.5 mol of ATP, and 1 mol of acetyl-CoA yields 10 mol of ATP. You should verify the correctness of these values.

TABLE 6-3. ATP Molecules Generated During the Complete Oxidation of One Molecule of Glucose by Glycolysis with the Malate-Aspartate Shuttle (Glycerol Phosphate Shuttle), Citric Acid Cycle Reactions, and Oxidative Phosphorylation

Process or Reaction	ATP Yield
Glycolysis (glucose → 2 pyruvate)	2
2 NADH from glyceraldehyde-3-phosphate dehydrogenase and malate (or glycerol phosphate) shuttle	5 (3)
Pyruvate dehydrogenase (2 NADH)	5
Isocitrate dehydrogenase (2 NADH)	5
α-Ketoglutarate dehydrogenase (2 NADH)	5
Succinate thiokinase (2 substrate level)	2
Succinate dehydrogenase	3
Malate dehydrogenase (2 NADH)	5
Total	**32 (30) ATP yield per hexose**
Individual yields	
NADH (1)	2.5
$FADH_2$ (1)	1.5
Acetyl-CoA (1)	10.0
Pyruvate (1)	12.5

Acetyl-CoA, acetylcoenzyme A; $FADH_2$, reduced flavin adenine dinucleotide; NADH, reduced nicotinamide-adenine dinucleotide.

ADDITIONAL ASPECTS OF CARBOHYDRATE METABOLISM

Before the catabolism of lipids and amino acids are considered, some additional aspects of carbohydrate metabolism are covered, including the pathways for glycogen formation (**glycogenesis**) and glycogen degradation (**glycogenolysis**). In the next sections, the pentose phosphate pathway and gluconeogenesis are examined. **Gluconeogenesis** refers to the pathway for glucose biosynthesis from pyruvate, lactate, citric acid cycle intermediates, amino acids, and glycerol.

Glycogenesis

Glycogen serves as a storage form of carbohydrate. It is a polymer of glucose residues and is found in all cells except mature erythrocytes. The major stores of glycogen occur in **liver** and **muscle**. Muscle glycogen is a source of fuel for muscle contraction. Glycogen is a branched, treelike molecule with a high molecular weight (up to 1 million daltons). The straight chain portions are composed of α-1,4-glycosidic bonds, and the branch points occur at α-1,6 bonds. Branches occur at approximately every tenth residue (Fig. 6-14). The sugar monomers, due to their high –OH content, are highly hydrated and extended in space; therefore, glycogen is not an optimal storage form for energy (as compared with triglyceride). Moreover, lipids have a higher energy potential than sugars.

Glycogen biosynthesis begins with **glucose 6-phosphate**. Phosphoglucomutase (PGM) catalyzes the isoergonic conversion of glucose 6-phosphate to glucose 1-phosphate. The next reaction is designed to produce an activated high-energy form of glucose for biosynthesis, which is UDPG. **UDPG pyrophosphorylase** catalyzes a reaction between uridine triphosphate (UTP) and glucose 1-phosphate to produce UDPG and inorganic pyrophosphate (PP_i). This reaction is isoergonic.

The only known metabolic fate of **PP_i** in humans is hydrolysis to yield two P_i molecules; the reaction is catalyzed by a ubiquitous and separate enzyme named **inorganic pyrophosphatase**. Hydrolysis of PP_i is exergonic and physiologically irreversible. The formation of PP_i and its hydrolysis represents one mechanism for pulling reactions forward. This is a rather general and noteworthy principle of metabolism. Many reactions are associated with the so-called **pyrophosphate split**, and invariably, the bioenergetics and principles are those enunciated here. The glycosidic bond between a sugar and

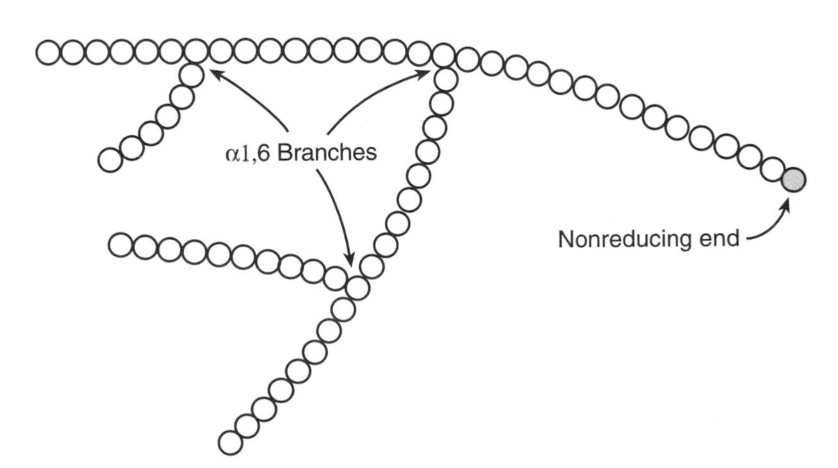

Figure 6-14. Branched structure of glycogen.

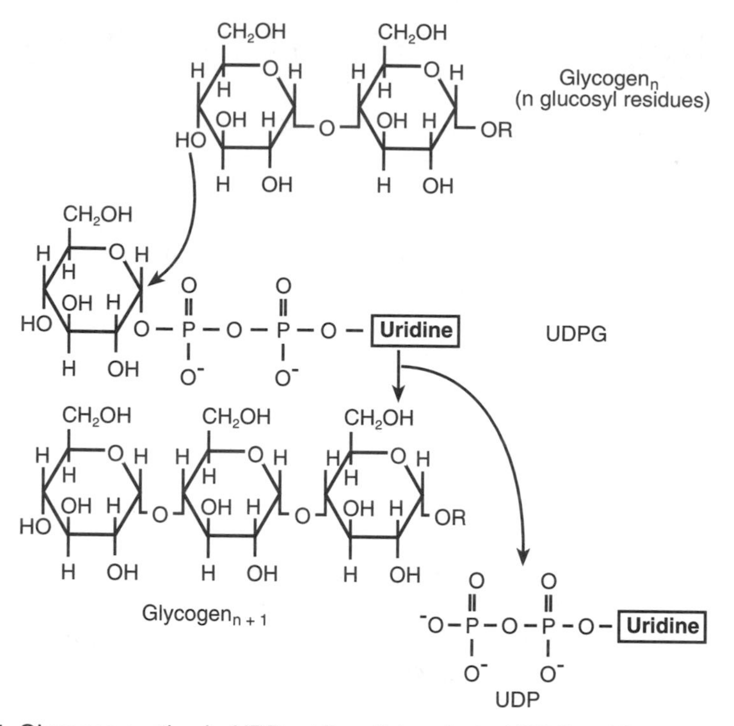

Figure 6-15. Glycogen synthesis. UDP, uridine diphosphate; UDPG, uridine diphosphoglucose.

pyrophosphate as found in UDPG is energy rich with a standard free energy of hydrolysis of –7 kcal/mol.

Two enzyme activities are required for **glycogen biosynthesis**. The first (glycogen synthase) is responsible for the formation of linear chains, and the second (branching enzyme) is responsible for the formation of branch points. **Glycogen synthase** catalyzes the reaction between glycogen$_n$ containing n glycosyl residues and UDPG to yield glycogen$_{n+1}$ and UDP ("n" refers to an arbitrary number of sugar moieties). The high-energy bond of UDPG is converted into a low-energy glycosidic bond, and the reaction is exergonic and physiologically irreversible (Fig. 6-15). After 12 to 16 glucosyl residues are added distal to a branch point, then a **branching enzyme** transfers a block of six or more residues to yield a new branch. Both ends of the branch can now be elongated in reactions catalyzed by glycogen synthase. Hepatic glycogen synthesis occurs postprandially from the glucose substrate transported by the hepatic portal vein.

Glycogenolysis

Two enzymes are necessary for glycogenolysis (the degradation or lysis of glycogen). **Glycogen phosphorylase** (usually called **phosphorylase**) catalyzes a reaction between P_i and glycogen to yield glucose 1-phosphate and glycogen$_{n-1}$ (Fig. 6-16). This phosphorolysis (lysis by phosphate) reaction occurs at α-1,4-glycosidic bonds and is modestly exergonic. Note that this is *not* a hydrolysis reaction. An additional noteworthy property of phosphorylase is that it contains pyridoxal phosphate as cofactor. Phosphorylase reactions occur until a glucose residue approximately four residues from a branch point is reached. Then a single protein with two enzymatic activities, called **debranching enzyme**, mediates the elimination of the branch. A glucosyltransferase activity moves three glucosyl residues as a block, leaving a single glucose in α-1,6 linkage at a branch point. The glycosyl group is added elsewhere to extend a straight chain with the α-1,4 bond (Fig. 6-17). Then the debrancher catalyzes the hydrolytic

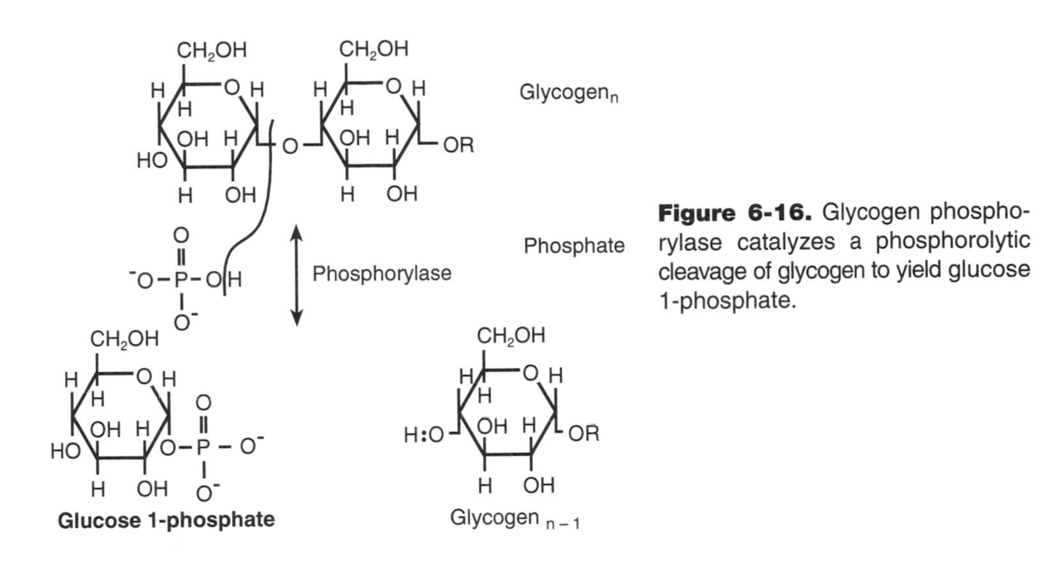

Figure 6-16. Glycogen phosphorylase catalyzes a phosphorolytic cleavage of glycogen to yield glucose 1-phosphate.

removal of glucose at the branch point to yield free glucose and the remainder of the glycogen molecule (Fig. 6-18).

Most of the glucosyl residues from glycogen are released as phosphate esters by the phosphorylase reaction. The resulting glucose 1-phosphate is converted into glucose 6-phosphate by phosphoglucomutase. Approximately 10% of the residues are hydrolytically released as free glucose by the debranching reaction. For metabolism, free glucose must be phosphorylated by ATP to form glucose 6-phosphate in a reaction catalyzed by hexokinase or glucokinase. Several inborn errors of metabolism, called **glycogen storage diseases**, are known. Their names and associated enzyme deficiencies are given in Table 6-4.

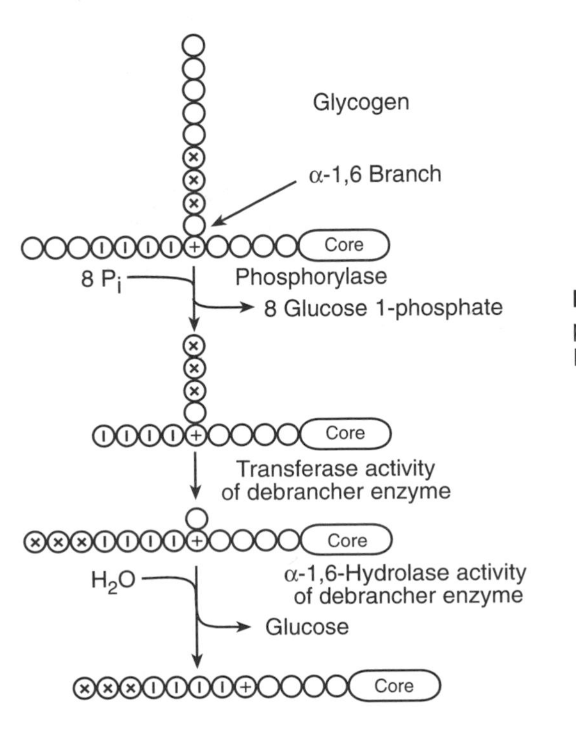

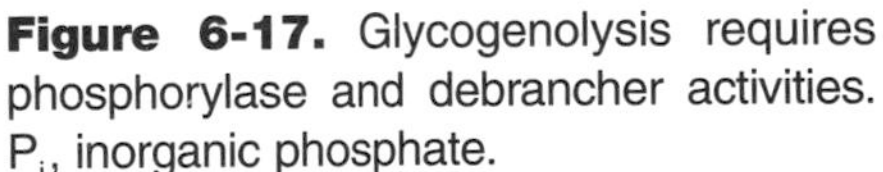
Figure 6-17. Glycogenolysis requires phosphorylase and debrancher activities. P_i, inorganic phosphate.

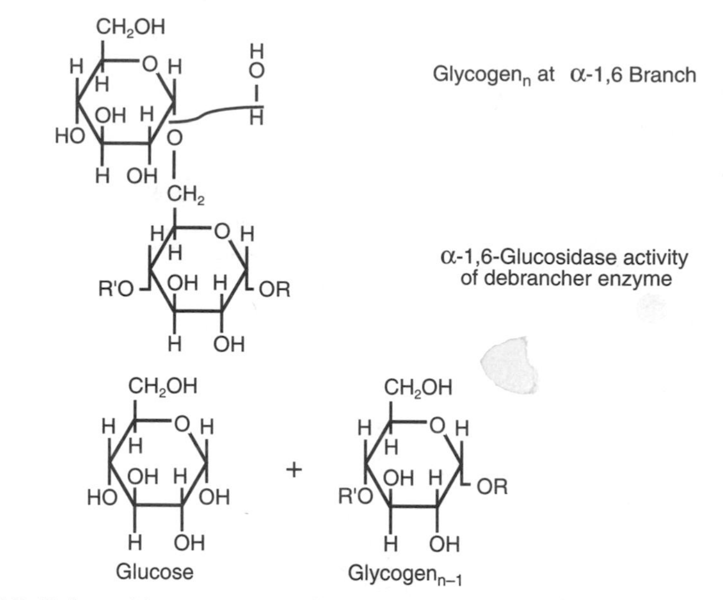

Figure 6-18. Debranching enzyme catalyzes a hydrolysis reaction, yielding free glucose.

Regulation of Glycogenesis and Glycogenolysis

The biosynthesis and degradation of glycogen are the result of distinct enzyme-catalyzed reactions. This allows for enhancement of the activity of one process with concomitant inhibition of the other. The mechanism of **hyperglycemia** is discussed in response to the secretion of glucagon. Glucagon activates its receptor found in the liver cell membrane and triggers a series of events known as a **second messenger cascade** (Fig. 6-19). The activated receptor interacts with a stimulatory guanine nucleotide-binding (G_s) protein. G_s is composed of three subunits (α, β, and γ). The α subunit exchanges GTP for GDP and $G_s\alpha$/GTP dissociates from the β/γ complex, and $G_s\alpha$/GTP interacts with adenylyl cyclase and activates it. Adenylyl cyclase catalyzes the formation of cyclic AMP and PP_i from ATP. Cyclic AMP then activates a cognate protein kinase (cyclic AMP–dependent protein kinase or protein kinase A). The following equation describes this activation:

TABLE 6-4. Glycogen Storage Diseases

Type	Name	Deficiency	Comments
I	von Gierke's	Glucose-6-phosphatase	Liver and kidney have increased glycogen of normal structure
II	Pompe's	α-1 → 4 Glucosidase	Lysosomal disease
III	Cori's	Debrancher	Highly branched glycogen
IV	Andersen's	Brancher	Sparsely branched glycogen
V	McArdle's	Muscle phosphorylase	Proved role of glycogen synthase in glycogenesis
VI	—	Liver phosphorylase	—

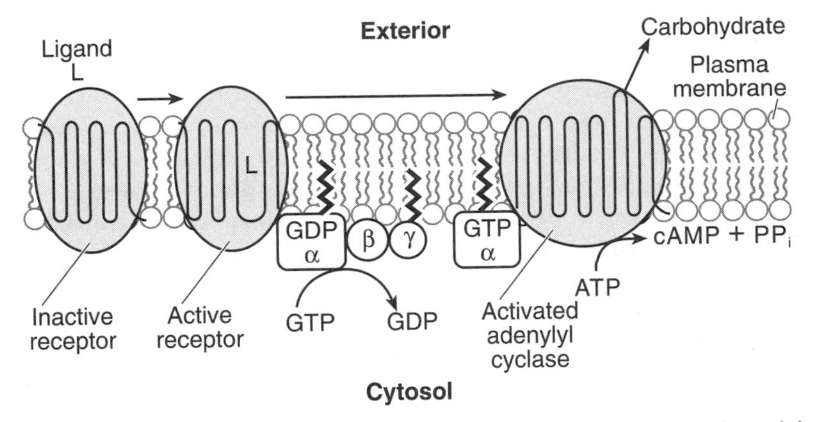

Figure 6-19. Regulation of adenylyl cyclase by G proteins. ATP, adenosine triphosphate; cAMP, cyclic adenosine monophosphate; GDP, guanosine diphosphate; GTP, guanosine triphosphate; PP_i, inorganic pyrophosphate.

$$\underset{\text{(less active)}}{R_2C_2 + 4\ \text{cyclic AMP}} \rightleftharpoons \underset{\text{(more active)}}{2C + R_2 - \text{cyclic AMP}_4}$$

R designates a regulatory subunit, and C designates a catalytic subunit.

First, consider how **cyclic AMP–dependent protein kinase** activates glycogenolysis, and then consider how this enzyme inhibits glycogenesis. The catalytic subunit of a cyclic AMP–dependent protein kinase catalyzes the phosphorylation of a second protein kinase called **phosphorylase kinase**. The activity of phosphorylase kinase is thereby increased. Phosphorylase kinase enzyme now catalyzes the phosphorylation of the enzyme (glycogen) phosphorylase. The phosphorylated enzyme, called **phosphorylase *a***, is the more active form. The unphosphorylated enzyme is called **phosphorylase *b*** and is less active. Phosphorylase *a* then catalyzes the degradation of glycogen. Consider the role of glycogenolysis *in vivo*. The major metabolite, glucose 1-phosphate, is converted into glucose 6-phosphate, which can in turn be hydrolyzed to glucose and released into the bloodstream. The enzyme catalyzing this reaction is glucose-6-phosphatase, which is present in liver and kidney. All other tissues lack glucose-6-phosphatase. The liver is the most important organ in releasing carbohydrate into the bloodstream as glucose. Other organs, except the kidney, are unable to do so because they lack glucose-6-phosphatase.

An enhancement of the activity of the degradative enzyme represents one side of the coin, and the mechanism that enables glucagon to decrease the rate of biosynthesis of liver glycogen is less intricate than that involved in the regulation of glycogenolysis. The regulatory scheme parallels that described previously in that an activation of cyclic AMP–dependent protein kinase occurs. The inhibition of the chief biosynthetic enzyme, namely **glycogen synthase**, is produced in a direct fashion without the intermediacy of another protein kinase. The activated cyclic AMP–dependent enzyme catalyzes the direct phosphorylation of glycogen synthase.

After phosphorylation, glycogen synthase is less active. The phosphorylated, less active form is dependent on **glucose 6-phosphate** for its activity and is called the **D-form**. The unphosphorylated form is independent of glucose 6-phosphate and is the **I-form**. As in the case of phosphorylase, regulation of glycogen synthase depends on a combination of covalent and allosteric effectors.

When the stimulus subsides, consider the processes that reestablish the initial state of the enzymes. Decreasing glucagon concentration decreases receptor occupation.

The **α-subunit of G_s** catalyzes the hydrolysis of GTP to GDP and P_i. **G_sα/GDP** then recombines with the βγ-complex and adenylyl cyclase activity returns to unstimulated or basal levels. A cyclic AMP phosphodiesterase catalyzes the hydrolysis and destruction of cyclic AMP by converting it to 5'-AMP. A decrease in cyclic AMP concentration favors the reassociation of the catalytic and regulatory subunits of cyclic AMP–dependent protein kinase, rendering the enzyme less active. Both phosphorylase kinase and glycogen synthase are regenerated from their phosphorylated forms after hydrolytic reactions catalyzed by phosphoprotein phosphatase–1. There is a family of protein phosphatase enzymes that participate in the many phosphorylation-dephosphorylation reactions in biochemistry, including phosphoprotein phosphatases –1, –2A, –2B, and –2C.

To summarize, two enzymes are required for glycogen biosynthesis, and two are required for its degradation. The synthetic enzymes are **glycogen synthase** and **brancher**; the catabolic enzymes are **phosphorylase** and **debrancher**. Glucose 1-phosphate is the product of phosphorylase, and free glucose results from the action of debrancher. Cyclic AMP and its cognate protein kinase enhance glycogenolysis by protein phosphorylation. Phosphorylase kinase is the target enzyme. Phosphorylase kinase then catalyzes the ATP-dependent phosphorylation and activation of phosphorylase. Cyclic AMP–dependent protein kinase catalyzes the direct phosphorylation of glycogen synthase, rendering it less active. Protein phosphorylation can change the activity of the target enzyme in the positive or negative direction—the result depends on the enzyme.

Other mechanisms can play a role in glycogen metabolism, and these mechanisms react to different physiologic stimuli and work through different effector systems. Elevated **5'-AMP** activates phosphorylase *b* allosterically. 5'-AMP serves as a signal that ATP levels are low. The **glucose 1-phosphate** resulting from phosphorylase activation serves as a fuel for ATP formation. Glucose per se also decreases phosphorylase activity in an allosteric fashion and thereby inhibits glycogenolysis. Higher blood glucose levels stimulate the release of insulin from the β-cells of the pancreas. **Insulin** promotes glucose transport from the extracellular space into muscle and adipose tissue; however, insulin has no direct effect on glucose transport into liver or brain cells. At least seven glucose translocases have been identified, and GLUT4 is the insulin responsive transporter. Insulin may increase glycogen synthase activity in liver cells by enzyme induction.

Chapter 15 discusses hormone function and regulation and how **insulin** works through a molecular mechanism distinct from the cyclic AMP–dependent second messenger cascade. **Glucagon** release, on the other hand, is enhanced at low blood glucose concentrations. Glucagon interacts with its specific receptor in the liver cell membrane, which in turn activates **adenylyl cyclase**. The receptors of many hormones interact with G-proteins that stimulate or inhibit adenylyl cyclase activity or that activate phospholipase C. These receptors, such as those for glucagon and norepinephrine, have seven transmembrane segments.

Gluconeogenesis

The supply of hepatic glycogen is limited, and other fuels must be used to maintain normal blood glucose levels after an overnight fast. **Gluconeogenesis** is the process responsible for converting lactate (produced by red blood cells, muscle, and other tissues), glycerol (produced from lipolysis or triglyceride catabolism), pyruvate, and intermediates of the tricarboxylic acid cycle (derived from amino acid catabolism) into glucose. Lactate is also produced by muscle in anaerobic conditions, and functions as part of the Cori cycle mentioned under Anaerobic Glycolysis. Muscle releases alanine into the blood, which serves as a substrate for gluconeogenesis.

Gluconeogenesis occurs in the **liver** and to a lesser extent in the **kidney**. The pathway is absent in muscle, heart, and brain tissue and in other organs. Although muscle and heart can use other metabolic fuels, such as free fatty acids and ketone bodies, the

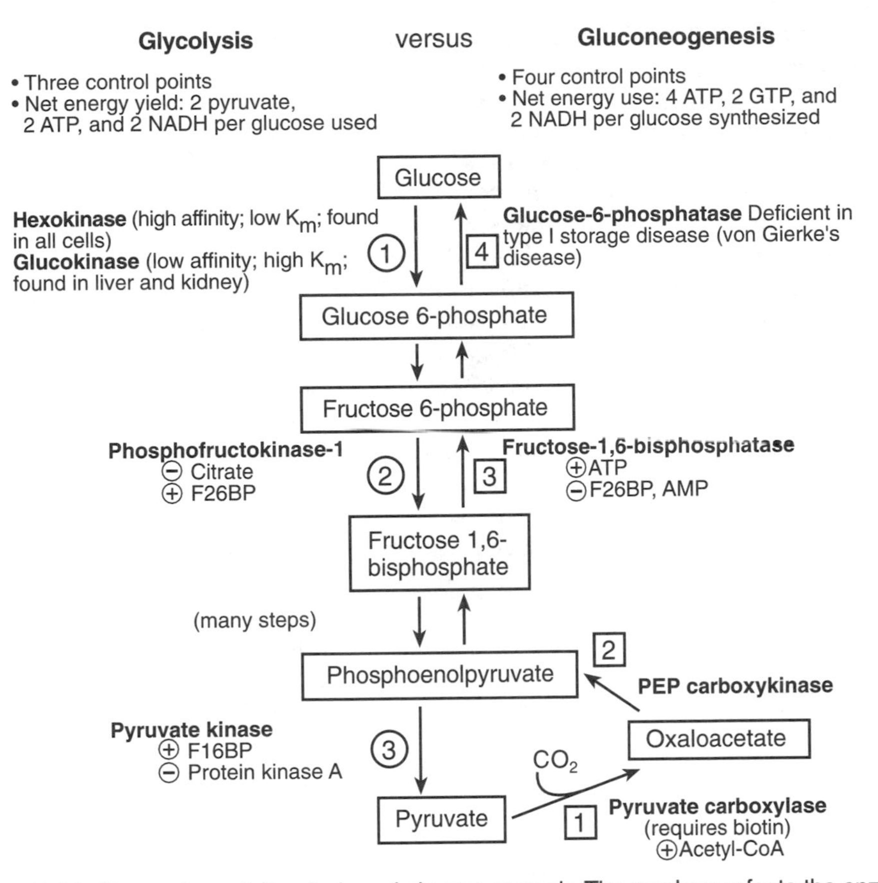

Figure 6-20. Comparison of glycolysis and gluconeogenesis. The numbers refer to the enzymes unique to glycolysis (three numbers in circles) and gluconeogenesis (four numbers in squares). Acetyl-CoA, acetylcoenzyme A; AMP, adenosine monophosphate; ATP, adenosine triphosphate; GTP, guanosine triphosphate; K_m, Michaelis constant; NADH, reduced nicotinamide adenine dinucleotide; PEP, peptidase.

brain is completely dependent on circulating glucose, and it is for this reason that adequate blood glucose levels must be maintained.

Consider first the **conversion of pyruvate to glucose**. As noted earlier, eight enzyme-catalyzed reactions are shared by glycolysis and gluconeogenesis. Three reactions of glycolysis are highly exergonic and biochemically irreversible. These are the reactions catalyzed by hexokinase, PFK, and pyruvate kinase. The process of gluconeogenesis can be more easily understood if we consider how these three irreversible reactions are bypassed (Fig. 6-20).

First, the **conversion of pyruvate to PEP** is examined. The very large negative free energy of hydrolysis of PEP (–14.8 kcal/mol) is an indication that PEP is very energy rich. The direct phosphorylation of pyruvate by ATP does not occur to a physiologically important extent (because the free energy of hydrolysis of ATP is only –7.3 kcal/mol). To overcome the thermodynamic barrier, nature has used a two-step pathway. **Pyruvate carboxylase** catalyzes a reaction between ATP, pyruvate, and bicarbonate to yield oxaloacetate, ADP, and P_i. The pyruvate carboxylase reaction occurs within the mitochondrion. **PEP carboxykinase** catalyzes a reaction between oxaloacetate and GTP to yield PEP, GDP, and P_i. Two high-energy bonds ($\Delta G^{\circ\prime}$ of hydrolysis of –7.3 kcal/mol) are expended to form the energy-rich linkage of PEP ($\Delta G^{\circ\prime}$ = –14.8 kcal/mol). Mitochondrial PEP is translocated into the cytosol in exchange for P_i.

An alternative route is operative for the conversion of pyruvate and Krebs cycle intermediates into phosphoenolpyruvate. **Pyruvate**, which is translocated into the mitochondrion from the cytosol, is converted to **oxaloacetate** by the pyruvate carboxylase reaction. As noted in the section Metabolite Transport across the Inner Mitochondrial Membrane, oxaloacetate per se cannot be transported across the inner mitochondrial membrane. Oxaloacetate is converted to **malate**, transported outside the mitochondrion, and oxidized to yield **oxaloacetate and NADH**. This is the route followed by pyruvate and Krebs cycle intermediates, which, unlike lactate, do not generate cytosolic NADH. Transport of malate, however, results in the translocation of reducing equivalents to the cytosol that can be used in gluconeogenesis. Oxaloacetate is converted to PEP in a reaction catalyzed by cytosolic PEP carboxykinase. In humans, the activities of mitochondrial and cytosolic PEP carboxykinase are approximately equal.

Next, the enzymes of glycolysis convert phosphoenolpyruvate to fructose 1,6-bisphosphate. Circumventing the PFK reaction, **fructose-1,6-bisphosphatase** catalyzes the hydrolysis of this compound to yield fructose 6-phosphate and P_i; this is an exergonic reaction and is unidirectional. After a reaction catalyzed by phosphohexose isomerase, **glucose-6-phosphatase** catalyzes the hydrolysis of its substrate to yield glucose and P_i; the reaction is unidirectional.

The stoichiometry for the **gluconeogenesis pathway** is given by the following formula:

$$\begin{array}{ccc} \text{2 Pyruvate} & & \text{Glucose} \\ + & & + \\ \text{2 NADH} + 2\,H^+ & & \text{2 NAD}^+ \\ + & & + \\ \text{2 GTP} & \rightarrow & \text{2 GDP} \\ + & & + \\ \text{2 ATP} & & \text{2 ADP} \\ + & & + \\ 4\,H_2O & & 4\,P_i \end{array}$$

Note that four high-energy bonds are expended and that 2 mol of reduced NADH are required.

Consider the **regulation of gluconeogenesis** (see Fig. 6-20). When glucose and insulin levels are low, there is an increase in catabolism of fatty acids in the liver. This is accompanied by an increase in the concentrations of citrate and acetyl-CoA. **Acetyl-CoA** is required for pyruvate carboxylase activity (the first step in gluconeogenesis). Acetyl-CoA also inhibits pyruvate dehydrogenase by activating a pyruvate dehydrogenase kinase. Furthermore, citrate inhibits PFK and decreases catabolism by glycolysis.

Fructose 2,6-bisphosphate is another allosteric regulator of glycolysis and gluconeogenesis; it interacts with PFK and fructose-1,6-bisphosphatase (see Fig. 6-20). Fructose 2,6-bisphosphate activates PFK and inhibits fructose-1,6-bisphosphatase in liver. Fructose 2,6-bisphosphate promotes glycolysis. Under conditions favoring gluconeogenesis, the concentration of fructose 2,6-bisphosphate declines. This decrease removes a stimulus for PFK and an inhibitor of fructose-1,6-bisphosphatase.

Fructose 2,6-bisphosphate is formed in a reaction involving fructose 6-phosphate and ATP; ADP is the other product (Fig. 6-21). Fructose 2,6-bisphosphate is degraded by hydrolysis to form fructose 6-phosphate and P_i. A single protein contains both kinase and phosphatase activities, which catalyze the formation and degradation of fructose 2,6-bisphosphate. The individual activities, however, are mutually exclusive—that is, when the kinase is on, the phosphatase is off, and vice versa. The phosphorylation of kinase/phosphatase by cyclic AMP–dependent pro-

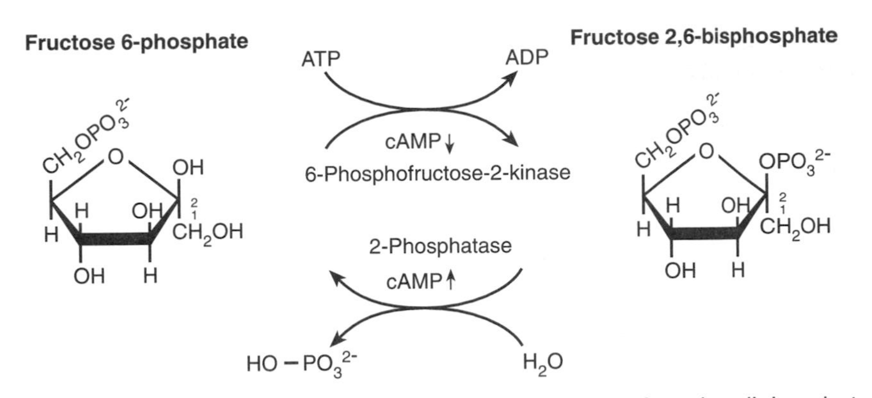

Figure 6-21. Metabolism of fructose 2,6-bisphosphate. ADP, adenosine diphosphate; ATP, adenosine triphosphate; cAMP, cyclic adenosine monophosphate.

tein kinase increases the phosphatase activity and decreases the fructose 6-phosphate 2-kinase activity.

Kinase/phosphatase phosphorylation occurs in conditions of low plasma glucose and insulin or high glucagon levels. A high glucagon/insulin ratio occurs in diabetes mellitus, and the chief mechanism responsible for the hyperglycemia of diabetes is enhanced gluconeogenesis. Agents that elevate cyclic AMP levels in the liver promote gluconeogenesis, an observation that can be used to rationalize the reciprocal effects and levels of fructose 2,6-bisphosphate.

The simultaneous operation of two opposing pathways, such as glycolysis and gluconeogenesis, is called a **futile cycle**. The net transformation of metabolites fails to occur, and if unregulated, substances are sequentially synthesized and degraded to no apparent advantage. Futile cycles are associated with the loss of ATP, illustrated as follows:

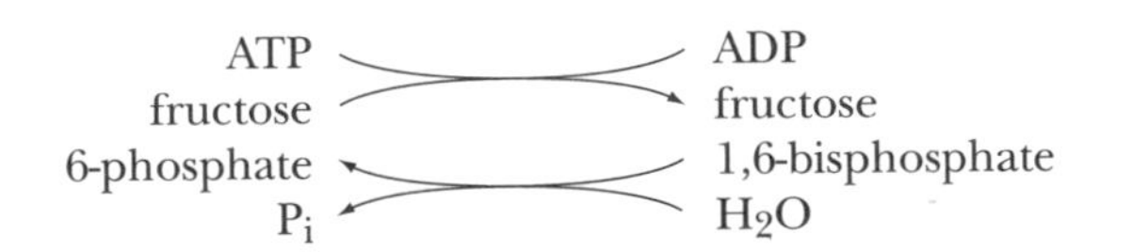

Regulatory mechanisms apparently operate to minimize futile cycles *in vivo.* This is very dramatically illustrated by the **glycolytic and gluconeogenic pathways** (see Fig. 6-20). Thinking once again about the metabolic Big Picture (see Fig. 1-1), when nutrients are high, glycolysis and energy use are activated. When nutrients are low, however, the liver shuts down glycolysis and activates gluconeogenesis so that this organ can synthesize and release glucose for use in the rest of the body. Remember that the liver and kidney are unique in their capacity to supply glucose *de novo* for the rest of the body; therefore, there is logic in the design of the regulatory mechanisms. Fructose 2,6-bisphosphate activates glycolysis and inhibits gluconeogenesis. Similarly, energy indices of nutrient environment (ATP and AMP concentrations) have reciprocal effects on the unique steps in glycolysis and gluconeogenesis.

Consider the feeder pathways for other substances that serve as substrates for gluconeogenesis. Lactate is converted to pyruvate by the **lactate dehydrogenase** reaction in one step, as considered previously. **Alanine** is converted into pyruvate by alanine aminotransferase in one step and is considered in Chapter 8. Alanine is one of the most important substrates for human gluconeogenesis and is mobilized from muscle. Amino acids, which can be converted into TCA cycle intermediates, also serve as substrates for gluconeogenesis. They can be converted to malate in the Krebs cycle, and the subsequent reactions of gluconeogenesis follow.

Gluconeogenesis is important in maintaining blood glucose levels. Rare **inborn errors of metabolism** can lead to hypoglycemia and lactic acidosis when enzymes of gluconeogenesis are deficient or absent. Such enzyme defects can involve pyruvate carboxylase, fructose-1,6-bisphosphatase, and glucose 6-phosphatase. A deficiency of glucose-6-phosphatase leads to type I glycogen storage disease, known as **von Gierke's disease**.

The β-oxidation of fatty acids containing an even number of carbon atoms (the most common case) yields **acetyl-CoA.** In humans, acetyl-CoA cannot lead to a net increase in glucose and other carbohydrates because of the unidirectional nature of the TCA cycle. After the reaction with oxaloacetate to yield citrate, two carbon atoms are eliminated during the conversion to malate or oxaloacetate (see Fig. 6-8). Because of the lack of a net increase in the number of carbon atoms in the resulting metabolite, acetyl-CoA cannot serve as a source for the net production of carbohydrate. This is the explanation for the aphorism that, in humans, fat (fatty acids) cannot be converted to carbohydrate.

Pentose Phosphate Pathway

In addition to the Embden-Meyerhof glycolytic pathway, the **Warburg-Dickens pentose phosphate pathway** represents another scheme for the metabolism of glucose 6-phosphate. The function of the pathway is twofold.

First, this pathway is responsible for the production of biosynthetic reducing equivalents such as **NADPH.** One of the noteworthy principles of human biochemistry is that NADPH is used in biosynthetic reactions, and NADH is produced in catabolic reactions. NADPH is important in fatty acid and steroid biosynthesis. The tissues and organs that exhibit high activities of the enzymes of the pentose phosphate pathway include the liver (an important organ for fatty acid and cholesterol biosynthesis), adrenal cortex (steroid hormone biosynthesis), and lactating mammary gland (lipid biosynthesis).

The second major function of the pentose phosphate pathway is the generation of **five-carbon sugars**. These occur in nucleotides such as ATP, NAD^+, $NADP^+$, FAD, CoA, RNA, and DNA. NADPH is produced concomitantly with pentose phosphates. The requirement for NADPH, however, is generally much greater than that of the pentose phosphates. The pentose phosphate pathway is the sole source of these important cellular reagents. Enzymes exist that convert the pentose phosphates to triose phosphate and glucose 6-phosphate to salvage the five-carbon sugar derivatives. The pentose phosphate pathway is a very flexible one and allows for the interconversion of many carbohydrate intermediates. The components of the pentose phosphate pathway and glycolysis are located in the cytosol (see Table 1-1).

The pentose phosphate pathway can be divided into two portions: The first is the **oxidative segment**, which is associated with NADPH production, and the second is the **nonoxidative segment**, which is associated with the interconversion of several pairs of sugar phosphates (Fig. 6-22). These range from trioses to heptoses. A simplified stoichiometry of the pathway is given:

$$\text{3 Glucose 6-phosphate} + 6\,NADP^+ \rightarrow \text{2 Fructose 6-phosphate} + 3\,CO_2 + \text{1 Glyceraldehyde 3-phosphate} + 6\,NADPH + 6\,H^+$$

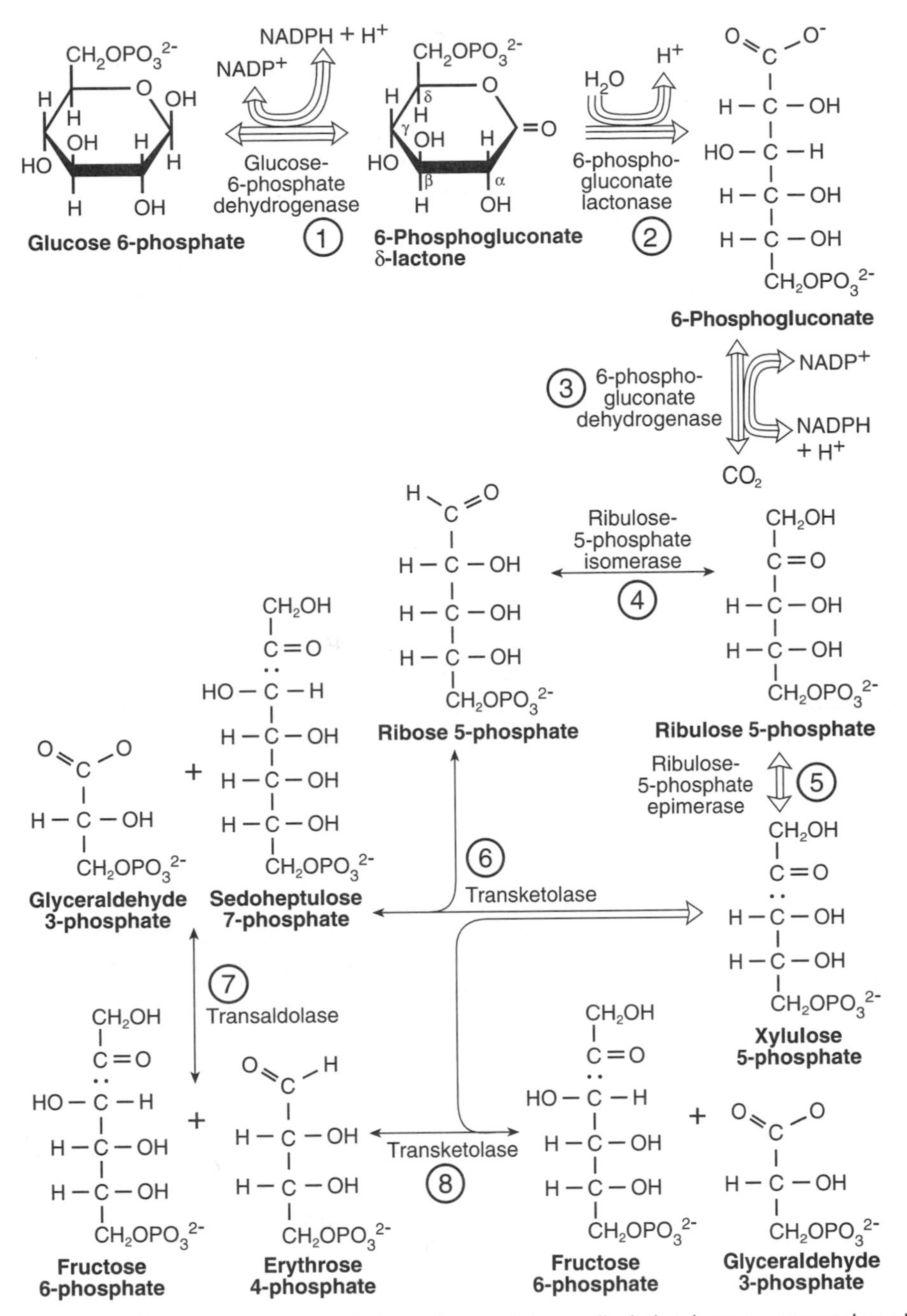

Figure 6-22. The pentose phosphate pathway (also called the hexose monophosphate shunt). The circled numbers represent the sequential reactions and flow of carbon atoms. $NADP^+$, oxidized form of nicotinamide adenine dinucleotide phosphate; NADPH, reduced nicotinamide adenine dinucleotide phosphate.

First, consider the oxidative segment. **Glucose-6-phosphate dehydrogenase** catalyzes a reaction between substrate and $NADP^+$ to form 6-phosphogluconolactone and NADPH + H^+ in a reversible fashion. Next, a specific **lactonase** catalyzes the hydrolysis of the lactone to yield 6-phosphogluconate in an irreversible hydrolytic reaction. **6-Phosphogluconate dehydrogenase** catalyzes an oxidative decarboxylation yielding **ribulose 5-phosphate**, carbon dioxide, NADPH, and H^+. Two mol of NADPH are generated per hexose phosphate. This concludes the oxidative segment.

The nonoxidative portion of the pentose phosphate pathway involves four enzymes: Two enzymes catalyze specific reactions, and two enzymes catalyze general reactions. The specific enzymes include a **3-epimerase**, which mediates the reversible interconversion of ribulose 5-phosphate and xylulose 5-phosphate. The second is a **keto isomerase**, which catalyzes the reversible interconversion of ribulose 5-phosphate (a ketopentose) and ribose 5-phosphate (an aldopentose). The two general enzymes are **transaldolase** and **transketolase**. Transaldolase contains an essential lysine residue that mediates the transfer of a three-carbon fragment from a donor to an acceptor. Transketolase contains **thiamine pyrophosphate** as an essential cofactor and mediates the transfer of a two-carbon ketonyl fragment to an appropriate acceptor. From the reactants and products of a reaction, you can deduce whether the enzyme is a transketolase (two-carbon transfer) or a transaldolase (three-carbon transfer). The association of vitamin deficiency and alcoholism has prompted an examination of the properties of transketolase in such individuals. Initial studies suggest that the activity of transketolase is altered in alcoholism, but additional research is required to verify this idea.

Next, consider the pathway for the transformation of 3 mol of **ribulose 5-phosphate** (15 carbons) to 2 mol of **fructose 6-phosphate** (12 carbons) and 1 mol of **glyceraldehyde 3-phosphate** (3 carbons). Two mol of ribulose 5-phosphate are transformed into xylulose 5-phosphate by the 3-epimerase, and 1 mol of ribulose 5-phosphate is transformed into ribose 5-phosphate by keto isomerase. Then **transketolase** transfers a two-carbon fragment from xylulose 5-phosphate to ribose 5-phosphate to form sedoheptulose 7-phosphate and glyceraldehyde 3-phosphate. **Transaldolase** then operates on these two substrates and transfers a three-carbon fragment from sedoheptulose 7-phosphate to glyceraldehyde 3-phosphate, yielding fructose 6-phosphate and erythrose 4-phosphate.

Fructose 6-phosphate can be metabolized by glycolysis or converted to glucose 6-phosphate and metabolized by the pentose phosphate pathway. Transketolase then catalyzes the transfer of a two-carbon fragment from xylulose 5-phosphate to erythrose 4-phosphate to yield fructose 6-phosphate and glyceraldehyde 3-phosphate. Both compounds are intermediates for glycolysis or gluconeogenesis and will be further metabolized, depending on metabolic need. In particular, the fructose 6-phosphate residues are readily converted to glucose 6-phosphate molecules that can reenter the pentose phosphate pathway or be diverted to the biosynthesis of glycogen.

Despite extensive investigation, knowledge of the mechanism of regulation of the pentose phosphate pathway is incomplete. Most evidence favors the $NADP^+$ to NADPH ratio as the most important parameter. A high ratio promotes pentose phosphate cycle activity at the glucose 6-phosphate dehydrogenase step.

The pentose phosphate pathway plays an important role in maintaining the mature erythrocyte. The pentose phosphate pathway provides NADPH for the reduction of oxidized glutathione. **Glutathione** is a tripeptide (γ-glutamylcysteinylglycine). Glutathione exists as the reduced (G-SH) and oxidized forms (G-S-S-G). **Glutathione reductase** converts the oxidized to the reduced form:

$$\begin{array}{ccc} \text{GSSG} & & \text{2 GSH} \\ + & \rightarrow & + \\ \text{NADPH} + \text{H}^+ & & \text{NADP}^+ \end{array}$$

Reduced glutathione is a substrate for **glutathione peroxidase** (selenium is a cofactor; see Table 2-1):

$$2\ \text{GSH} + H_2O_2 \rightarrow \text{GSSG} + 2\ H_2O$$

Glutathione peroxidase plays an important role in destroying H_2O_2 in the erythrocyte. Recall that the mature erythrocyte lacks membranous cellular organelles, such as peroxisomes, which are found in nucleated cells. Catalase in peroxisomes mediates the conversion of hydrogen peroxide to water and molecular oxygen ($H_2O_2 \rightarrow H_2O + \frac{1}{2} O_2$). Glutathione peroxidase functions in the red blood cell to destroy hydrogen peroxide as catalase functions in other cells. A deficiency of glucose-6-phosphate dehydrogenase is associated with drug-induced hemolytic anemia produced by chloroquine and primaquine (antimalarial agents) and other substances. Hemolysis may be related to deficient NADPH production for the glutathione peroxidase reaction.

Chapter 7

Lipid Metabolism

FATTY ACID METABOLISM

The important classes of lipids in humans include triglycerides (triacylglycerols), phospholipids, and steroids. **Triglyceride** serves as the main storage form of metabolic fuel in humans and is the most concentrated form of metabolic energy (9 kcal/g) (Fig. 7-1). Moreover, triglyceride, which is stored in an anhydrous state, represents an average of 19% of the total body mass (the percentage varies considerably among individuals from a minimum of approximately 9%). It can subserve humans for weeks or months of starvation. In contrast, stored **carbohydrate** is depleted in a day or so and must be replenished by gluconeogenesis. Glycogen is extensively hydrated, which increases its bulk. A gram of glycogen tissue contains an equivalent amount of bound water. Glycogen does not represent as efficient an energy storage form as does triglyceride. Amino acids and proteins are not stored to any appreciable extent in humans.

The structure of the **fatty acids** (number of carbons and degree of unsaturation; Table 7-1) dramatically affects the physical characteristics of the compounds as well as the triglycerides formed from them; for instance, the longer the side chain, the higher the melting temperature. This is also true of saturated versus unsaturated fatty acids. An 18-carbon fatty acid containing a single double bond (oleic acid) has a melting temperature more than 50°C lower than that of a saturated fatty acid of the same size (palmitate). This is due to the inability of the kinked fatty acid to pack into regular arrays and can have a dramatic effect on the fluid nature of a given membrane depending on its fatty acid composition.

Fatty acids stored in adipose tissue are released from triglycerides by hydrolysis reactions catalyzed by triglyceride lipase. This enzyme is also called **hormone-sensitive lipase** because it responds to circulating epinephrine by undergoing phosphorylation by protein kinase A and concomitant activation (see the second messenger cascade model presented in Fig. 6-19). Hormone-sensitive lipase, moreover, is inhibited by the action of insulin. The liberated fatty acids, bound to albumin, are transported in the circulation, and the fatty acids are taken up by most organs or tissues (except the brain).

Fatty Acid Activation

After entering the cells, the fatty acids must be derivatized as thioesters with coenzyme A (CoA) before metabolism. A family of **fatty acyl-coenzyme A synthetases** and a single **pyrophosphatase** catalyze the following reactions in the cytosol:

$$\text{Fatty acid} + \text{ATP} + \text{coenzyme A} \rightleftharpoons \text{fatty acyl-CoA} + \text{AMP} + PP_i$$

$$PP_i + H_2O \rightarrow 2\ P_i$$

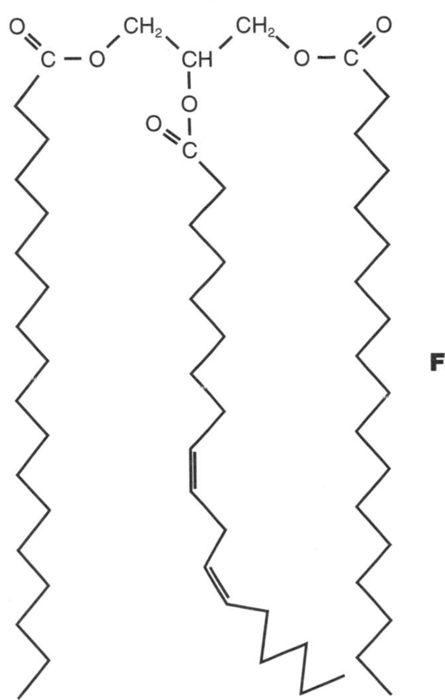

Figure 7-1. Structure of a triacylglycerol, or triglyceride.

TABLE 7-1. Some Naturally Occurring Fatty Acids

Carbon Atoms	Structure	Systematic Name	Common Name (Acid)	Melting Point (°C)
Saturated fatty acids				
12	$CH_3(CH_2)_{10}COOH$	*n*-Dodecanoic	Lauric	44.2
14	$CH_3(CH_2)_{12}COOH$	*n*-Tetradecanoic	Myristic	53.9
16	$CH_3(CH_2)_{14}COOH$	*n*-Hexadecanoic	Palmitic	63.1
18	$CH_3(CH_2)_{16}COOH$	*n*-Octadecanoic	Stearic	69.6
20	$CH_3(CH_2)_{18}COOH$	*n*-Eicosanoic	Arachidic	76.5
24	$CH_3(CH_2)_{22}COOH$	*n*-Tetracosanoic	Lignoceric	86.0
Unsaturated fatty acids				
16	$CH_3(CH_2)_5CH{=}CH(CH_2)_7COOH$	—	Palmitoleic	–0.5
18	$CH_3(CH_2)_7CH{=}CH(CH_2)_7COOH$	—	Oleic	13.4
18	$CH_3(CH_2)_4CH{=}CHCH_2CH{=}$ $CH(CH_2)_7COOH$	—	Linoleic	–5.0
18	$CH_3CH_2CH{=}CHCH_2CH{=}$ $CHCH_2CH{=}CH(CH_2)_7COOH$	—	Linolenic	–11.0
20	$CH_3(CH_2)_4CH{=}CHCH_2CH{=}$ $CHCH_2CH{=}CHCH_2CH{=}$ $CH(CH_2)_3COOH$	—	Arachidonic	–49.5

The first part of the reaction involves the formation of an intermediate **fatty acyladenylate and inorganic pyrophosphate (PP_i)**. In the second part of the reaction, CoA displaces adenylate to yield **fatty acyl-CoA and adenosine monophosphate (AMP)**. The family of fatty acyl-CoA synthetases differs in chain length specificity. Long-chain (16 carbon atoms or more), intermediate-chain (six to 14 carbon atoms), and short-chain (two to four carbon atoms) specific enzymes exist. Thioesters are energy-rich bonds. To pull the reaction forward, pyrophosphatase (a separate enzyme) catalyzes the hydrolysis of PP_i to yield two inorganic phosphate (P_i) molecules in an exergonic and unidirectional reaction.

As noted previously, CoA and its derivatives do not pass through the inner mitochondrial membrane. To effect the transfer of fatty acids into the mitochondrial matrix (the main site of fatty acid oxidation; see Table 1-1), the formation of fatty acylcarnitine is required. This isoergonic process is catalyzed by **carnitine acyltransferase I**:

$$\text{Fatty acyl-CoA} + \text{carnitine} \rightleftharpoons \text{fatty acylcarnitine} + \text{coenzyme A}$$

The fatty **acylcarnitine** is transported into mitochondria in exchange for carnitine. Once inside the mitochondrion, carnitine acyltransferase II catalyzes the reverse reaction, yielding fatty acyl-CoA and free carnitine. **Carnitine** ($[CH_3]_3N^+CH_2CH[OH]CH_2COO^-$) forms an ester linkage through its β-OH group with the carboxyl group of the fatty acid.

β-Oxidation

The conversion of fatty acyl-CoA to acetyl-CoA requires the action of four enzymes. The stoichiometry for the conversion of stearoyl-CoA (C_{18}) to 9 mol of acetyl-CoA is shown in the following equation:

$$\text{Stearoyl-CoA} + 8\,\text{FAD} + 8\,\text{NAD}^+ + 8\,\text{CoA} \rightarrow 9\,\text{acetyl-CoA} + 8\,\text{FADH}_2 + 8\,\text{NADH} + 8\,\text{H}^+$$

The cyclic pathway for β-oxidation is illustrated in Fig. 7-2. An **acyl-CoA dehydrogenase** catalyzes an isoergonic oxidation to yield *trans*-enoyl-CoA and reduced flavin adenine dinucleotide ($FADH_2$). The reducing equivalents are donated to the electron transport chain at the level of coenzyme Q and provide energy for the formation of 1.5 mol of adenosine triphosphate (ATP) per mol of $FADH_2$. Next, **enoyl-CoA hydratase** catalyzes an isoergonic hydration yielding an L-β-hydroxyacyl–CoA. The resulting compound is oxidized by nicotinamide-adenine dinucleotide (NAD^+) in a reaction catalyzed by **hydroxyacyl-CoA dehydrogenase** to give β-ketoacyl-CoA, reduced nicotinamide-adenine dinucleotide (NADH), and H^+. Oxidation of 1.0 mol NADH by the respiratory chain yields 2.5 mol of ATP. Next, **β-ketothiolase** (thiolase) catalyzes an exergonic cleavage or thiolysis by CoA to yield acetyl-CoA and a fatty acyl-CoA lacking the two carbon atoms. For stearoyl-CoA, the β-oxidation spiral occurs eight times.

This series of reactions in β-oxidation can be more easily remembered and understood by comparison with three analogous reactions in the Krebs cycle. The pattern of the first oxidation in β-oxidation by **flavin adenine dinucleotide (FAD)** is analogous to the succinate dehydrogenase reaction. The hydration reaction in β-oxidation parallels that catalyzed by **fumarase**. Finally, the oxidation of an alcohol by **NAD^+** to form a ketone parallels that of the malate dehydrogenase reaction. Note that the three reactions in the citric acid cycle and these three reactions in β-oxidation occur with similar chemistry. This is an example of the conservation of form and function in nature.

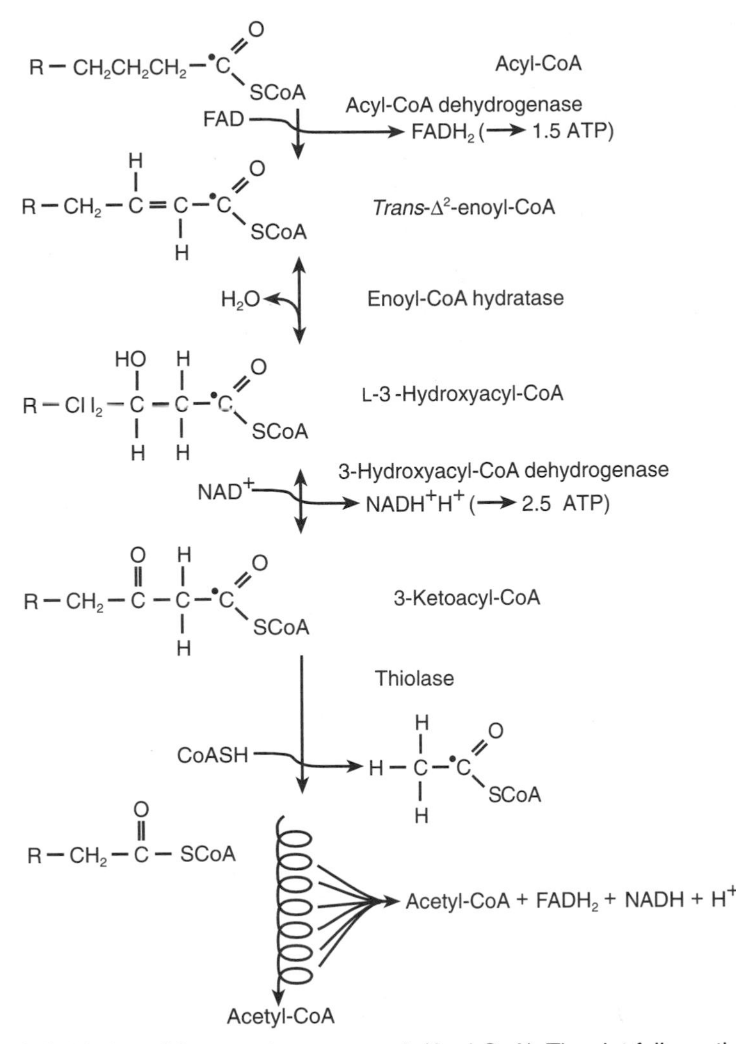

Figure 7-2. β-Oxidation of fatty acyl-coenzyme A (Acyl-CoA). The dot follows the fate of the terminal carbon in the acyl-CoA molecule. ATP, adenosine triphosphate; CoA, coenzyme A; CoASH, uncombined coenzyme A; FAD, flavin adenine dinucleotide; $FADH_2$, reduced flavin adenine dinucleotide; NAD^+, oxidized form of nicotinamide adenine dinucleotide; NADH, reduced form of nicotinamide adenine dinucleotide; SCoA, derivatized coenzyme A.

Next, consider the number of mol of **ATP** formed from adenosine diphosphate (ADP) and P_i as a result of the β-oxidation of the 18-carbon stearic acid and reactions of the Krebs cycle and oxidative phosphorylation. The value is shown in Table 7-2. The net production of ATP takes into account the requirement that the fatty acid must be converted to a CoA derivative with the expenditure of two high-energy bonds (ATP + 2 $H_2O \rightarrow$ AMP + 2 P_i).

Peroxisomes also possess a pathway for β-oxidation. There are several differences in the peroxisomal and mitochondrial processes. Peroxisomes can initiate the oxidation of fatty acids that are longer than 18-carbon atoms, but mitochondria cannot; moreover, carnitine is not required for the translocation of fatty acids into the peroxisome. The first oxidation step in the peroxisome differs from that of the mitochondrion in that oxygen (not FAD) is the oxidant. The products include the *trans* α,β-unsaturated **fatty acyl-CoA** and **hydrogen peroxide**. Hydrogen peroxide is degraded to oxygen and water in a reaction that is catalyzed by catalase. After the fatty acid is degraded to an acyl-CoA less

TABLE 7-2. ATP Yield from the Complete Oxidation of Stearic Acid (C18)

Process	ATP Resulting	Totals
8 $FADH_2$	8.0 × 1.5	12
8 NADH + 8 H^+	8.0 × 2.5	20
9 Acetyl-CoA	9.0 × 10.0	+90
		122
ATP required for stearyl-CoA formation*		−2
Net ATP production		120

$ATP \rightarrow AMP + 2\ P_i$

Acetyl-CoA, acetylcoenzyme A; AMP, adenosine monophosphate; ATP, adenosine triphosphate; $FADH_2$, reduced flavin adenine dinucleotide; NADH, reduced nicotinamide adenine dinucleotide; P_i, inorganic phosphate; stearyl-CoA, stearyl-coenzyme A.

than 18-carbon atoms, the metabolite is translocated into the mitochondrion (as the carnitine derivative), and complete β-oxidation occurs.

Metabolism of Propionyl-CoA

A very small percentage of fatty acids in the human diet contain an odd number of carbon atoms. These fatty acids undergo β-oxidation and yield propionyl-CoA (propionate = C_3) and acetyl-CoA (acetate = C_2) after the final thiolytic cleavage. **Propionyl-CoA** is also produced during the catabolism of several amino acids (valine, isoleucine, and methionine). The pathway for the metabolism of propionyl-CoA is quantitatively more important for amino acid metabolism than for the metabolism of the uncommon odd-chain fatty acids. Propionyl-CoA metabolism requires the participation of vitamin B_{12}, and this is a distinctive property.

Propionyl-CoA carboxylase catalyzes a reaction between substrate, bicarbonate, and ATP to yield D-methylmalonyl-CoA, ADP, and P_i (Fig. 7-3). This ATP-dependent carboxylation reaction, like several others, involves a biotin prosthetic group. An **epimerase** catalyzes an isoergonic conversion of D-methylmalonyl-CoA to the L-isomer. **Methylmalonyl-CoA mutase** (a vitamin B_{12}–dependent enzyme) catalyzes the conversion of L-methylmalonyl-CoA to succinyl-CoA. The carboxylic acid is transferred to the carbon marked with the asterisk to yield the final product in Fig. 7-3. Succinyl-CoA is an intermediate in the Krebs cycle and is metabolized by familiar reactions. The association of **vitamin B_{12}** with **methylmalonyl-CoA mutase** is noteworthy.

Metabolism of Unsaturated Fatty Acids

The metabolism of fatty acids with double bonds beginning at even-numbered carbon atoms, such as the Δ^{12} position of linoleate, requires two additional enzyme activities not required for saturated fatty acid metabolism: **2,4-dienoyl-CoA reductase** and **enoyl-CoA isomerase**. A Δ^2-*trans*, Δ^4-*cis* double bond results from the action of acyl-CoA dehydrogenase on linoleate. 2,4-Dienoyl-CoA reductase catalyzes a reaction of the substrate with **reduced nicotinamide adenine dinucleotide phosphate (NADPH)** to form a Δ^3-*trans* derivative. Note that this is an unusual case in which NADPH is required for catabolism. Enoyl-CoA isomerase catalyzes the con-

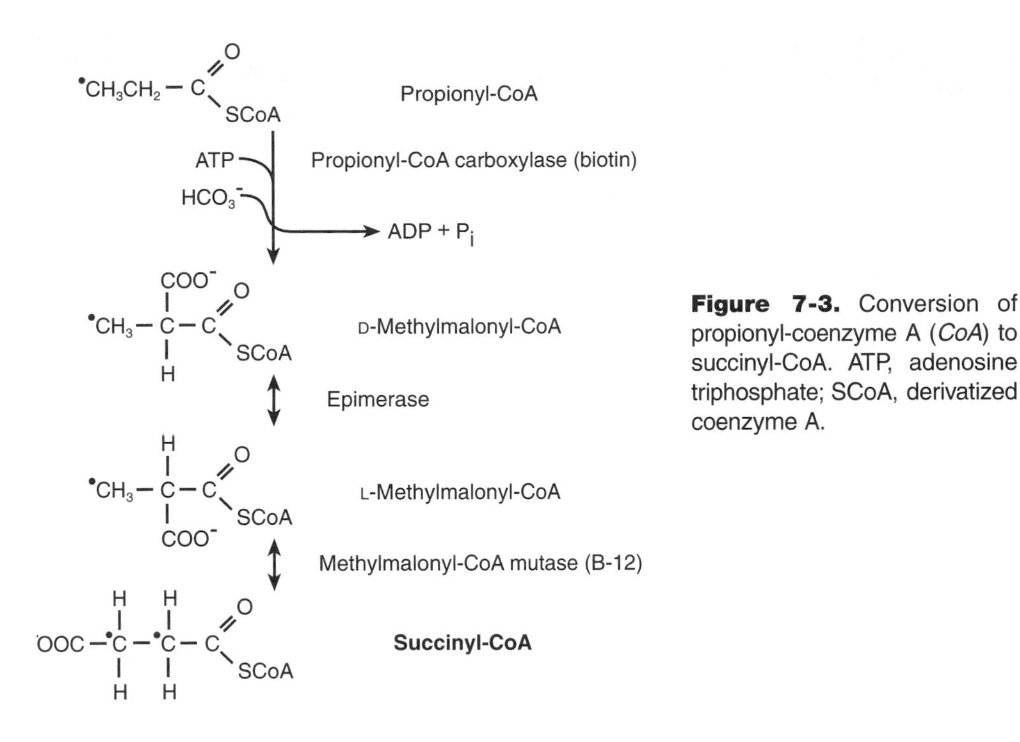

Figure 7-3. Conversion of propionyl-coenzyme A (*CoA*) to succinyl-CoA. ATP, adenosine triphosphate; SCoA, derivatized coenzyme A.

version of the Δ^3-*trans* derivative to the Δ^2-*trans* derivative, a substrate for the hydratase of β-oxidation.

The metabolism of fatty acids with double bonds beginning at odd-numbered carbon atoms, such as the Δ^9 position of linoleate, requires **3,5-dienoyl-CoA isomerase**, a third enzyme. A Δ^2-*trans*, Δ^4-*cis* metabolite of linoleate is acted on by enoyl-CoA isomerase to form a Δ^3-*trans*, Δ^5-*cis* derivative. This substrate for 3,5-dienoyl-CoA isomerase is converted to a Δ^2-*trans*, Δ^4-*trans* metabolite, which is in turn a substrate for 2,4-dienoyl-CoA reductase. The resulting Δ^3-*trans* compound is a substrate for enoyl-CoA isomerase, and β-oxidation can proceed.

Metabolic Defects in Fatty Acid Oxidation

There are several **acyl-CoA dehydrogenases** that differ in the chain length of their substrates, and metabolic defects in several of these have been characterized. The enzymes consist of very-long-chain (C_{20} to C_{14}), long-chain (C_{18} to C_{12}), medium-chain (C_{12} to C_4), and short-chain (C_6 to C_4) acyl-CoA dehydrogenases.

Deficiency of the **medium-chain enzyme** is the most common **inborn error of metabolism** involving these enzymes. Defects in muscle carnitine palmitoyltransferase II, electron transfer flavoprotein, long-chain 3-hydroxyacyl-CoA dehydrogenase, the mitochondrial carnitine/acylcarnitine translocase, plasma membrane carnitine translocase, and muscle short-chain 3-hydroxyacyl-CoA dehydrogenase also exist. When acyl-CoAs cannot be metabolized, these compounds accumulate. This accumulation leads to a decrease in CoA levels and to a decrease in carnitine levels as acyl-CoAs and acyl-carnitines build up (leading to a depletion of free CoA and carnitine). The acyl-containing compounds undergo hydrolysis, and some of them are metabolized by β-oxidation to produce increased amounts of dicarboxylic acids, which accumulate in tissues and are excreted in the urine.

The presence of **dicarboxylic acid** metabolites in the urine can be helpful in diagnosing and deciphering enzyme defects. Fatty acid oxidation yields acetyl-CoA, a substrate for ketone body synthesis. Defects in fatty acid oxidation are not accompanied by ketosis. Acidosis without ketosis is commonly observed in the diseases involving defects in fatty acid oxidation.

TABLE 7-3. Comparison of Fatty Acid Oxidation and Synthesis

Fatty Acid	Oxidation	Synthesis *de novo*
Acetyl-CoA required	+	+
Malonyl-CoA required	–	+
NADH	+	–
NADPH	+ (unsaturated fatty acids)	+
CoA	+	+
ACP (acyl carrier protein)	–	+
Mitochondria	+	–
Cytosol	–	+
Multienzyme complex	–	+
L-Hydroxyacyl intermediate	+	–
D-Hydroxyacyl intermediate	–	+

Acetyl-CoA, acetylcoenzyme A; CoA, coenzyme A; Malonyl-CoA, malonylcoenzyme A; NADH, reduced nicotinamide adenine dinucleotide; NADPH, reduced nicotinamide adenine dinucleotide phosphate; +, positive; –, negative.

Fatty Acid Biosynthesis

Although fatty acid biosynthesis can be considered to be the reverse of β-oxidation, fatty acid anabolic and catabolic pathways do not share common reversible steps, unlike gluconeogenesis and glycolysis. Rather, different substrates and different molecular machinery are used (Table 7-3). The reactions of fatty acid synthesis occur in the cytosol.

The reactions of fatty acid biosynthesis take place on a **fatty acid synthase multienzyme complex**. A priming reaction using acetyl-CoA initiates the process. Then, malonyl-CoA adds successive two-carbon fragments to the primer. After each addition, four reactions convert a β-ketoacyl compound to the reduced acyl derivative. NADPH serves as a reductant in keeping with its role as a central component in anabolic pathways. The fatty acid synthase contains **acyl carrier protein** (ACP). This is a small protein containing covalently bound **4'-phosphopantetheine**. The latter constitutes part of the molecular structure of CoA, and ACP can be considered to represent protein-bound CoA. ACP carries covalently linked intermediates from the various active sites necessary for biosynthesis. The –SH group of ACP is called the **central thiol**. An enzymic cysteine constitutes a **peripheral thiol**. After the 16-carbon palmitoyl group is formed, free palmitate is released by hydrolysis.

Consider the pathway for palmitate biosynthesis. Acetyl-CoA initiates the process by reacting with the multienzyme complex to yield an acetyl-enzyme thioester intermediate involving the peripheral thiol group. This acetyl group contributes carbon atoms 15 and 16 of the 16-carbon palmitate. They are at the omega (ω) end of the fatty acid—that is, the farthest removed from the carboxyl group. The other 14-carbon atoms are contributed by malonyl-CoA. The formation of malonyl-CoA, as catalyzed by **acetyl-CoA carboxylase**, is illustrated by the following reaction:

$$\text{ATP} + HCO_3^- + \text{acetyl-CoA} \rightleftharpoons \text{malonyl-CoA} + \text{ADP} + P_i$$

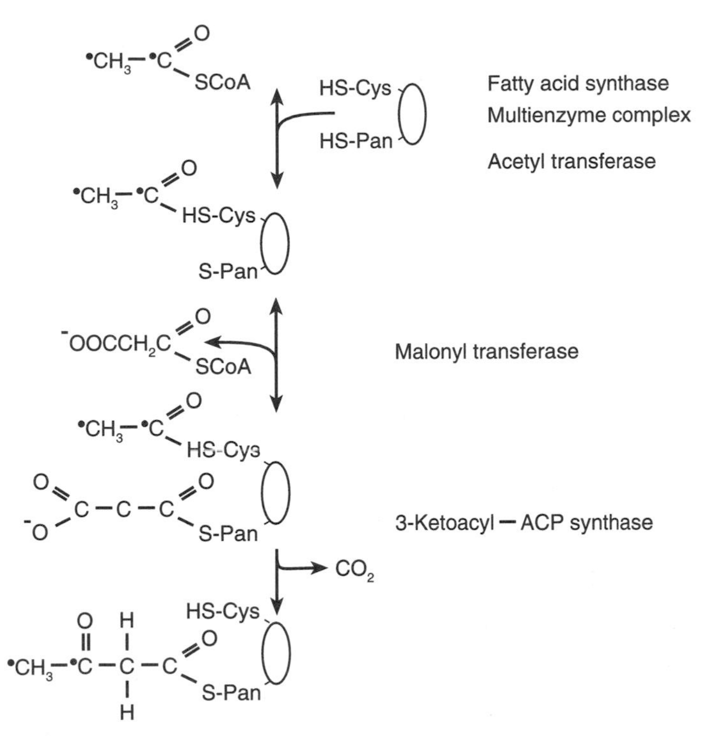

Figure 7-4. Pathway for fatty acid biosynthesis. The dots follow the fates of the initiating acetyl-CoA carbons. ACP, acyl carrier protein; HS-Cys, peripheral cysteinyl thiol; Pan, central pantetheinyl thiol; SCoA, derivatized coenzyme A.

Acetyl-CoA carboxylase is the rate-limiting step in fatty acid biosynthesis and is activated by citrate. Malonyl-CoA reacts with the acetyl-fatty acid synthase complex to yield CoA and the malonyl group bound to the central thiol of ACP (Fig. 7-4).

The sequential series of reactions leading to fatty acid synthesis begins as the α-carbon of malonyl-ACP attacks the carbonyl group of acetyl-CoA in a condensation reaction catalyzed by **3-ketoacyl synthase** (see Fig. 7-4). A β-ketoacyl group is attached to the central thiol as the peripheral thiol is freed, and carbon dioxide is displaced. This decarboxylation renders the process exergonic. The exergonic decarboxylation accounts for the formation and use of the malonyl group; the carboxyl group activates the tail of the acetyl group within the malonyl CoA for this condensation reaction. A **3-ketoacyl reductase** catalyzes a reaction with NADPH + H^+ to form the D-β-hydroxyacyl–ACP and $NADP^+$. A **dehydratase** catalyzes the elimination of water, yielding a 2,3-unsaturated acyl-ACP. An **enoyl reductase** catalyzes a reaction with NADPH + H^+ to yield the acyl-ACP. The acyl group, now elongated by two carbons, is transferred to the peripheral thiol. Next, malonyl-CoA reacts with the acyl-enzyme, and the malonyl group is linked to the central thiol. This is followed by another condensation reaction resulting in the formation of a β-ketoacyl group elongated by two carbon atoms, and carbon dioxide is the other product. The sequence of reactions is repeated as a **reduction, dehydration**, and **second reduction** occur. The acyl group is transferred to the peripheral thiol.

A total of seven condensation reactions is required to produce the **palmitoyl enzyme**. A **thioesterase** catalyzes the hydrolytic cleavage and release of palmitate, and the free enzyme is regenerated. That fatty acids longer than palmitate cannot be synthesized by the complex is probably related to a limitation in the size of one of the active sites of the complex. In a figurative sense, the fatty acid synthase complex counts to 16 carbons and releases palmitate.

Fatty acid metabolism is very stringently regulated. As alluded to earlier, the control of fatty acid synthesis occurs at the synthesis of malonyl-CoA with **acetyl-CoA carboxylase**

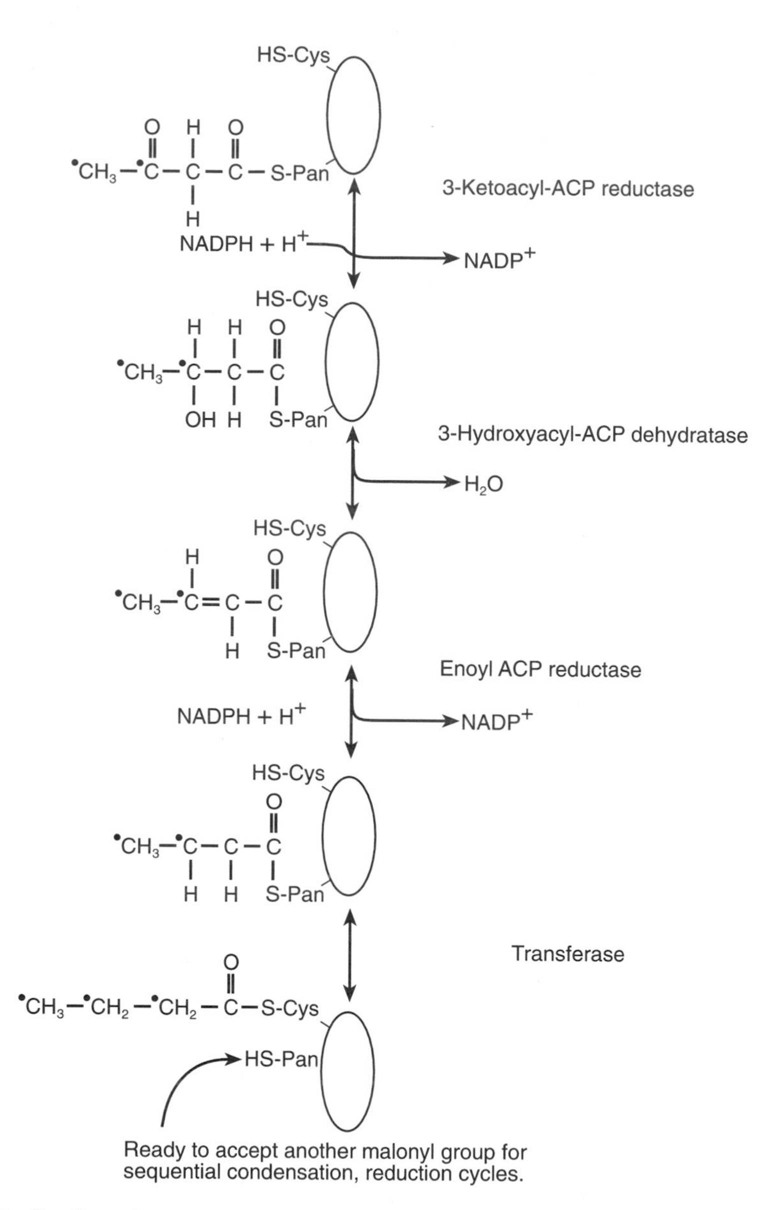

Figure 7-4. *Continued*

serving as the gatekeeper. This should be intuitively obvious. Looking once again at the metabolic Big Picture (see Fig. 1-1), you can see that acetyl-CoA occupies a crossroads in the metabolic road map. The cell must decide whether to send this metabolite into the Krebs tricarboxylic acid (TCA) cycle yielding of NADH and ATP, or to divert it to the biosynthesis of fatty acids and to cholesterol and steroids. As with most regulatory steps, this decision is based on the energy and lipid status of the cell (Fig. 7-5). If fatty acids are high, palmitoyl-CoA acts as a sensor and shuts off further fatty acid synthesis. If energy levels are low (i.e., there are high AMP concentrations), the carboxylase is inhibited. Conversely, citrate serves as an indicator that energy stores are high and diverts acetyl-CoA to fatty acid synthesis by stimulating the carboxylase. Malonyl-CoA inhibits carnitine acyltransferase I and prevents the translocation of fatty acids into the mitochondrion, thereby inhibiting β-oxidation. **Insulin** serves as a signal for high glucose and diverts acetyl-CoA to fatty acid synthesis; insulin induces acetyl-CoA carboxylase by increasing transcription of its gene and increases total enzyme activity.

A comparison of **fatty acid oxidation** and **synthesis** is given in Table 7-3. The separate pathways and intracellular localization (see Table 1-1) permit independent reg-

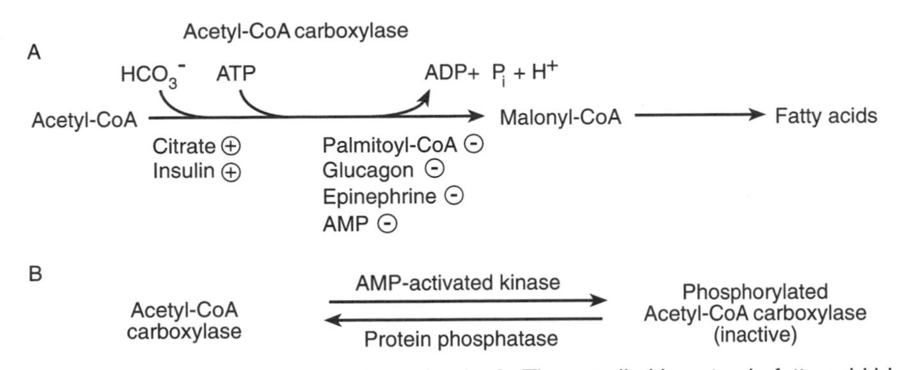

Figure 7-5. Regulation of fatty acid biosynthesis. **A:** The rate-limiting step in fatty acid biosynthesis is the conversion of acetylcoenzyme A (*CoA*) into the building block malonyl-CoA. This reaction is catalyzed by acetyl-CoA carboxylase, an enzyme heavily regulated by feedback inhibitors and allosteric modulators tied to the energy needs of the body. **B:** Adenosine monophosphate (*AMP*; an indication of reduced energy stores and therefore *decreased* need for fatty acid synthesis) activates a protein kinase that inactivates acetyl-CoA carboxylase. ADP, adenosine diphosphate; ATP, adenosine triphosphate; P_i, inorganic phosphate.

ulation of the processes. Although carbon dioxide as bicarbonate is required for malonyl-CoA formation, the added carbonyl group is released during the condensation reactions, and bicarbonate is not a direct source of the carbon atoms in fatty acids. All the carbon atoms are derived from **acetyl-CoA**: The two on the ω-end are derived directly from **acetyl-CoA**, and the others are derived from **malonyl-CoA**.

Provision of NADPH and Acetyl-CoA

The source of NADPH and acetyl-CoA used as substrates for palmitate biosynthesis varies. In the liver, which is the predominant triglyceride-synthesizing organ in humans, most of the NADPH is derived from the pentose phosphate pathway. In adipocytes, approximately half the NADPH results from the reaction catalyzed by **malic enzyme** (NADP-malate dehydrogenase):

$$\text{Malate} + \text{NADP}^+ \rightarrow \text{pyruvate} + \text{carbon dioxide} + \text{NADPH} + \text{H}^+$$

The producers and consumers of NADPH occur within the cytosol, so that no barrier to effective use occurs.

The provision of **cytosolic acetyl-CoA**, however, requires transport across the mitochondrial membrane. Glucose serves as the source of carbon atoms for much of fatty acid biosynthesis in liver. Glucose is converted to pyruvate by glycolysis. **Pyruvate** is transported into the mitochondrion by its specific translocase and is converted to acetyl-CoA in a reaction catalyzed by pyruvate dehydrogenase. CoA and its derivatives are not directly transportable through the inner membrane. The transport of acetyl groups occurs via citrate. **Citrate** is formed by the usual Krebs cycle reaction involving acetyl-CoA, oxaloacetate, and water. Citrate is transported through the inner mitochondrial membrane in exchange for malate. Once in the cytosol, **ATP-citrate lyase** (citrate cleavage enzyme) catalyzes a reaction with substrate, CoA, and ATP to yield acetyl-CoA, oxaloacetate, ADP, and P_i. **Acetyl-CoA** then functions as a precursor for fatty acid formation. **Oxaloacetate** is reduced to malate and can be transported into the mitochondrion in exchange for citrate.

The provision of cytosolic acetyl-CoA by mitochondrial citrate provides a likely explanation for the regulation of the Krebs cycle at the isocitrate dehydrogenase reaction. This allows for the biosynthesis of citrate and still permits the regulation of the Krebs cycle by ADP at a distal step—that is, at the isocitrate dehydrogenase reaction.

Also note that cytosolic citrate activates the **acetyl-CoA carboxylase** reaction (the rate-limiting reaction in fatty acid biosynthesis). Citrate serves both as a precursor and feed-forward activator of fatty acid biosynthesis.

TRIGLYCERIDE BIOSYNTHESIS

The precursors for triglyceride, or **triacylglycerol**, synthesis include 3 mol of fatty acyl-CoA and glycerol phosphate. The palmitate produced by synthesis *de novo* is thioesterified as palmitoyl-CoA in a reaction catalyzed by fatty acyl-CoA synthetase. Fatty acid, ATP, and CoA are reactants, and fatty acyl-CoA, AMP, and PP_i are products. **Palmitoyl-CoA** may also be elongated, desaturated, or both before incorporation into triglyceride. Dietary **fatty acids**, which are derived from triglyceride, are a major precursor of stored triglyceride. Glycerol phosphate can be obtained by the reduction of dihydroxyacetone phosphate catalyzed by **glycerol phosphate dehydrogenase** or by the ATP-dependent phosphorylation of glycerol as catalyzed by **glycerol kinase**. The fatty acyl-CoA groups are energy rich and activated forms of fatty acids, which are logical donors in biosynthetic reactions.

The following four reactions yield triglyceride. **Glycerol phosphate acyltransferase** catalyzes the first reaction between activated fatty acyl-CoA, which is activated like a warhead, and glycerol 3-phosphate to yield 1-acylglycerol 3-phosphate. An **acyltransferase** catalyzes the addition of the next acyl group from fatty acyl-CoA (usually unsaturated) to yield 1,2-diacylglycerol 3-phosphate (also called **phosphatidate**), as shown in Fig. 7-6. Next, a **phosphatidate phosphatase** catalyzes a hydrolysis to yield

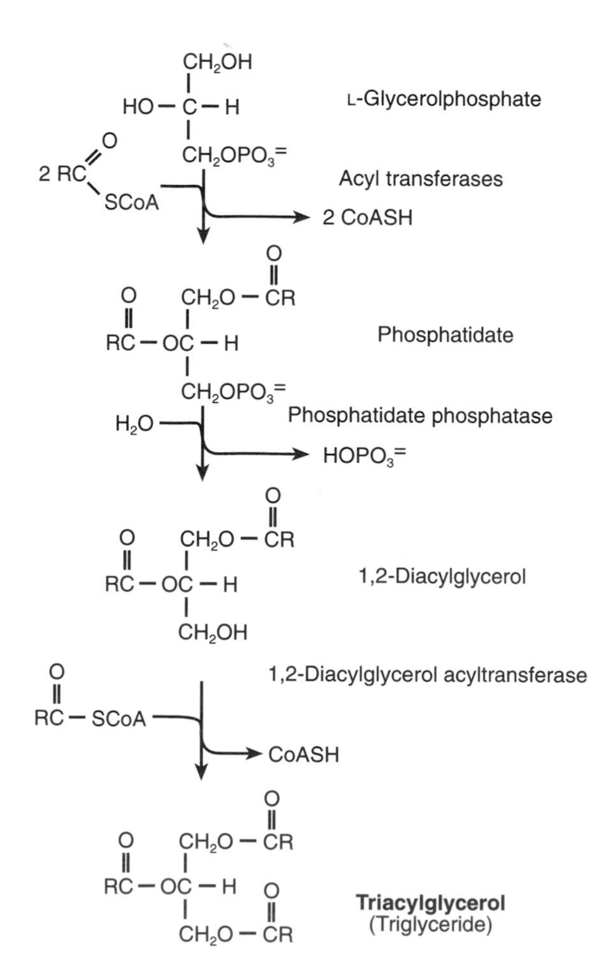

Figure 7-6. Biosynthesis of triacylglycerol. CoASH, uncombined coenzyme A; R, various fatty acid side chains; SCoA, derivatized coenzyme A.

1,2-diacylglycerol and P_i. Finally, **1,2-diacylglycerol acyltransferase** catalyzes a reaction of diacylglycerol with fatty acyl-CoA to yield triglyceride and CoA.

Each of the four reactions is highly exergonic. The fatty acid thioesterified to CoA constitutes a high-energy and activated form of the fatty acid. Activated monomers serve as energetically favorable precursors in condensation reactions.

PHOSPHOLIPID METABOLISM

Phosphoglycerides

Important phospholipids include **phosphatidylcholine (lecithin)**, **phosphatidylserine**, and **phosphatidylethanolamine**. The latter two compounds can be converted into phosphatidylcholine. The pathway for phosphatidylcholine formation begins with free choline and 1,2-diacylglycerol. This is called the *salvage pathway* because preformed choline is salvaged or reused; in contrast to the *de novo* pathway, in which choline is formed from phosphatidylethanolamine.

In the first step of the salvage pathway for **phosphatidylcholine** biosynthesis, choline is phosphorylated by ATP to yield phosphocholine and ADP in a reaction catalyzed by **choline kinase** (Fig. 7-7). Phosphocholine must now be converted to an acti-

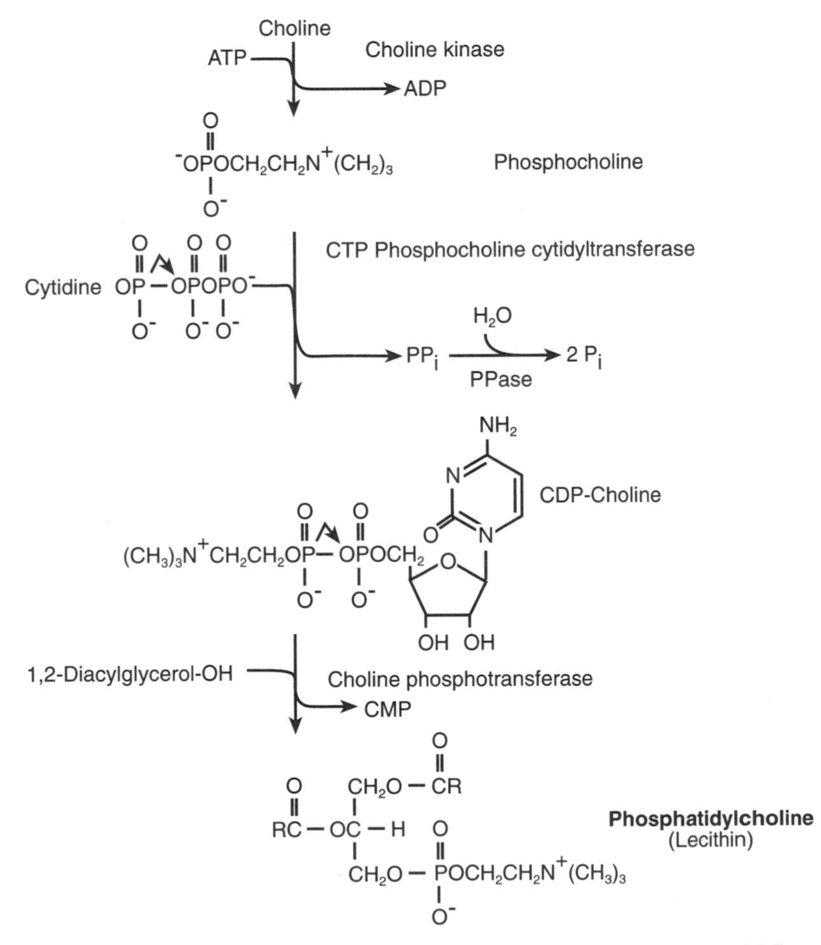

Figure 7-7. Phosphatidylcholine biosynthesis by the salvage pathway. ADP, adenosine diphosphate; ATP, adenosine triphosphate; CDP, cytidine diphosphate; CMP, cytidine monophosphate; CTP, cytidine triphosphate. P_i, inorganic phosphate; PPase, pyrophosphatase; PP_i, inorganic pyrophosphate.

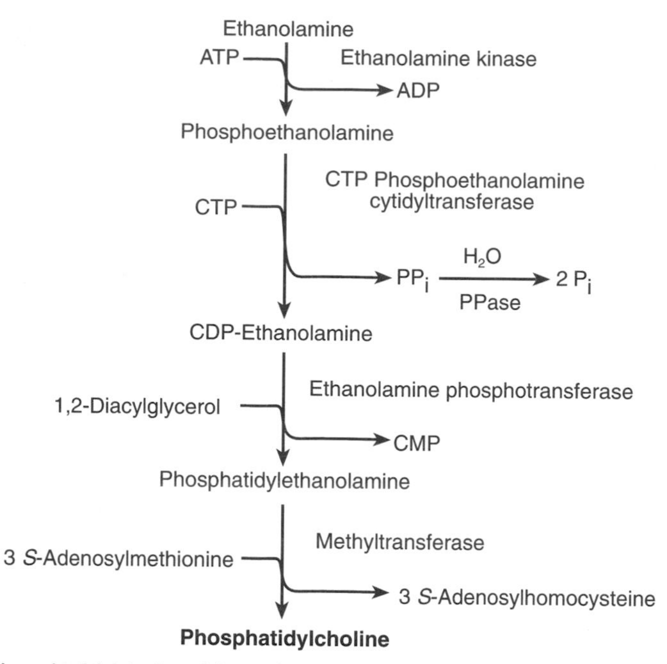

Figure 7-8. Phosphatidylcholine biosynthesis *de novo*. ADP, adenosine diphosphate; ATP, adenosine triphosphate; CDP, cytidine diphosphate; CMP, cytidine monophosphate; CTP, cytidine triphosphate; PPase, pyrophosphatase; P_i, inorganic phosphate; PP_i, inorganic pyrophosphate.

vated, or energy-rich, form before combining with an acceptor molecule. The activated form is cytidine diphosphate (CDP)–choline. The diphosphate linkage is that of an acid anhydride and is energy rich with a $\Delta G°'$ of hydrolysis of –7.2 kcal/mol. The activation process involves a reaction of cytidine triphosphate (CTP) and phosphocholine to yield CDP-choline and PP_i and is catalyzed by **CTP-phosphocholine cytidyltransferase**. The number of high-energy bonds is the same in the reactants and products, and this is an isoergonic reaction. The process is coupled to the hydrolysis of PP_i catalyzed by **pyrophosphatase**, which serves to pull the reaction forward. **CDP-choline** reacts with 1,2-diacylglycerol to yield phosphatidylcholine and cytidine monophosphate (CMP) in a reaction catalyzed by a **choline phosphotransferase**.

The pathway for **phosphatidylethanolamine** biosynthesis is analogous to that for phosphatidylcholine biosynthesis by the salvage pathway. Ethanolamine is phosphorylated by ATP to form ethanolamine phosphate (Fig. 7-8). The latter reacts with CTP to form CDP-ethanolamine, which then reacts with 1,2-diacylglycerol to form phosphatidylethanolamine. Phosphatidylethanolamine can be converted into phosphatidylcholine by three successive transmethylation reactions involving *S*-adenosylmethionine (see Fig. 7-8). Phosphatidylethanolamine and phosphatidylcholine occur in the lipid bilayer of cell membranes.

Another cytidine derivative important in lipid metabolism is CDP-diacylglycerol. **CDP-diacylglycerol** results from a reaction involving phosphatidate and CTP and yields CDP-diacylglycerol and PP_i. CDP-diacylglycerol reacts with inositol (a hexitol) to form phosphatidylinositol and CMP. **Phosphatidylinositol** is phosphorylated by ATP to form ADP and phosphatidylinositol 4-phosphate (PIP). The latter is phosphorylated by ATP to form ADP and phosphatidylinositol 4,5-bisphosphate (PIP_2). PIP_2 is an important precursor for two intracellular second messengers: **inositol 1,4,5-trisphosphate (IP_3)** and **diglyceride**. The physiologic roles of these agents are considered later.

SPHINGOLIPID BIOSYNTHESIS

This group of phospholipids contains a complex amino alcohol named *sphingosine*; sphingolipids lack glycerol. **Sphingosine** is formed in a three-step process from palmitoyl-CoA and serine. In a reaction catalyzed by a pyridoxal phosphate–dependent enzyme, the 16-carbon palmitoyl-CoA reacts with the 3-carbon serine to yield the 18-carbon dihydrosphingosine (3-ketosphinganine), CO_2 (from serine), and CoA (Fig. 7-9). Dihydrosphingosine reacts with acyl-CoA to form CoA and *N*-acyldihydrosphingosine. The latter is oxidized to yield *N*-acylsphingosine, or ceramide.

Ceramide (*N*-acylsphingosine) is a key intermediate in sphingolipid biosynthesis. Ceramide is converted to **sphingomyelin** (*O*-phosphocholine ceramide) after a reaction with CDP-choline. Ceramide is also converted to **cerebrosides** and **gangliosides**. The alcohol group of ceramide reacts with an activated uridine diphosphate (UDP)-sugar to yield cerebroside and UDP (Fig. 7-10). The addition of several (up to five) sugars to ceramide yields a family of gangliosides. These are glycolipids (sugars attached to lipid).

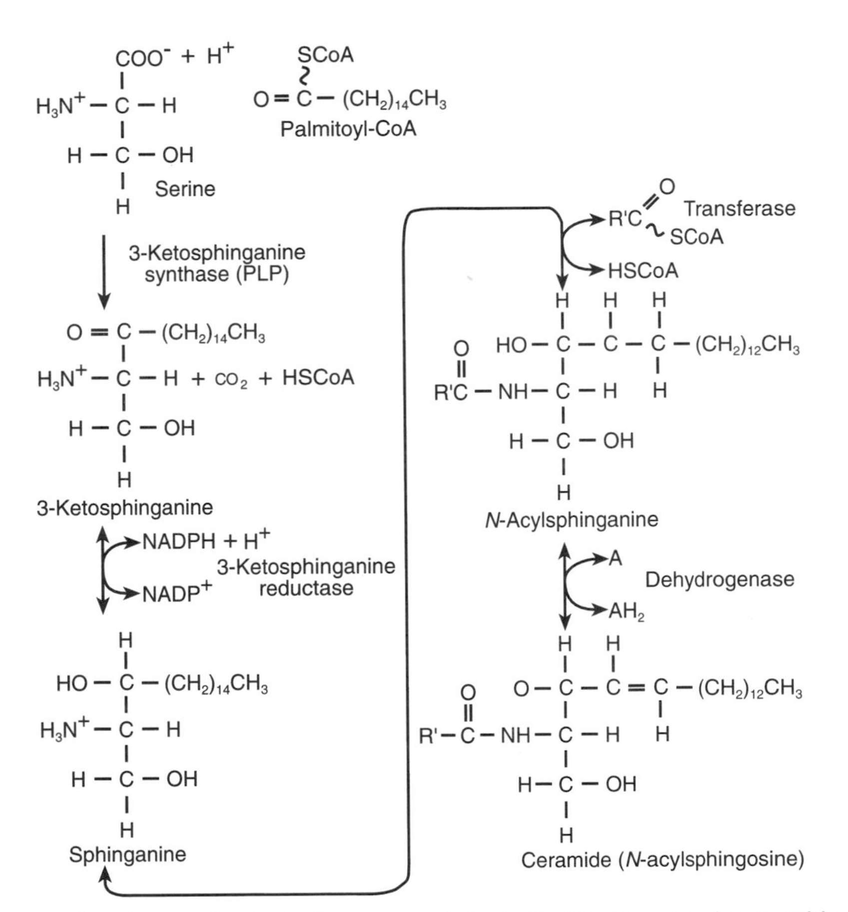

Figure 7-9. Conversion of palmitoyl-coenzyme A (*palmitoyl-CoA*) and serine to sphingosine. HSCoA, uncombined coenzyme A; $NADP^+$, oxidized nicotinamide adenine dinucleotide phosphate; $NADPH^+$, reduced nicotinamide adenine dinucleotide phosphate; SCoA, derivatized coenzyme A.

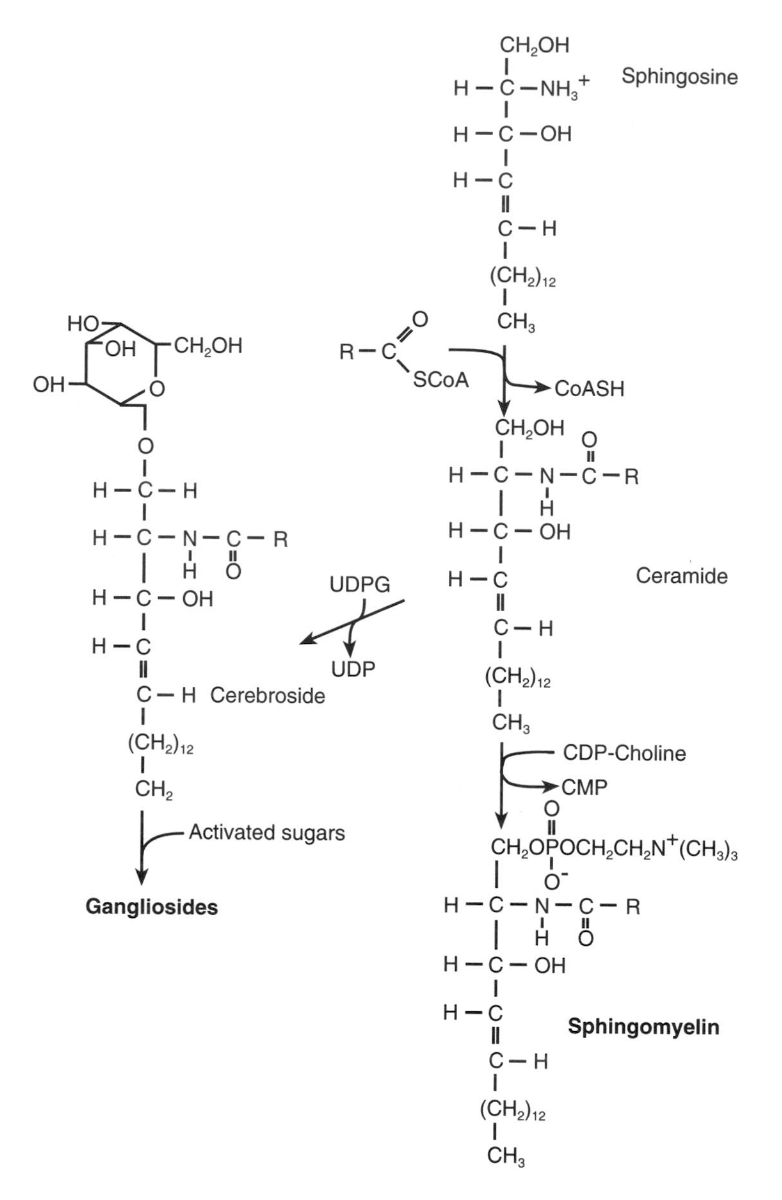

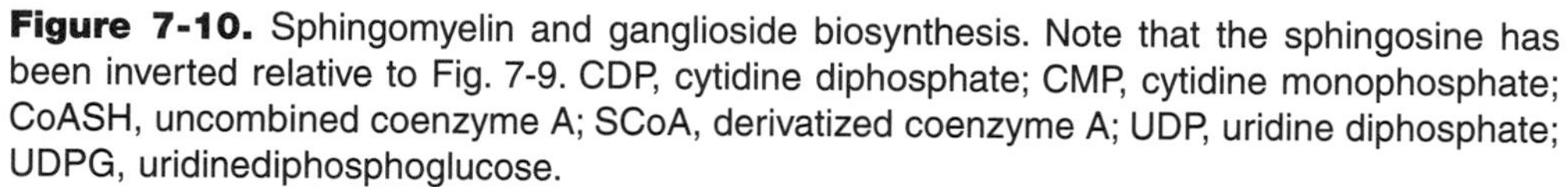
Figure 7-10. Sphingomyelin and ganglioside biosynthesis. Note that the sphingosine has been inverted relative to Fig. 7-9. CDP, cytidine diphosphate; CMP, cytidine monophosphate; CoASH, uncombined coenzyme A; SCoA, derivatized coenzyme A; UDP, uridine diphosphate; UDPG, uridinediphosphoglucose.

Phospholipid Degradation

The specificity of several phospholipases is indicated in Fig. 7-11. Two phospholipases have been implicated in several regulatory processes. Phospholipase A_2, for example, catalyzes the hydrolysis of the fatty acid at C^2 of the derivatized glycerol. Arachidonate and other polyunsaturated fatty acids are often attached to C^2. Arachidonate is converted into biologically active eicosanoids [prostaglandins (PGs), prostacyclins (PGIs), thromboxanes (TXs), and leukotrienes (LTs)]. Phospholipase A_2 is the rate-limiting enzyme in eicosanoid biosynthesis. Phospholipase C catalyzes the removal of a phosphate at position 3. Important products derived from phosphatidylinositol 4,5-bisphosphate include IP_3 and diglyceride.

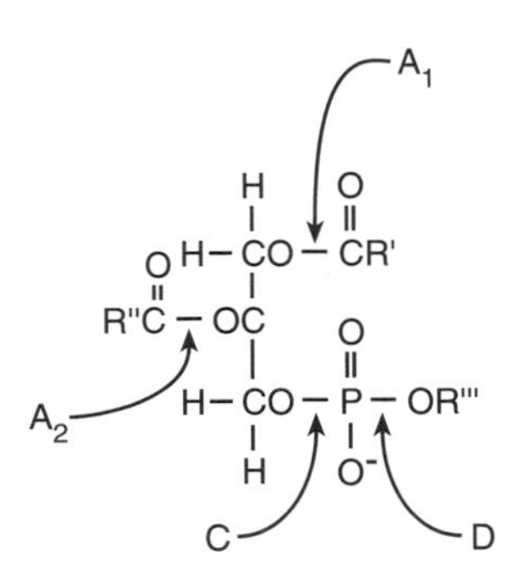

Figure 7-11. Bonds hydrolyzed by phospholipases A_1, A_2, C, and D.

TABLE 7-4. Enzyme Deficiencies in Selected Sphingolipidoses

Disease	Enzyme	Site of Deficient Reaction (Denoted by /)
Gaucher's	β-Glucosidase	Cer/Glc
Krabbe's	β-Galactosidase	Cer/Gal
Niemann-Pick	Sphingomyelinase	Cer/P-Choline
Metachromatic leukodystrophy	Arylsulfatase A	Ger-Gal/OSO_3^-
Tay-Sachs	Hexosaminidase A	Cer-Glc-Gal(NeuAc)/GalNac

Cer, ceramide; Gal, galactose; GalNac, N*-acetylgalactosamine; Glc, glucose; NeuAc,* N*-acetylneuraminic.*

Sphingolipid Degradation

A number of diseases, called **sphingolipidoses**, are caused by the deficiency of a lysosomal enzyme that catalyzes the hydrolysis of sphingolipids. Selected sphingolipidoses are listed in Table 7-4. In the absence of hydrolytic activity, the sphingolipid that cannot be degraded accumulates and produces the characteristic pathology. The diseases have been known much longer than the existence and functions of lysosomes. Basic research has provided an intellectual and scientific framework for understanding these disorders. It is possible to diagnose affected individuals *in utero*. Diagnosis is based on enzyme activity measurements or DNA analysis. The design of curative or effective palliative treatment remains for the future.

KETONE BODY METABOLISM

Ketone bodies are a group of three related compounds, including **acetoacetate**, **β-hydroxybutyrate**, and **acetone** (Fig. 7-12). The condition of ketosis occurs during carbohydrate deprivation and starvation. A more severe form of ketosis is that of diabetic ketoacidosis. Ketosis occurs in humans during the extensive mobilization of fatty acids. Ketone bodies are synthesized but not used by the liver. They are transported to extrahepatic tissues where they are metabolized to provide reducing equivalents for oxidative phosphorylation. Although a contradiction in terms, ketone bodies can be considered to be water-soluble lipids.

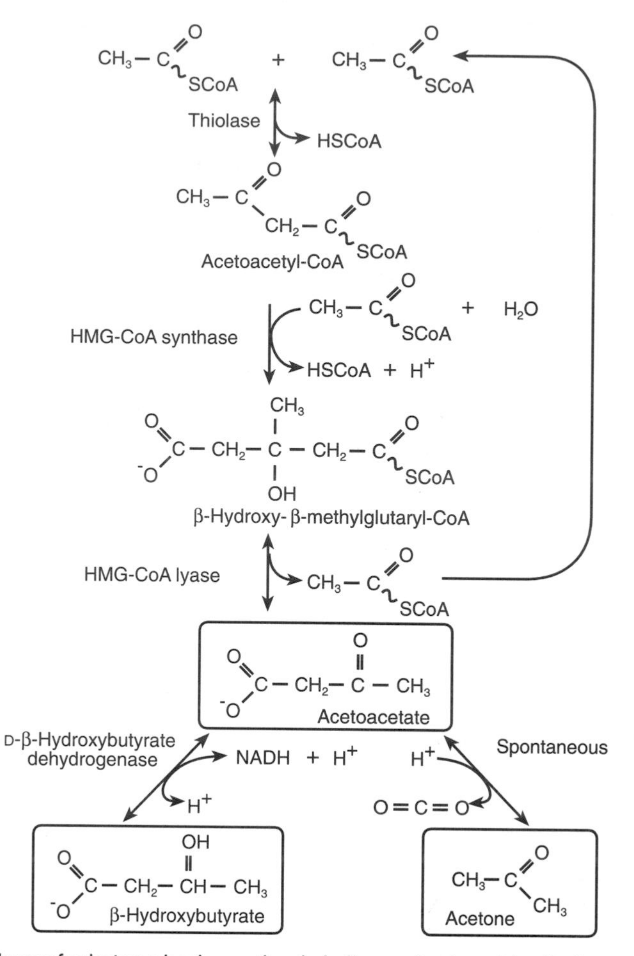

Figure 7-12. Pathway for ketone body synthesis in liver mitochondria. CoA, coenzyme A; HMG-CoA, hydroxymethylglutaryl-coenzyme A; HSCoA, unmodified coenzyme A; SCoA, derivatized coenzyme A.

Like fatty acids, ketone bodies are derived from acetyl-CoA. Two mol of acetyl-CoA condense to yield acetoacetyl-CoA and CoA. This process constitutes the reversal of the **thiolase** reaction of β-oxidation. Acetoacetyl-CoA is also produced by β-oxidation per se. You might envisage the synthesis of acetoacetate by hydrolysis of its CoA thioester. This pathway, however, either is not prominent or is nonexistent. The absence of such a hydrolytic enzyme prevents the loss of acetoacetyl-CoA formed during the course of β-oxidation of fatty acids. Acetyl-CoA condenses with acetoacetyl-CoA, followed by hydrolysis of one thioester bond to yield CoA and 3-hydroxy-3-methylglutaryl-CoA (HMG-CoA) in a highly exergonic reaction catalyzed by **HMG-CoA synthase**. HMG-CoA is a substrate for **HMG-CoA lyase**, which catalyzes the nonhydrolytic conversion of substrate to acetoacetate and acetyl-CoA. HMG-CoA synthase and lyase occur in liver mitochondria. This localization accounts for the *exclusive production of ketone bodies by the liver.* This pathway is active in conditions that favor fatty acid oxidation.

A portion of acetoacetate is reduced to form β-hydroxybutyrate by β-hydroxybutyrate dehydrogenase. **β-Hydroxybutyrate dehydrogenase** is found in liver and extrahepatic tissues. The reaction is reversible and functions *in vivo* in both directions depending on metabolic need. Acetoacetate undergoes a nonenzymatic decarboxylation reaction to yield carbon dioxide and acetone. Acetone constitutes a metabolic dead end and is excreted in the urine or exhaled by the lungs. It does not form to a significant extent during human fasting. Small amounts form, however, during **diabetic ketoacidosis**. A suggestive diagnosis of ketoacidosis can be made by smelling acetone in the breath of a comatose patient. It is reminiscent of the odor associated with fruity chewing gums.

β-Hydroxybutyrate and acetoacetate are transported from the liver to extrahepatic organs where they are taken up by cells. The β-hydroxybutyrate is oxidized to acetoacetate. Acetoacetate is derivatized in mitochondria in a reaction catalyzed by **succinyl-CoA:acetoacetyl-CoA transferase**; the CoA is transferred from succinyl-CoA, yielding succinate and acetoacetyl-CoA. The latter is a substrate for thiolase and yields 2 mol of acetyl-CoA. The HMG-CoA synthase and lyase in the liver explain ketone body synthesis there. Liver tissue lacks the CoA transferase, which is present in extrahepatic tissues. The distribution of the enzymes explains why ketone bodies are not catabolized in the liver but are catabolized in other tissues. Moreover, there are no mechanisms in mammals for converting acetyl-CoA to glucose (i.e., it is nonglucogenic). Ketone body formation therefore appears to have evolved to make use of acetyl-CoA in times of low glucose or nutrients.

CHOLESTEROL METABOLISM

Cholesterol is an essential constituent of human cell membranes, and depriving mammalian cells of cholesterol results in cell death. **Cholesterol** makes up approximately 40% of the Golgi and plasma membranes by mass, but it is present to a lesser extent in the endoplasmic reticulum and nuclear membranes. The structure and numbering system for cholesterol, a C_{27} compound, are shown in Fig. 7-13. There are two sources of cholesterol in humans: (1) ***de novo* synthesis**, and (2) **diet** (plants are unable to synthesize cholesterol and dietary cholesterol is obtained from animal products). Approximately 1.0 g per day is synthesized *de novo*, and 0.3 g per day is absorbed from the gut.

Cholesterol is converted into steroid hormones, including estradiol, progesterone, testosterone, cortisol, and aldosterone. Cholesterol is converted by the liver into **bile salts**, which are important in lipid digestion in and absorption from the intestine. Bile salts, which include cholate and deoxycholate, are quantitatively the most important metabolites of cholesterol. These negatively charged metabolites are secreted in the bile, and more than 95% of them are reabsorbed in the terminal ileum (enterohepatic circulation). One strategy to decrease body stores of cholesterol involves the administration of nonabsorbable ion exchange resins by mouth. These resins bind bile salts, prevent their reabsorption, and are excreted in feces.

Animals lack the ability to degrade the steroid nucleus, or ring system (see Fig. 7-13), to carbon dioxide and water. Conversion of cholesterol into bile acids and excretion into the feces represent the quantitatively important route of its elimination. Conversion to steroid hormones, although important biologically, constitutes a quantitatively minor route of metabolism.

The early steps of **cholesterol biosynthesis** are important and are considered in detail. Cholesterol is derived entirely from acetyl-CoA. Acetyl-CoA is converted into the six-carbon HMG-CoA (Fig. 7-14) as previously described for ketone body formation (see Fig. 7-12), except that the reactions occur in the cytosol (see Table 1-1) and not in mitochondria, as in the case of ketone bodies. The synthesis of HMG-CoA by **HMG-CoA synthases** is an important regulatory step. Then **HMG-CoA reductase** catalyzes a reaction between HMG-CoA and two molecules of NADPH to form two $NADP^+$ and **mevalonate** (see Fig. 7-14). HMG-CoA reductase is also a pacemaker enzyme, and it is the target enzyme of a variety of drugs used in the treatment of **hypercholesterolemia**. Mevalonate undergoes three successive phosphorylation

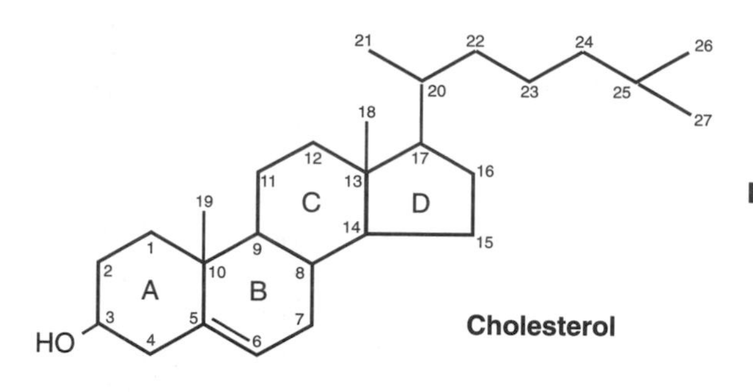

Figure 7-13. Structure of cholesterol.

reactions by ATP, yielding mevalonate phosphate, mevalonate pyrophosphate, and mevalonate-3-phospho-5-pyrophosphate (Fig. 7-15). The latter compound undergoes decarboxylation and elimination of P_i to yield **isopentenyl-PP$_i$**. Isopentenyl-PP$_i$ undergoes an isomerization reaction to form **dimethylallyl pyrophosphate**.

The conversion of these isoprenoid compounds to cholesterol is outlined in Fig. 7-15. Two 5-carbon fragments (isopentenyl-PP$_i$ and dimethylallyl-PP$_i$) condense to form the 10-carbon **geranyl-PP$_i$**. The latter reacts with the five-carbon isopentenyl-PP$_i$ to yield the 15-carbon **farnesyl-PP$_i$**. Two of these react to form the 30-carbon **squalene**. In a reaction involving molecular oxygen, squalene is converted into the four-membered ring system of steroids in the form of the 30-carbon **lanosterol**. Formate is released in an oxygen-dependent reaction, and this step is followed by two decarboxylations, resulting in the 27-carbon **zymosterol**. Zymosterol, in turn, is converted to cholesterol, which also contains 27 carbon atoms.

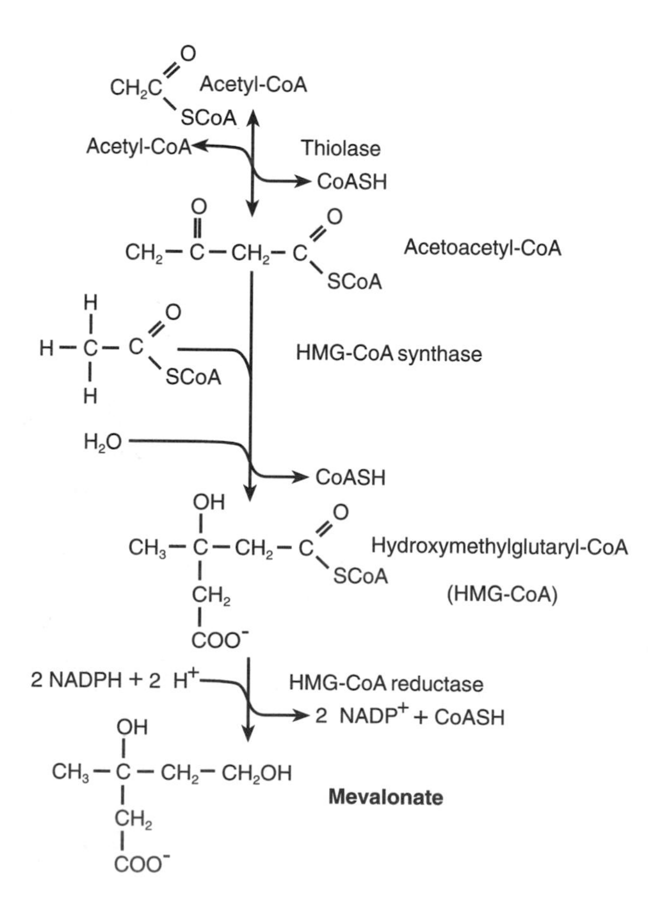

Figure 7-14. First stage of cholesterol biosynthesis: conversion of acetylcoenzyme A (*acetyl-CoA*) to mevalonate. CoASH, uncombined coenzyme A; $NADP^+$, oxidized nicotinamide adenine dinucleotide phosphate; NADPH, reduced nicotinamide adenine dinucleotide phosphate; SCoA, derivatized coenzyme A.

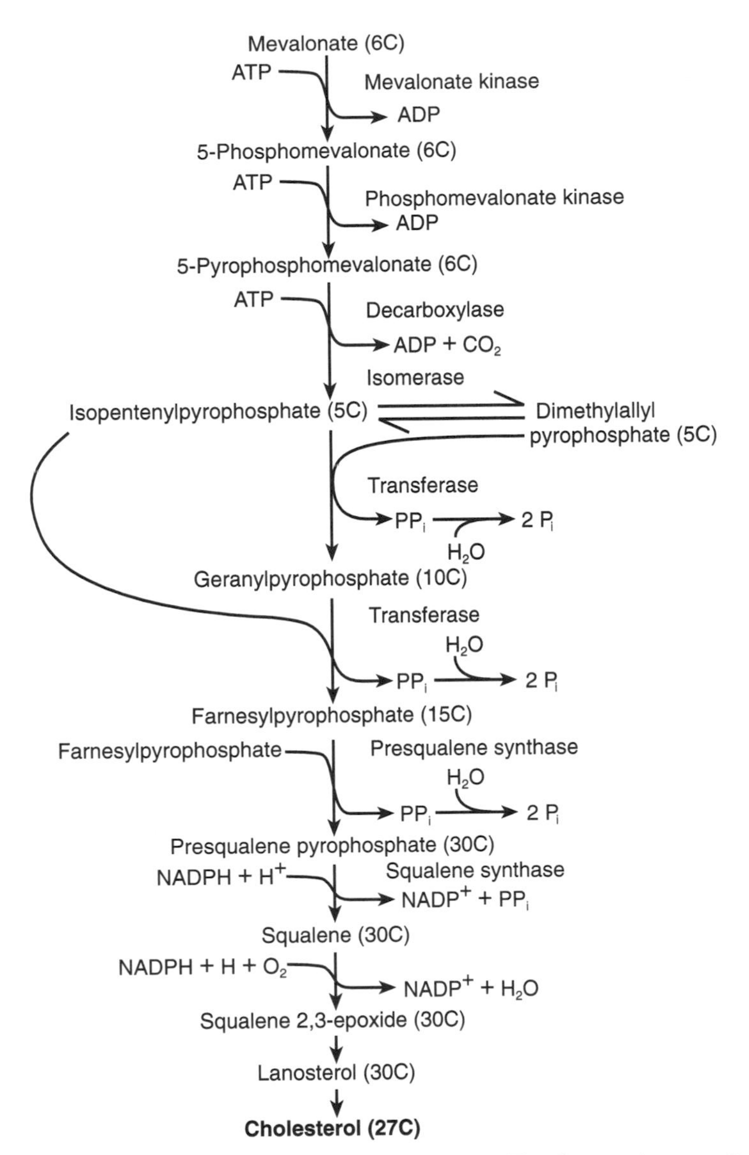

Figure 7-15. Second stage of cholesterol biosynthesis. The figures in parentheses denote the number of carbon atoms in each intermediate. ADP, adenosine diphosphate; ATP, adenosine triphosphate; $NADP^+$, oxidized nicotinamide adenine dinucleotide phosphate; NADPH, reduced nicotinamide adenine dinucleotide phosphate; P_i, inorganic phosphate; PP_i, inorganic pyrophosphate.

EICOSANOID METABOLISM

Arachidonate (5,8,11,14 eicosatetraenoate) gives rise to eicosanoids (20-carbon acids), including PGs, PGIs, TXs, and LTs. Humans cannot synthesize arachidonate *de novo.* Arachidonate is derived from linoleate. Linoleate is termed an **essential fatty acid** to reflect the inability of humans to synthesize this required compound.

The eicosanoids have multiple and diverse effects. **TXs**, for example, are synthesized in platelets and produce vasoconstriction and platelet aggregation. **PGIs** are produced by blood vessel walls and inhibit platelet aggregation. **PGs** increase cyclic AMP levels in some cells (platelets, the anterior pituitary, and lung) but lower cyclic AMP levels in others (renal tubules and adipose cells). **LTs** are produced in leukocytes,

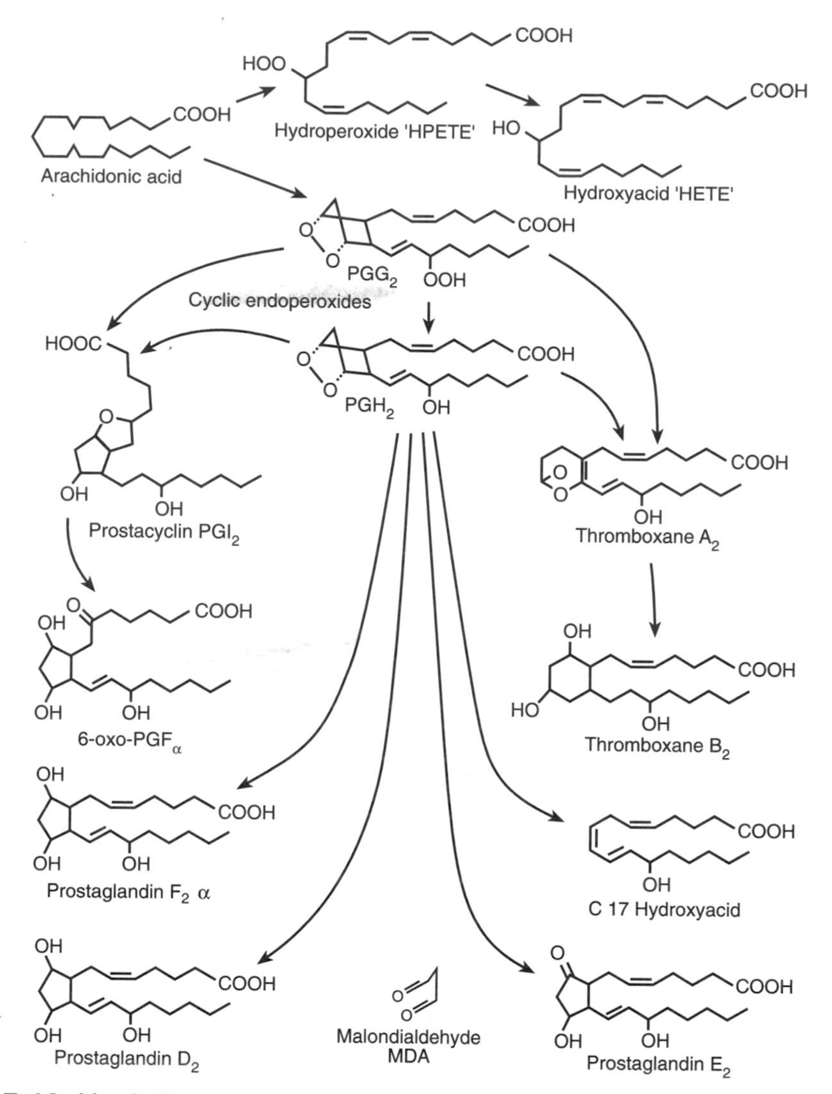

Figure 7-16. Metabolic conversions of arachidonic acid. HETE, hydroxyeicosatetraenoate; HPETE, hydroperoxyeicosatetraenoate; MDA, malondialdehyde; PGF, prostaglandin F; PGG_2, prostaglandin G_2; PGH_2, prostaglandin H_2; PGI_2, prostaglandin I_2.

platelets, and macrophages, and they attract and activate leukocytes. They also play a role in inflammation and hypersensitivity reactions, including asthma. Eicosanoids are synthesized on demand (they are not stored) and are released and act locally in an autocrine or paracrine fashion. These compounds have an evanescent existence. Only the general pathway for biosynthesis is considered here.

Arachidonate is derived from phospholipids in the plasma membrane by hydrolysis, as catalyzed by **phospholipase A_2** (see Fig. 7-12). This is the rate-limiting enzyme in the pathway. Phospholipase A_2 activity is stimulated by angiotensin II, bradykinin, epinephrine, and thrombin under specific conditions. Phospholipase A_2 stimulation involves a cell surface receptor and the action of a specific G-protein. Phospholipase A_2 is inhibited by anti-inflammatory corticosteroids. Arachidonate exists at a branch point in metabolism. Along with the action of cyclooxygenase, it reacts with 2 mol of oxygen to form a prostanoid termed prostaglandin G_2 (PGG_2). PGG_2 is converted to one of the PGs, PGIs, or TXs by specific enzymes (Fig. 7-16). **Cyclooxygenase** is inhibited by aspirin and indomethacin, and this is a noteworthy property.

Arachidonate, in a reaction catalyzed by **5-lipoxygenase**, is converted to 5-hydroperoxyeicosatetraenoate (**5-HPETE**) (see Fig. 7-16). This is converted to 5-hydroxyeicosatetraenoate (**5-HETE**) or to leukotrienes. Leukotrienes are potent constrictors of bronchial smooth muscle. Leukotriene A_4 forms a covalent adduct via a thioether linkage with glutathione to yield leukotriene C_4. Glutamate is hydrolytically removed, yielding leukotriene D_4, and glycine is hydrolytically removed to yield leukotriene E_4. Sophisticated chemical techniques, including gas chromatography or mass spectrometry, are required to elucidate the reactions outlined in Fig. 7-16.

LIPID TRANSPORT AND LIPOPROTEINS

Glucose, amino acids, and ketone bodies are soluble in water. In contrast, fatty acids, triglycerides, and cholesterol esters are sparingly soluble in water. There are two general processes for mediating the transport of lipids. **Free fatty acids** are transported as a complex with albumin. **Albumin** has a low capacity for binding and transporting fatty acids and an even lower capacity for binding and transporting cholesterol.

A second system with a larger capacity for triglyceride and cholesterol is composed of four classes of lipoproteins. The lipoproteins are made up of a hydrophilic phospholipid monolayer exterior, specific proteins (**apolipoproteins**) associated with the phospholipid surface, and an apolar lipid core. Lipoproteins mediate the transport of lipids in the extracellular compartments between the intestine, liver, adipose tissue, and other peripheral organs. Lipoprotein solubility and transport are dependent on the hydrophilic nature of the particle exterior.

Free Fatty Acid Transport

Adipose tissue is a storage depot for triglyceride. A **hormone-sensitive lipase** catalyzes the hydrolysis of two of the three fatty acids of triglyceride. A second enzyme catalyzes the removal of the third fatty acid. The fatty acids are released into the circulation, where they are transported as a complex with albumin. Nearly all cells except the brain and erythrocytes take up fatty acid. Fatty acids furnish between one-fourth and one-half of the required metabolic energy in fasting conditions. The plasma content of underivatized fatty acid (bound to albumin) is approximately 0.5 mmol postabsorptively and 0.8 mmol after an overnight fast.

Triglyceride and Cholesterol Transport

There are four major classes of lipoproteins. They are (in order of decreasing percentage of lipid and increasing density) **chylomicrons**, very-low-density lipoproteins (**VLDLs**), low-density lipoproteins (**LDLs**), and high-density lipoproteins (**HDLs**). The major function of chylomicrons is to transport triglyceride and cholesterol derived from the diet from the intestine to other tissues (Table 7-5). Chylomicrons are synthesized by the intestine but derive some apolipoproteins from HDL. Lipoprotein lipase in the capillary endothelium throughout the body catalyzes the hydrolysis of triglyceride to fatty acids, which are then taken up by cells of the tissues or organs. Chylomicron remnants return apolipoproteins to HDL, and the remnants are taken up and metabolized by the liver.

The function of VLDL is to transport triglyceride from liver to extrahepatic tissues. **Lipoprotein lipase** catalyzes the hydrolysis and release of fatty acids from the core triglyceride. VLDL is synthesized by the liver and also receives apolipoproteins from HDL. After transporting a portion of its triglyceride and after returning specific apolipoproteins to HDL, VLDL is converted into intermediate-density lipoprotein and then LDL by the liver.

TABLE 7-5. Composition and Function of the Major Classes of Lipoproteins

Particle	Function	Origin	Fate	Apolipoproteins
Chylomicron	Transports dietary triglyceride and cholesterol from gut to other tissues	Gut and HDL	Liver and HDL	Apo B-48 Apo CII, CIII Apo E
VLDL	Transports triglyceride from liver to other tissues	Liver and HDL	Converted into LDL	Apo B-100 Apo CI, CII, CIII Apo E
LDL	Transports cholesterol from liver to other tissues (bad cholesterol)	Formed from VLDL by liver	Taken up by target cells	Apo B-100
HDL	Transports cholesterol to liver from other tissues (good cholesterol)	From chylomicrons and liver	Liver	Apo AI, AII Apo B Apo CI, CII, CIII Apo D Apo E

Apo, apolipoprotein; HDL, high-density lipoprotein; LDL, low-density lipoprotein; VLDL, very-low-density lipoprotein.

The main function of **LDL** is to transport cholesterol from the liver to extrahepatic tissues. This process has been implicated as a potential factor in atherogenesis, and sometimes LDL is called **"bad" cholesterol**. The LDL complex is recognized by specific plasma membrane receptors in most cells. LDL is taken up by the cell by receptor-mediated endocytosis and is delivered to the lysosomes. The latter degrade the apolipoproteins and hydrolyze cholesterol ester to cholesterol. Cholesterol acts to decrease the synthesis of HMG-CoA synthase, HMG-CoA reductase, and the LDL receptor and thereby decreases intracellular cholesterol levels. The regulation of metabolic pathways at more than one step is called the dispersive control of metabolism. The LDL receptor is nonfunctional in **familial hypercholesterolemia**, an autosomal dominant disease.

The main function of **HDL** is to transport cholesterol from extrahepatic tissues to the liver. HDL is sometimes called **"good" cholesterol** because it removes cholesterol from peripheral tissues. HDL is synthesized in the liver and donates and receives components from chylomicrons and VLDL. VLDL is degraded by the liver. These properties are summarized in Table 7-5.

The proteins that constitute the various classes of lipoprotein particles are endowed with properties important in secretion, cell recognition, and activation of participating enzymes. Apolipoprotein AI, for example, is an activator of lecithin-cholesterol acyltransferase (LCAT). LCAT converts cholesterol and lecithin on the surface of HDL to cholesterol ester and lysolecithin. The cholesterol ester is incorporated into the interior of HDL, and lysolecithin is transferred to albumin. **Apolipoprotein CII** activates lipoprotein lipase, which catalyzes triglyceride hydrolysis. Chylomicrons and VLDL serve as substrates for this lipase. Most of the released free fatty acid is transported into the tissues that contain lipoprotein lipase. **Apolipoprotein B-100** is recognized by the LDL receptor. Apolipoprotein E is recognized by the LDL receptor and by the chylomicron remnant receptor.

Chapter 8

Amino Acid Metabolism and Urea Biosynthesis

METABOLIC CLASSIFICATION OF AMINO ACIDS

Oxidation of the carbon skeletons of amino acids accounts for approximately 15% of the metabolic energy of humans. The amino groups are converted to **ammonia** and **urea**. The metabolism of some amino acids involves complex pathways and is covered only briefly. Amino acids are derived from the breakdown of body proteins and of dietary protein. Approximately 85% of the amino acids resulting from the breakdown of endogenous body proteins are reused for protein synthesis. The breakdown of **proteins** into amino acids and their reuse for protein synthesis is called **turnover**. Based on these considerations, it should be obvious that under normal circumstances, proteins and amino acids are not a preferred energy source.

The metabolism of amino acids is divided into three categories. Amino acids are designated as **glycogenic** if they lead to carbohydrate formation, **ketogenic** if they lead to ketone body formation, and both glycogenic and ketogenic if they lead to increases in both types of compounds. This classification was derived from experiments performed by administering each amino acid to animals and determining whether there was an increase in blood glucose (glycogenic amino acid), circulating ketone bodies (ketogenic amino acid), or both. Ketogenic amino acids are catabolized to acetyl-coenzyme A (acetyl-CoA), acetoacetyl-CoA, or both. As discussed in Chapter 7, acetyl-CoA and, by inference, acetoacetyl-CoA (the product of the condensation of two acetyl-CoA molecules), cannot be used to synthesize glucose. The general classification of amino acids in this fashion is given in Table 8-1. Note that **leucine** is the sole amino acid that is ketogenic only. (Lysine administration leads to an elevation of both glucose and ketone bodies, but the metabolic pathway given in textbooks indicates that it is ketogenic only—indicating that our knowledge of lysine catabolism is incomplete.) The aromatic amino acids lysine and isoleucine are both glycogenic and ketogenic, and the remainder are glycogenic only.

TRANSAMINATION REACTIONS

Transamination reactions, which are isoergonic, are pivotal for both the degradation and biosynthesis of the majority of the genetically encoded amino acids. The following equation is representative of all the reactions.

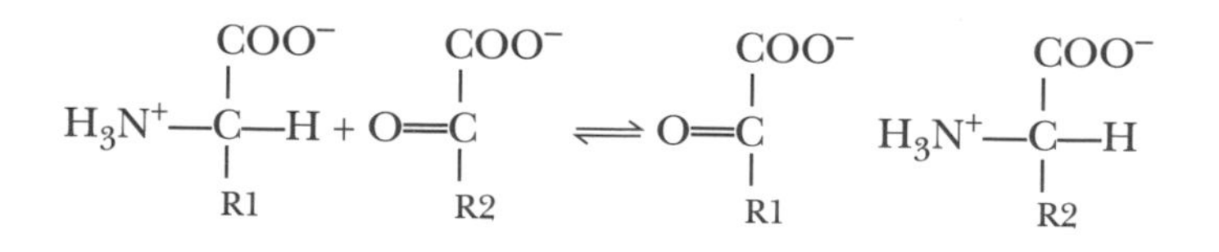

Several enzymes catalyze this reaction; **pyridoxal phosphate** (a derivative of vitamin B_6) is the cofactor. **Transaminases** generally demonstrate a preference for one of the pairs of amino group donors and acceptors and exhibit a varying latitude for the reciprocal cognate pair. The involvement of glutamate and α-ketoglutarate as substrate is prevalent and pivotal. The ability of most amino acids to donate their amino groups to α-ketoglutarate to form glutamate is responsible in part for the central role of glutamate in nitrogen metabolism. Glutamate is also a common amino group donor in the formation of many other amino acids.

The transaminase enzymes occur in the cytosol and mitochondria of all human cells. Aminotransferases also occur in human blood plasma in low but measurable amounts. An increase in serum transaminases occurs during **hepatitis**, in other liver diseases, and accompanying the tissue necrosis due to a myocardial infarction. Laboratories routinely measure **SGOT** (serum glutamate oxaloacetate transaminase) and **SGPT** (serum glutamate pyruvate transaminase) for a variety of diagnostic purposes.

Glutamate is the predominant source of free ammonium ion. **Glutamate dehydrogenase** catalyzes a reversible reaction involving the following components:

$$\text{Glutamate} + NAD(P)^+ + H_2O \rightleftharpoons \alpha\text{-ketoglutarate} + NAD(P)H + NH_4^+$$

Both the oxidized form of nicotinamide adenine dinucleotide (NAD^+) and the oxidized form of nicotinamide adenine dinucleotide phosphate ($NADP^+$) are substrates for this enzyme. The glutamate dehydrogenase reaction is amphibolic and can function in biosynthesis or degradation depending on physiologic need.

TABLE 8-1. Classification of Glycogenic and Ketogenic Amino Acids

Glycogenic Only	Both Glycogenic and Ketogenic	Ketogenic Only
Gly	Ile	Leu
Ala	Lys	
Val	Phe	
Asp	Tyr	
Asn	Trp	
Glu		
Gln		
Arg		
His		
Ser		
Thr		
Cys		
Met		
Pro		

UREA BIOSYNTHESIS

Quantitatively, the predominant excretory product of nitrogen metabolism in humans is urea. A human consuming approximately 100.0 g of protein daily excretes approximately 16.5 g of nitrogen per day. Approximately 80% to 90% of this is in the form of urinary urea. A small proportion is excreted as uric acid and free ammonium ion. Approximately 5% of nitrogen is eliminated in organic form in the feces. Given that free ammonium ion (NH_4^+) is toxic in humans, nature has evolved a complex and expensive means for eliminating it in the form of urea. Urea is synthesized in the liver (and to a lesser extent in the kidney) but not in other tissues or organs.

The precursors of urea are shown in Fig. 8-1. One nitrogen of urea is derived from **ammonium ion**, and the second is derived from **aspartate**. The carbonyl group is derived from carbon dioxide (as bicarbonate). Amino groups are funneled into glutamate (see Fig. 8-1). Glutamate is oxidized, and one precursor (ammonia) is formed; glutamate donates its amino group to oxaloacetate to yield the second precursor (aspartate). Again, note the central role of glutamate in this entire process.

Urea biosynthesis involves a cyclic metabolic pathway called the *Krebs urea cycle*, or *urea cycle*. This was the first cyclic pathway to be discovered, and the urea cycle represents the intellectual cornerstone for the elucidation of the citric acid cycle.

The urea cycle pathway involves a complex interplay of mitochondrial and cytosolic reactions. The stoichiometry of the overall process is given as follows:

$$NH_3 + HCO_3^- \text{ asparate} + 3\,ATP + H_2O \rightarrow 2\,ADP + 2\,P_i + AMP + PP_i + \text{fumarate} + \text{urea}$$

The first step in the urea cycle involves the formation of activated carbamate as **carbamoyl phosphate**. Its formation requires the expenditure of two high-energy bonds from two adenosine triphosphate (ATP) molecules and is catalyzed by carbamoyl-phosphate synthetase I, a mitochondrial enzyme. This enzyme requir

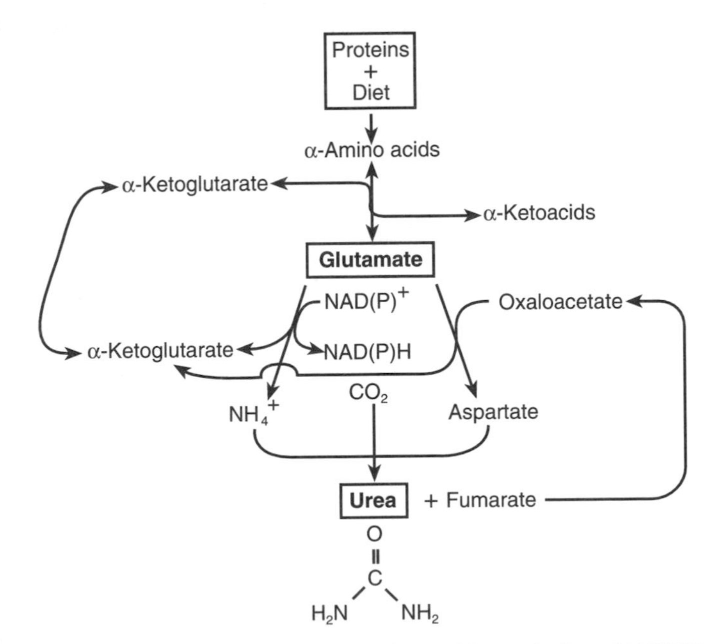

Figure 8-1. Glutamate and its central role in amino acid metabolism. NAD(P)+, oxidized form of nicotinamide adenine dinucleotide (with or without phosphate); NAD(P)H, reduced nicotinamide adenine dinucleotide (with or without phosphate).

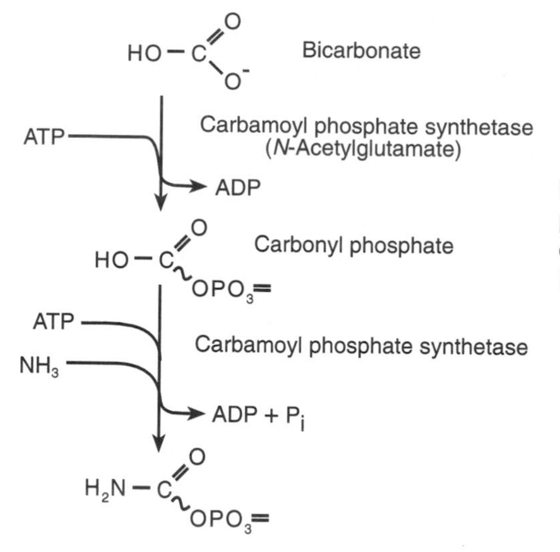

Figure 8-2. Synthesis of carbamoyl phosphate (active carbonate). ADP, adenosine 5'-diphosphate; ATP, adenosine triphosphate; P_i, inorganic phosphate.

es ***N*-acetylglutamate** as an allosteric activator. Carbamoyl-phosphate synthetase I is distinct from the cytosolic enzyme that is involved in pyrimidine formation, carbamoyl-phosphate synthetase II (Fig. 8-2). Activated carbamate reacts with ornithine in a reaction catalyzed by **ornithine transcarbamoylase** to yield citrulline. **Citrulline** is transported into the cytosol (in exchange for ornithine) before the subsequent reactions of urea formation (Fig. 8-3). Aspartate, ATP, and citrulline react to form adenosine monophosphate (AMP), inorganic pyrophosphate (PP_i), and argininosuccinate in a reaction catalyzed by **argininosuccinate synthetase**. PP_i is hydrolyzed by pyrophosphatase to pull the reaction forward. Next, **argininosuccinase** catalyzes a lyase (not hydrolase) reaction to yield arginine and fumarate. **Arginase** catalyzes the hydrolysis of arginine to form urea and ornithine. Ornithine is transported into mitochondria in exchange for citrulline.

From the stoichiometry of the cycle, note that three ATP molecules and four energy-rich bonds are expended [2 ATP → 2 adenosine 5'-diphosphate (ADP) + 2 inorganic phosphate (P_i) (2), and ATP → AMP + 2 P_i (2)]. The overall process is

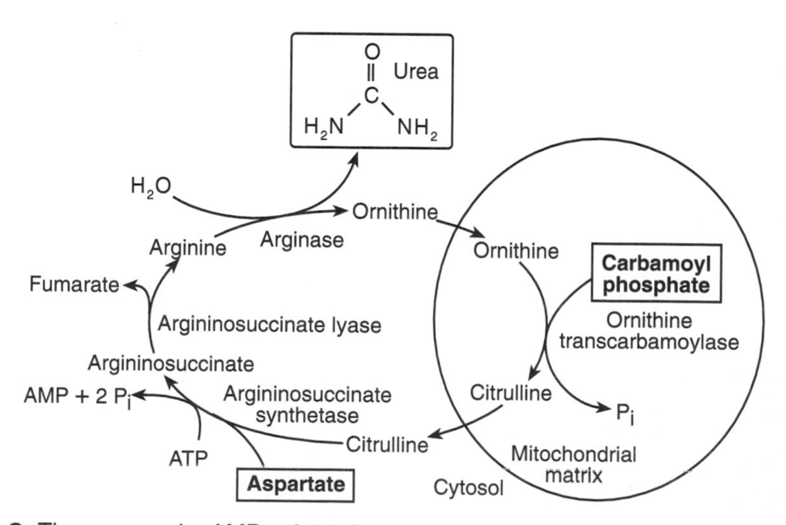

Figure 8-3. The urea cycle. AMP, adenosine monophosphate; ATP, adenosine triphosphate; P_i, inorganic phosphate.

TABLE 8-2. Inborn Errors of Urea Cycle and Amino Acid Metabolism

Name	Enzyme Deficiency	Metabolism Affected
Argininosuccinaturia	Argininosuccinase	Urea cycle
Citrullinemia	Argininosuccinate synthetase	Urea cycle
Hyperammonemia I	Ornithine transcarbamoylase	Urea cycle
Hyperammonemia II	Carbamoyl-phosphate synthetase	Urea cycle
Alkaptonuria	Homogentisate oxidase	Phe, Tyr
Argininemia	Arginase	Arg
Cystathioninuria	Cystathionase	Cys, Met, Ser
Histidinemia	Histidine ammonia lyase	His
Homocystinuria	Cystathionine synthase	Cys, Met, Ser
Maple syrup urine disease	Branched-chain ketoacid dehydrogenase	Val, Leu, Ile
Phenylketonuria	Phenylalanine hydroxylase	Phe

highly exergonic and represents an expensive means for "fixing" nitrogen as urea, then excreting it in the urine.

Renal disease is often associated with an elevation of **blood urea nitrogen** (BUN). In severe liver disease, there is an elevation of blood ammonia. One mechanism proposed for the toxicity of ammonia involves the depletion of mitochondrial citric acid cycle intermediates by converting α-ketoglutarate to glutamate in the reaction catalyzed by glutamate dehydrogenase. Krebs citric acid cycle function and aerobic metabolism are especially important in the brain, and ammonium toxicity leads to hepatic encephalopathy with confusion, stupor, or even coma and death. **Inborn errors of urea cycle metabolism** have been reported that involve each of the urea cycle enzymes (Table 8-2). Defects in the urea cycle are unusual in that these inborn errors of metabolism are not associated with metabolic acidosis.

CATABOLISM OF THE CARBON SKELETONS OF THE AMINO ACIDS

Amino acids that give rise to pyruvate and intermediates of the citric acid cycle (**α-ketoglutarate**, **succinyl-CoA**, **fumarate**, and **oxaloacetate**) are glycogenic. These compounds can furnish mitochondrial oxaloacetate and malate, substrates for gluconeogenesis. In contrast, those amino acids that directly yield acetyl-CoA or acetoacetyl-CoA cannot support gluconeogenesis. These compounds yield ketone bodies. Several amino acids exhibit multiple pathways of metabolism, but are not considered here.

Catabolism to Pyruvate

Four amino acids (**alanine**, **glycine**, **serine**, and **cysteine**) are converted to pyruvate. **Alanine** undergoes a transamination reaction and yields pyruvate directly (compare their structures). **Glycine** reacts with N^5,N^{10}-methylenetetrahydrofolate and water to yield **serine** and tetrahydrofolate. Serine dehydratase catalyzes the pyridoxal phosphate–dependent elimination of ammonia with a rearrangement to yield pyruvate. Note that the initial steps in the catabolism of threonine, glycine, and serine do not involve transaminations. The main pathway for **cysteine** metabolism in humans involves oxidation of the thiol group to sulfonate ($R–SO_2^-$), transamination, and then hydrolysis (releasing $HO–SO_2^-$), yielding pyruvate.

Catabolism to α-Ketoglutarate

Five amino acids (proline, arginine, histidine, glutamine, and glutamate) are metabolized to α-ketoglutarate. **Proline** is dehydrogenated (NAD^+), forming a double bond in the ring between the α-carbon and nitrogen (Δ1-pyrroline-5-carboxylate), and this compound is hydrolyzed between the α-carbon and nitrogen atoms to form L-glutamate semialdehyde. A dehydrogenase (NAD^+) oxidizes the aldehyde on C^5 to yield glutamate. **Glutamate** is converted into α-ketoglutarate by transamination or by dehydrogenation (glutamate dehydrogenase). **Arginine** is hydrolyzed to urea and ornithine. Ornithine undergoes a transamination reaction to form L-glutamate semialdehyde.

Histidine metabolism is too complex to consider fully. Histidine ammonia lyase catalyzes the elimination of ammonia to form urocanate (first isolated in canine urine). Urocanate undergoes two successive hydrolysis reactions to yield *N*-formimino-L-glutamate (FIGLU). This reacts with tetrahydrofolate to form L-glutamate and N^5-formiminotetrahydrofolate. In **folic acid deficiency**, FIGLU is excreted in the urine and forms the basis of a diagnostic test after a large dose of histidine. Glutaminase catalyzes the hydrolysis of the amide group of **glutamine** to yield glutamate and ammonia. It is noteworthy that the glutaminase reaction in the kidney is responsible for the generation of most of the ammonium ion excreted in the urine. Glutamate is converted into α-ketoglutarate in one step by transamination or dehydrogenation.

CATABOLISM TO SUCCINYL-CoA

Valine, isoleucine, threonine, and methionine are metabolized to succinyl-CoA. The pathway from **valine** to succinyl-CoA consists of eight steps and is not covered in its entirety. The first step is a transamination to form α-ketoisovalerate. This undergoes an oxidative decarboxylation involving NAD^+ and CoA. The reaction is catalyzed by a **branched-chain keto acid dehydrogenase**. The reaction is analogous to the pyruvate dehydrogenase and α-ketoglutarate dehydrogenase reactions (with three enzymes and five cofactors). Subsequent steps eventually yield methylmalonyl-CoA, which is metabolized in a vitamin B_{12}–dependent pathway (that was considered previously in the metabolism of fatty acids) with an odd number of carbon atoms (see Fig. 7-3). Methylmalonyl-CoA is converted to succinyl-CoA in this process.

Isoleucine is first transaminated to form α-keto-ß-methylvalerate, which undergoes an oxidative decarboxylation by the same branched-chain keto acid dehydrogenase mentioned earlier. After several more steps, a thiolytic cleavage produces propionyl-CoA and acetyl-CoA. Propionyl-CoA is converted to succinyl-CoA by the B_{12}–dependent pathway (see Fig. 7-3). Propionyl-CoA is glycogenic and acetyl-CoA is ketogenic, accounting for the classification of isoleucine in Table 8-1. **Threonine** undergoes a dehydration-deamination reaction catalyzed by serine/threonine dehydratase to yield α-ketobutyrate. Serine/threonine dehydratase contains pyridoxal phosphate as cofactor. α-Ketobutyrate undergoes a NAD^+-dependent oxidative decarboxylation reaction to form propionyl-CoA, CO_2, and nicotinamide-adenine dinucleotide (NADH) and H^+. α-Ketobutyrate dehydrogenase is made up of three different proteins and requires the same five cofactors as pyruvate dehydrogenase. Propionyl-CoA is converted into succinyl-CoA (see Fig. 7-3).

The metabolism of **methionine** is complex. The reactions of its derivative, *S*-adenosylmethionine, are considered in some detail in Chapter 9. ***S*-Adenosylmethionine** is an important methyl donor in several reactions. The conversion of phosphatidylethanolamine to phosphatidylcholine, for example, was considered previously (see Fig. 7-7). **Homocysteine**, which forms after transmethylation and hydrolysis, is a four-carbon homologue of cysteine with the thiol group on the γ-carbon. Homocysteine reacts with serine to yield **cystathionine** ($HOOCCH[NH_2]CH_2CH_2SCH_2CH[NH_2]COOH$) and water. **Cystathionine** is cleaved in a lyase reaction catalyzed by cystathionase to form cysteine, ammonia, and α-ketobutyrate; the latter is derived from the four carbons of methionine after hydrolysis of the amino group. **α-Ketobutyrate** undergoes an oxidative decarboxylation as catalyzed by its dehydrogenase to yield **propionyl-CoA**. The latter is converted to succinyl-CoA by the vitamin B_{12}–dependent pathway (see Fig. 7-3). Valine, isoleucine, threonine, and methionine constitute the succinyl-CoA family. Methionine is an essential amino acid, and cysteine, which can be derived from methionine and serine, is nonessential. Provision of cysteine decreases the requirement for methionine, and this process is called "sparing."

Catabolism to Fumarate and Acetoacetyl-CoA

Phenylalanine and tyrosine are converted to fumarate (glycogenic) and acetoacetyl-CoA (ketogenic) and are thus both glycogenic and ketogenic. **Phenylalanine** (essential) is converted into tyrosine (nonessential) in a reaction catalyzed by **phenylalanine hydroxylase**, which occurs only in the liver. **Tetrahydrobiopterin** is the reductant, and molecular oxygen is required. The products include tyrosine, dihydrobiopterin, and water. The hereditary deficiency of phenylalanine hydroxylase is associated with **phenylketonuria** (PKU), one of the more common inborn errors of metabolism. Tyrosine undergoes a transamination reaction with α-ketoglutarate to yield 4-hydroxyphenylpyruvate and glutamate; the enzyme catalyzing this reaction is **tyrosine aminotransferase**. In a reaction involving molecular oxygen, 4-hydroxyphenylpyruvate is decarboxylated, hydroxylated, and rearranged to produce **homogentisate**. Homogentisate undergoes a reaction with oxygen catalyzed by **homogentisate oxidase**, which opens the aromatic ring yielding **maleylacetoacetate**. This undergoes an isomerization to **fumarylacetoacetate**. A lyase then catalyzes the formation of **fumarate** (glycogenic) and **acetoacetate** (ketogenic). Acetoacetate is transported from the liver for metabolism in extrahepatic tissues as a ketone body.

Catabolism to Oxaloacetate

Asparagine and aspartate are metabolized to oxaloacetate. Asparaginase catalyzes the hydrolytic removal of the amide group of **asparagine** to yield ammonia and aspartate. **Aspartate** undergoes transamination and yields oxaloacetate. The complete oxidation of any Krebs cycle metabolite is not as trivial as might be expected.

Recall that the Krebs cycle is catalytic and results in the catabolism of **acetyl-CoA**; the Krebs cycle intermediates are regenerated and not catabolized. The oxidation of Krebs cycle intermediates requires their conversion first to pyruvate and then acetyl-CoA. To achieve this requirement, Krebs cycle metabolites are converted to malate. A mitochondrial malic enzyme mediates the reaction of malate with $NADP^+$ to pyruvate, CO_2, and reduced nicotinamide adenine dinucleotide phosphate (NADPH). **Pyruvate** is converted to acetyl-CoA, the stoichiometric substrate of the Krebs cycle, in a reaction catalyzed by **pyruvate dehydrogenase**, also a mitochondrial enzyme. The yield of

ATP from the oxidation of NADPH is equivalent to that of NADH (2.5 mol of ATP per mol of NADH or NADPH).

Catabolism to Acetoacetyl-CoA

Leucine, lysine, and tryptophan are converted to acetoacetyl-CoA. **Leucine** is converted to α-ketoisocaproate by transamination. This substance undergoes an oxidative decarboxylation by the branched-chain keto acid dehydrogenase and yields isovaleryl-CoA. (Note that branched-chain keto acid dehydrogenase operates on valine, isoleucine, and leucine metabolites). **Isovaleryl-CoA dehydrogenase** catalyzes the next reaction in the pathway. A deficiency of this enzyme results in the accumulation of substrate, which reacts with glycine to form isovalerylglycine and is excreted in the urine.

The next enzyme in the pathway, **methylcrotonyl-CoA carboxylase**, uses biotin as a cofactor and ATP as cosubstrate. With inadequate methylcrotonyl-CoA carboxylase activity, 3-hydroxyisovalerate accumulates. A hydration reaction yields 3-hydroxy-3-methylglutaryl-CoA, and **HMG-CoA lyase** catalyzes its conversion to acetyl-CoA and acetoacetate. It is noteworthy that leucine is the only genetically encoded amino acid that is entirely ketogenic (see Table 8-1).

Catabolism to α-Ketoadipate

Lysine metabolism is complex and incompletely understood. α-Ketoadipate (a six-carbon dicarboxylic acid) is an intermediate metabolite. It is converted in several steps to acetyl-CoA (ketogenic). Studies in animals indicate that lysine is both glycogenic and ketogenic, and how lysine metabolites lead to the formation of net glucose is unknown. Because of this uncertainty, lysine is classified in some texts as ketogenic only, but this classification does not agree with metabolic studies *in vivo.*

Tryptophan metabolism is complex. Tryptophan, moreover, is metabolized in humans to nicotinic acid, a vitamin, and this is a noteworthy property. The first reaction in tryptophan catabolism involves its reaction with both atoms of an oxygen molecule, which opens the five-membered ring of imidazole and yields *N*-formyl kynurenine. After the hydrolytic cleavage of formate, **L-kynurenine** results. This undergoes a hydroxylation (O_2 and NADPH are reactants) and is followed by the elimination of alanine (glycogenic) and the formation of **3-hydroxyanthranylate**, which has two possible metabolic fates. The intermediate can be converted into α-ketoadipate, also a lysine metabolite. 3-Hydroxyanthranylate is also converted into quinolinate, which can be converted into **nicotinic acid ribose phosphate**. Nicotinic acid is a vitamin in humans because this pathway fails to provide physiologically adequate amounts of nicotinic acid. The formation of quinolinate from tryptophan decreases the requirement for niacin in humans.

INBORN ERRORS OF AMINO ACID METABOLISM

There are a large number of inborn errors in the metabolism of amino acids. The most common is PKU, and its incidence is 1 in 10,000 live births. Classic **PKU** is due to a relative deficiency of phenylalanine hydroxylase (see Table 8-2). Because phenylalanine cannot be degraded, alternative metabolites accumulate. These include phenylpyruvate (the phenylketone excreted in urine), phenylacetate, and phenylacetoacetate. Although several commercial methods are available as screening tests for this disorder, the definitive diagnosis requires a determination of plasma phenylalanine levels. The treatment includes a **diet deficient in phenylalanine**. Variants of PKU exist. One is due to a deficiency of dihydropteridine reductase activity. This enzyme is responsible for regenerating tetrahydrobiopterin from dihydrobiopterin.

Alkaptonuria is another disorder of phenylalanine and tyrosine metabolism. Alkaptonuria is attributable to a deficiency in **homogentisate oxidase**. Homogentisate is excreted in the urine, which turns dark on standing owing to auto-oxidation of the aromatic derivative. The biochemistry of alkaptonuria serves as a prototype of such genetic diseases. Alkaptonuria is a benign disorder in contrast to the morbidity associated with PKU. **Tyrosine aminotransferase**, which is a hepatic enzyme, is the rate-limiting enzyme for tyrosine catabolism, and its deficiency leads to **tyrosinemia**.

Maple syrup urine disease is a rare genetic disorder that is associated with deficiency of the branched-chain keto acid dehydrogenase involved in the metabolism of valine, leucine, and isoleucine (see Table 8-2). There are elevated levels of each of these three amino acids and their corresponding α-keto acids in plasma. The disease received its name from the odor of urine (and sweat) in affected individuals, which is reminiscent of maple syrup. **Homocystinuria** is due to a defect in cystathionine β-synthase, a vitamin B_6–dependent enzyme. Approximately half of the affected individuals respond to large doses (500 mg/day) of pyridoxine. Elevated circulating homocysteine is a risk factor for atherosclerosis; the relationship of homocysteine and atherosclerosis is an enigma. Administration of vitamin B_6 and folate are possible preventive treatments for atherosclerosis that act by decreasing homocysteine.

AMINO ACID BIOSYNTHESIS

There are 20 genetically encoded amino acids (see Figs. 2-2 to 2-8). If one of these amino acids is present in inadequate amounts, then protein synthesis is correspondingly diminished. **Nonessential amino acids** can be produced from endogenous metabolites. In contrast, **essential amino acids** cannot be derived from endogenous metabolites in required amounts, if at all, and must be obtained from the diet. A relative deficiency of an essential amino acid impairs protein synthesis and leads to a **negative nitrogen balance** (nitrogen excretion exceeds nitrogen intake). In healthy adults, a **nitrogen balance** exists (intake equals excretion). In growing children, in whom nitrogen intake exceeds excretion, a **positive nitrogen balance** occurs. Negative nitrogen balance results from a variety of nonphysiologic conditions, including infections, burns, and postsurgical stress. Negative nitrogen balance results in part from increased glucocorticoid secretion from the adrenal cortex.

Experiments with normal adult volunteers indicate that 8 of the 20 amino acids are essential and must be provided in the diet. Moreover, it is thought that infants require arginine for optimal growth. When volunteers are fed a diet lacking histidine, they do not go into negative nitrogen balance. The pathway for histidine synthesis, which is intricate in bacteria and plants, has not been demonstrated in humans. Human muscle, however, contains **carnosine**, a dipeptide made of **histidine** and **β-alanine**. During starvation, the levels of carnosine decrease. It is postulated that carnosine masks the essential nature of histidine. A mnemonic for the essential amino acids is the acronym **PVT TIM *HA*LL** ("private Tim Hall"). The letters correspond to **p**henylalanine, **v**aline, **t**hreonine, **t**ryptophan, **i**soleucine, **m**ethionine, **h**istidine, **a**rginine, **l**eucine, and **l**ysine. If you remember that tyrosine is nonessential because it is derived from phenylalanine (essential), then the mnemonic is less ambiguous. "HA" is italicized to signify that arginine is required in infants and that histidine is postulated to be essential.

This section outlines the pathways for the synthesis of the **nonessential amino acids**. The pathways for the biosynthesis of essential amino acids that occur in plants and microorganisms are intricate and are not considered. Glycine is formed by multiple pathways, including the pathway from serine by way of the hydroxymethyltransferase reaction involving tetrahydrofolate. Alanine is formed by transamination of

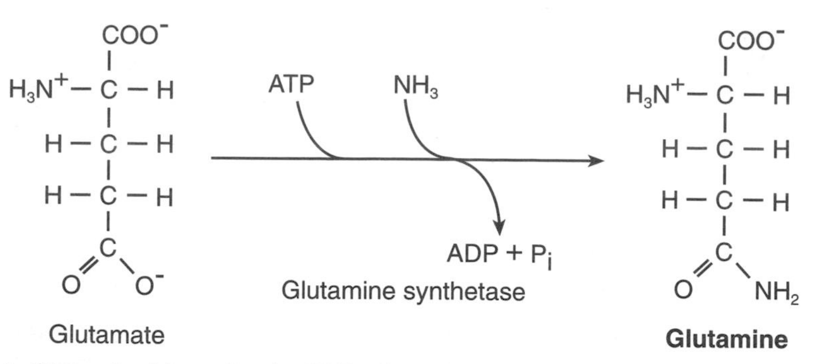

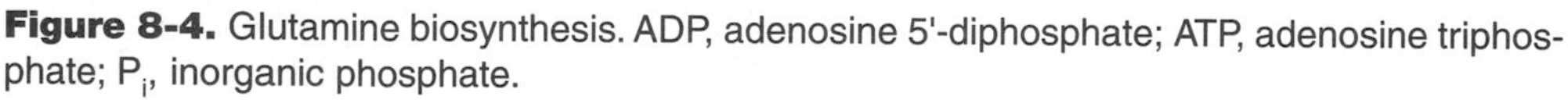
Figure 8-4. Glutamine biosynthesis. ADP, adenosine 5'-diphosphate; ATP, adenosine triphosphate; P_i, inorganic phosphate.

pyruvate. Glutamate and aspartate are formed by transamination of α-ketoglutarate and oxaloacetate, respectively. Glutamate is also formed from α-ketoglutarate through the action of glutamate dehydrogenase. This important reaction converts free ammonia to an organic amino function. Glutamate then serves as an amino donor in a variety of reactions.

Oxaloacetate, which can be converted to glucose by gluconeogenesis or to aspartate via transamination, is a product of the **pyruvate carboxylase** reaction. A hereditary deficiency of pyruvate carboxylase is accompanied by lactic acidosis and hyperammonemia. The lactic acidosis is due to the buildup of pyruvate, which is then reduced to form lactic acid. The **hyperammonemia** is due to a relative deficiency of aspartate, which is a nitrogen donor in the Krebs urea cycle. **Urea cycle defects** are accompanied by hyperammonemia in the absence of acidosis.

The biosynthesis of glutamine and asparagine requires ATP. In glutamine synthesis, ADP and P_i result. In asparagine synthesis, AMP and PP_i are formed. **Glutamine synthetase** catalyzes a reaction between ATP, glutamate, and ammonia to yield glutamine, ADP, and P_i (Fig. 8-4). γ-Phosphoglutamate is an activated acylphosphate intermediate that undergoes a reaction with ammonia.

In the **asparagine synthetase** reaction, aspartate, the amido group of glutamine, water, and ATP react to form asparagine, glutamate, AMP, and PP_i. The amido group of glutamine serves the important function of nitrogen donor. The Michaelis constant of the human enzyme for ammonia is very high, and ammonia is not a physiologically important donor in this reaction.

The main pathway for serine synthesis from 3-phosphoglycerate (an intermediate in glycolysis) involves its NAD^+-dependent dehydrogenation to phosphohydroxypyruvate. Phosphohydroxypyruvate undergoes a transamination reaction to form *O*-phosphoserine, which then undergoes hydrolysis to form serine and P_i. Although phosphoserine is found in proteins, the amino acid is added to the nascent polypeptide chain as a seryl group. A posttranslational phosphorylation catalyzed by protein kinases mediates the phosphorylation of the protein-serine to yield the phosphoseryl residue.

Proline is derived from glutamate. **Glutamate** is phosphorylated by ATP, and the derivative is reduced by NADPH. A dehydration with ring formation produces Δ1-pyrroline-5-carboxylate, which is reduced by NADPH to form proline. **Cysteine** is derived from methionine (essential) and serine (nonessential). **Methionine** is converted to homocysteine, which forms an adduct with serine called **cystathionine**, and cystathionine is cleaved during a lyase reaction to form cysteine and homoserine. **Tyrosine** is formed from phenylalanine (essential) in the phenylalanine hydroxylase reaction.

Chapter 9

One-Carbon and Nucleotide Metabolism

ONE-CARBON METABOLISM

The reactions of *S*-adenosylmethionine, biotin, and folate derivatives are important in the transfer of one-carbon groups in metabolism. The methyl group represents the most reduced form of carbon, and the carboxylate represents the most oxidized form of carbon (Table 9-1). ***S*-Adenosylmethionine** is important in many methyl-transfer reactions. **Biotin**, covalently linked to proteins, also serves as an intermediary of one-carbon carboxyl-transfer reactions. **Derivatives of tetrahydrofolate** (THF) play a role in a variety of one-carbon transfers of varying oxidation states. One-carbon groups transferred by THF include methyl, methylene, hydroxymethyl, methenyl, formyl, and formimino groups (see Table 9-1).

S-Adenosylmethionine Metabolism

Consider the biochemistry of *S*-adenosylmethionine (CH_3S^+[5'-adenosyl]CH_2CH_2CH[NH_2]COOH), which is formed in a very unusual reaction between adenosine triphosphate (ATP), methionine, and water (Fig. 9-1). In this reaction, **triphosphate** is displaced and is hydrolyzed to inorganic phosphate (P_i) and inorganic pyrophosphate (PP_i) before release from the enzyme active site. PP_i is then hydrolyzed to two P_i molecules by a separate **pyrophosphatase** activity. Two high-energy bonds and a low-energy bond are consumed during the process.

The standard free energy of hydrolysis of the methyl group from *S*-adenosylmethionine is –7 kcal/mol; thus, this compound is of the high-energy variety. *S*-Adenosylmethionine represents one form of an activated methyl group. It transfers its methyl group to a variety of acceptors to yield *S*-adenosylhomocysteine and a methylated compound. Methylated products include choline, creatine, epinephrine, capped mitochondrial RNA, and 5-methylcytosine in DNA.

Biotin and Carboxylation Reactions

Next, consider the role of biotin in ATP-dependent carboxylation reactions. Biotin forms a covalently linked prosthetic group in **propionyl-coenzyme A (propionyl-CoA) carboxylase**, **pyruvate carboxylase**, and **acetyl-CoA carboxylase**. The substrate bicarbonate presumably forms an activated carbonyl phosphate (acid anhydride) intermediate and adenosine 5'-diphosphate (ADP) in the first part of the reaction. The carbonyl phosphate reacts with the biotin prosthetic group to form an activated carbonyl-biotinyl group. The activated carboxyl group is transferred to an acceptor, such

TABLE 9-1. Oxidation States of One-Carbon Groups

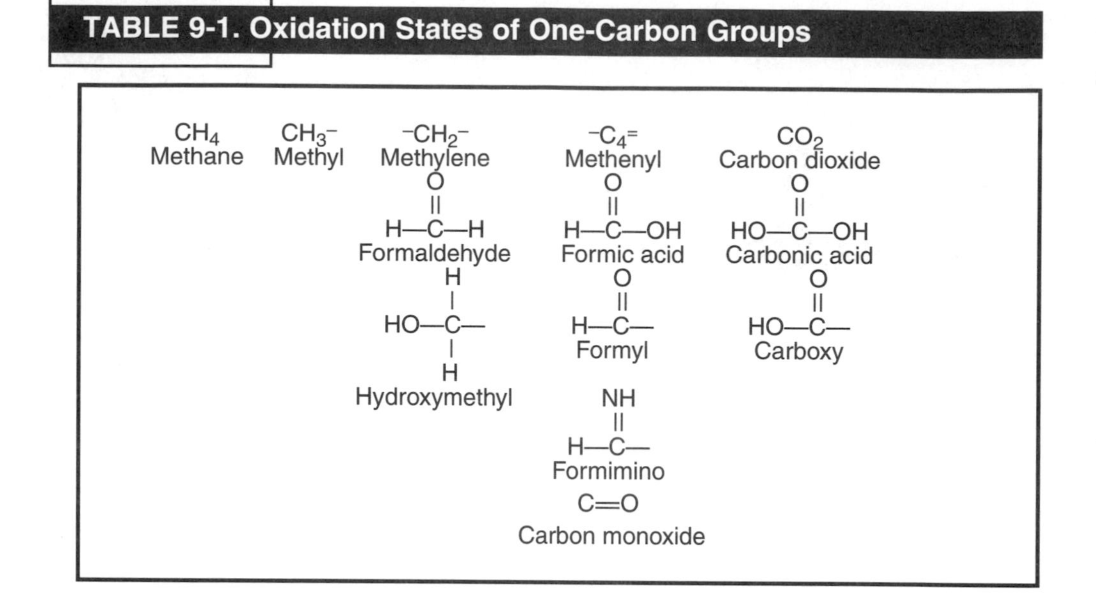

as propionyl-CoA, pyruvate, or acetyl-CoA, depending on the reaction, to produce the carboxylated compound and regenerated enzyme.

Folate Metabolism

Tetrahydrofolates are made up of tetrahydropterin, *p*-aminobenzoate, and glutamate (see Fig. 9-1). The glutamyl group is covalently linked to additional glutamate residues. The one-carbon groups are linked to N^5 or N^{10}, or to both nitrogen atoms. Serine is the major donor of one-carbon groups in humans in a reaction catalyzed by serine

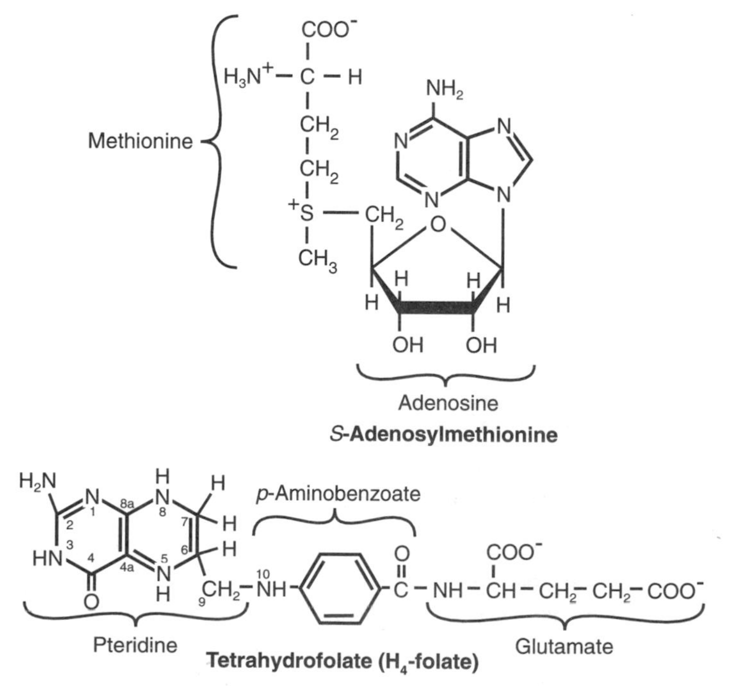

Figure 9-1. Structures of tetrahydrofolate and *S*-adenosylmethionine.

hydroxymethyltransferase. The reaction involves serine and THF and yields N^5,N^{10}-methylene-THF, glycine, and water.

Methylene-THF can be reduced by nicotinamide adenine dinucleotide (NADH) to yield N^5-methyl-THF and the oxidized form of nicotinamide adenine dinucleotide (NAD^+) in a unidirectional reaction. N^5-methyl-THF can transfer its methyl group to homocysteine in a vitamin B_{12}–dependent reaction catalyzed by methyl-THF–homocysteine methyltransferase to form methionine. With a deficiency of vitamin B_{12}, methyl-THF accumulates, because it can neither be converted back to methylene-THF nor react with homocysteine. This "methyl trap" hypothesis explains some of the metabolic derangements associated with pernicious anemia and vitamin B_{12} deficiency. Methionine is the precursor of *S*-adenosylmethionine, an important methyl donor, as noted previously. Methylene-THF also serves as methyl donor and reductant for **thymidylate** synthesis (see Pyrimidine Biosynthesis). Thymidylate is incorporated into DNA. N^5,N^{10}-Methylene-THF can be oxidized by the oxidized form of nicotinamide adenine dinucleotide phosphate ($NADP^+$) to yield N^5,N^{10}-methenyl-THF. This serves as a one-carbon donor in purine biosynthesis. N^5,N^{10}-methenyl-THF can also be formed from formimino-THF.

NUCLEOTIDE METABOLISM

Nucleotide Structures

Purine and pyrimidine nucleotides are nitrogen-containing, aromatic compounds with many important biological functions. Nucleotides serve as carriers of metabolic energy (ATP) and as coenzymes in oxidation-reduction reactions (NAD^+, $NADP^+$). Nucleotides serve as metabolic second messengers [e.g., cyclic adenosine monophosphate (cAMP)] and are components of DNA and RNA.

The structures of the common purine and pyrimidine bases are shown in Fig. 9-2. These compounds are called *bases* because they contain nitrogen, which accepts protons. In physiologic conditions, however, these nitrogen atoms rarely bear a frank positive charge as does the ammonium ion. The bases are aromatic and planar. **Adenine** and **guanine** occur in both DNA and RNA. **Cytosine** and **thymine** occur in DNA; **cytosine** and **uracil** occur in RNA. **Nucleosides** consist of a base and sugar (Fig. 9-3). The sugar is **ribose** in RNA and in the common coenzymes. The sugar is **2-deoxyribose** in DNA. **Nucleotides** are **phosphorylated nucleosides** and consist of a base, sugar, and

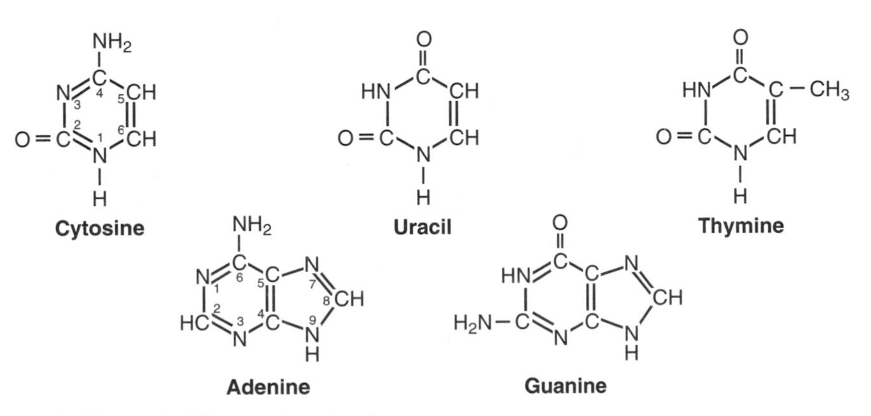

Figure 9-2. The pyrimidine and purine bases.

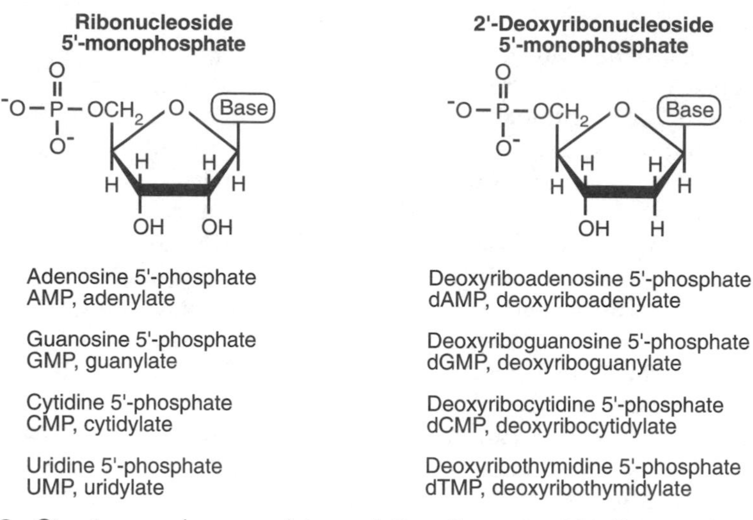

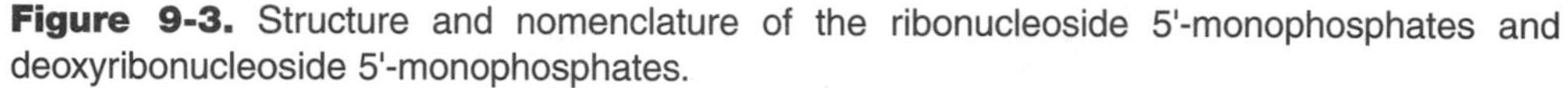

Figure 9-3. Structure and nomenclature of the ribonucleoside 5'-monophosphates and deoxyribonucleoside 5'-monophosphates.

phosphate. AMP is a nucleo**side** monophosphate, mistakenly called **nucleotide monophosphate**. The term **nucleotide** signifies the presence of phosphate. Table 9-2 lists the bases and the corresponding nucleosides and nucleotides.

Pyrimidine Biosynthesis

Nucleotide biosynthesis can be divided into two categories: simple and complex. Pyrimidine biosynthesis is simple when compared with purine biosynthesis. First, pyrimidine formation is discussed, and next the pathway for purine biosynthesis. The bases of the pyrimidines are derived from two precursors, namely, aspartate and carbamoyl phosphate (Fig. 9-4). The pathway for pyrimidine biosynthesis consists of four portions:

TABLE 9-2. Nomenclature of the Common Purines and Pyrimidines

Base (Abbreviation)	Nucleoside; Base-Sugar	Nucleotide; Nucleoside Monophosphate (Abbreviation)	Nucleoside Triphosphate (Abbreviation)
Purine			
Adenine (A)	Adenosine	Adenosine monophosphate (AMP)	Adenosine triphosphate (ATP)
Guanine (G)	Guanosine	Guanosine monophosphate (GMP)	Guanosine triphosphate (GTP)
Hypoxanthine (H)	Inosine	Inosine monophosphate (IMP)	Inosine triphosphate (ITP)
Xanthine (X)	Xanthosine	Xanthosine monophosphate (XMP)	Xanthosine triphosphate (XTP)
Pyrimidine			
Cytosine (C)	Cytidine	Cytidine monophosphate (CMP)	Cytidine triphosphate (CTP)
Utracil (U)	Uridine	Uridine monophosphate (UMP)	Uridine triphosphate (UTP)
Thymine (T)	Thymidine*	Thymidine* monophosphate (TMP)	Thymidine* triphosphate (TTP)

**Refers to the deoxyribose derivative; the unusual ribose derivative is called ribothymidine.*

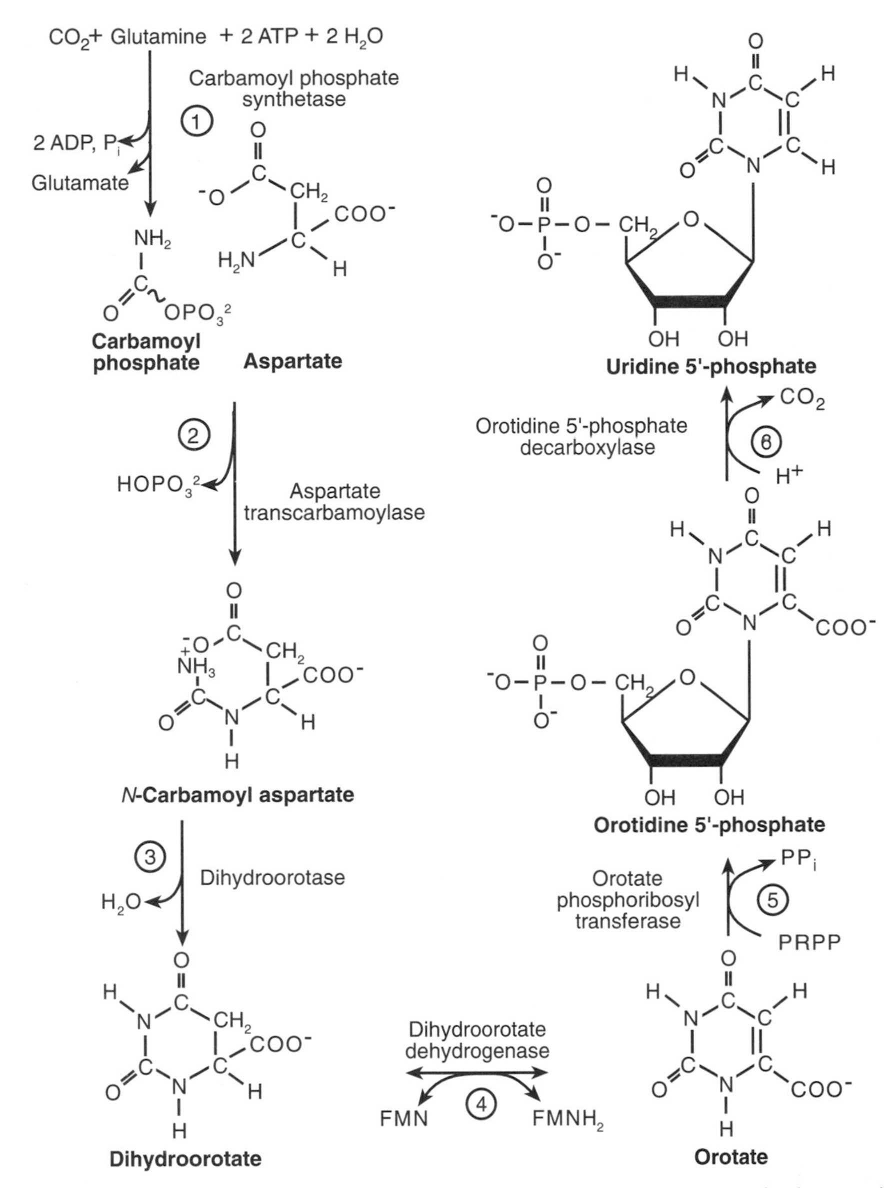

Figure 9-4. Pathway for pyrimidine biosynthesis. The CAD complex, a single protein, catalyzes reactions *1–3*. A single enzyme catalyzes the fourth reaction, whereas the UMP synthase complex catalyzes reactions *5* and *6*. ADP, adenosine diphosphate; ATP, adenosine triphosphate; FMN, flavin mononucleotide; $FMNH_2$, reduced form of flavin mononucleotide; PP_i, inorganic pyrophosphate; PRPP, phosphoribosylpyrophosphate.

(1) formation of carbamoyl phosphate, (2) formation of the pyrimidine ring, (3) addition of pentose phosphate, and (4) formation of the various pyrimidine derivatives.

The formation of active carbamate in humans is catalyzed by **carbamoyl-phosphate synthetase II**. This enzyme differs in three ways from the enzyme that participates in urea biosynthesis. The pyrimidine enzyme (1) is found in the cytosol (see Table 1-1) of all nucleated cells (and not just in liver and kidney mitochondria), (2) uses glutamine as the amido group donor, and (3) is not regulated by *N*-acetylglutamate. The reaction catalyzed by carbamoyl-phosphate synthetase II is given by the following chemical equation:

$$2\ ATP + HCO_3^- + glutamine \rightarrow 2\ ADP + P_i + glutamate + carbamoyl\ phosphate$$

Aspartate transcarbamoylase catalyzes the reaction between aspartate and carbamoyl phosphate to yield *N*-carbamoyl aspartate (see Fig. 9-4). Dihydroorotate then catalyzes a condensation reaction with the concomitant removal of the elements found in water to produce dihydroorotate. Dihydroorotate is oxidized by NAD^+ in a dihydroorotate dehydrogenase–catalyzed reaction to form orotate. Dihydroorotate dehydrogenase is a flavoprotein found on the outer face of the inner mitochondrial membrane.

Next, the reaction responsible for adding ribose-phosphate to the pyrimidine ring is considered. The donor is phosphoribosylpyrophosphate (PRPP), and we consider the pathway for PRPP formation. The reactants for PRPP formation include ribose 5-phosphate (from the pentose phosphate pathway) and ATP, and the enzyme is **PRPP synthetase**. The pyrophosphoryl group is transferred from ATP to the hemiacetal oxygen on C^1; PRPP and AMP are the products.

PRPP represents active phosphoribose; the standard free energy of hydrolysis of PRPP to ribose 5-phosphate and PP_i is approximately –7 kcal/mol. PRPP reacts with orotate to yield orotidine monophosphate (OMP) and PP_i. The latter is hydrolyzed to two P_i molecules by a separate pyrophosphatase that helps to pull the reaction forward. OMP is decarboxylated in a pyridoxal phosphate–dependent, or vitamin B_6–dependent, fashion to form uridine monophosphate (UMP) and CO_2. Uridylate kinase catalyzes an isoergonic reaction of UMP and ATP to yield uridine diphosphate (UDP) and ADP.

UDP is a key intermediate in pyrimidine metabolism (see Fig. 9-4). Nucleoside diphosphokinase catalyzes an isoergonic reaction between UDP and ATP to form uridine triphosphate (UTP) and ADP. UTP is a precursor for RNA biosynthesis and numerous uridine diphosphate sugar compounds used in carbohydrate metabolism. UTP is also converted to cytidine triphosphate (CTP). **CTP synthetase** catalyzes a reaction between UTP, ATP (an energy source), glutamine (an amido donor), and water to yield CTP, ADP, P_i, and glutamate. CTP is a precursor of RNA and numerous cytidine diphosphate (CDP) derivatives important in lipid biosynthesis such as CDP-choline (see Figs. 7-7 and 7-8).

UDP is also a precursor of thymidylate. Ribonucleotide reductase catalyzes the reduction of UDP to deoxyuridine diphosphate. **Ribonucleotide reductase**, or **ribonucleoside diphosphate reductase**, catalyzes a reaction between nucleoside **diphosphates** and reduced thioredoxin to form the corresponding deoxynucleoside diphosphate and oxidized thioredoxin (Fig. 9-5). The latter is decreased by reduced nicotinamide adenine dinucleotide phosphate (NADPH) and H^+ in a reaction catalyzed by **thioredoxin**

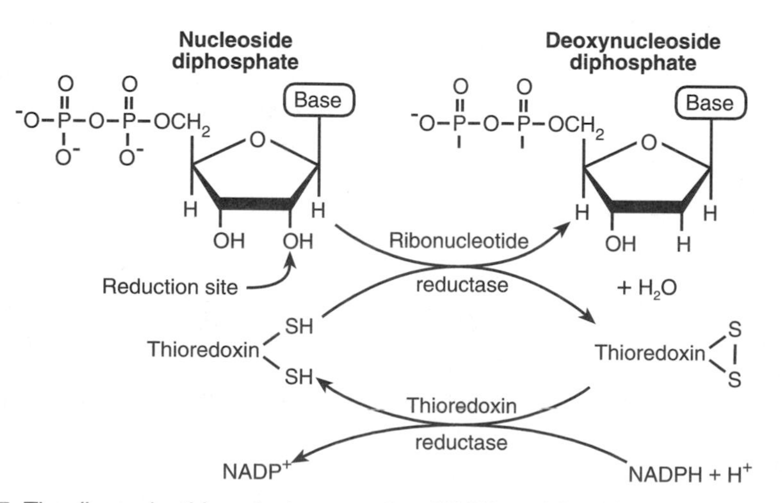

Figure 9-5. The ribonucleotide reductase reaction. $NADP^+$, oxidized form of nicotinamide adenine dinucleotide phosphate; NADPH, reduced nicotinamide adenine dinucleotide phosphate.

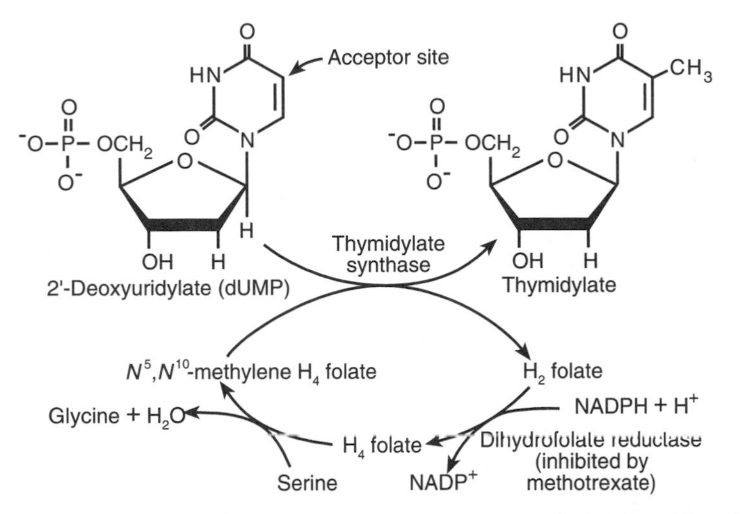

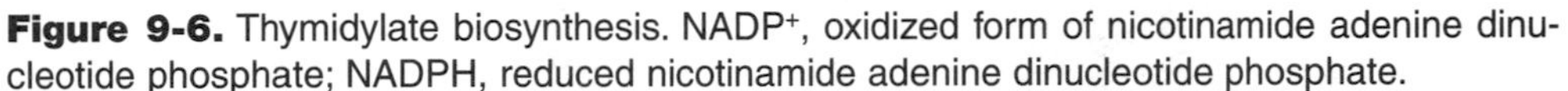
Figure 9-6. Thymidylate biosynthesis. $NADP^+$, oxidized form of nicotinamide adenine dinucleotide phosphate; NADPH, reduced nicotinamide adenine dinucleotide phosphate.

reductase. Ribonucleotide reductase is extremely important because of its obligatory participation in the synthesis of DNA precursors. Uridine nucleotides, however, are not genuine constituents of DNA. Deoxyuridine diphosphate is converted to deoxyuridine triphosphate (dUTP) by phosphorylation in a reaction catalyzed by nucleoside diphosphate kinase, and then dUTP is hydrolyzed to deoxyuridine monophosphate (dUMP) and PP_i in a reaction catalyzed by deoxyuridine triphosphatase.

dUMP is a substrate for the important **thymidylate synthase** reaction (Fig. 9-6). dUMP reacts with N^5,N^{10}-methylene-THF to yield deoxythymidine monophosphate (thymidylate monophosphate) and dihydrofolate. In this reaction, N^5,N^{10}-methylene-THF serves as a one-carbon donor and as a reductant. This sequence of reactions is important in understanding the mechanism of action of antifolates such as methotrexate.

Methotrexate is a drug used in the treatment of children with **acute lymphocytic leukemia** and a number of other neoplastic disorders. Methotrexate is a structural analogue of dihydrofolate and binds avidly to **dihydrofolate reductase**. Methotrexate inhibits the conversion of dihydrofolate to THF, thereby diminishing the levels of the fully reduced and active form of folate. The therapeutic effectiveness of this and other drugs is dependent on the differential sensitivity of normal and tumorigenic cells. Although neoplastic tissues may initially be sensitive to the action of methotrexate, resistance may develop. One form of resistance is related to overproduction of the enzyme dihydrofolate reductase. Thymidylate synthase is a target for fluorouracil, an anticancer agent.

Purine Biosynthesis

In contrast to the simple pathway for pyrimidines, that for purines is intricate, and only the initial steps are considered in detail. The purine ring is built on the ribose phosphate backbone. This contrasts with pyrimidine formation in which the sugar phosphate is added after formation of the ring (see Fig. 9-4). The source of atoms found in the purine ring, illustrated for inosine monophosphate (IMP), is indicated in Fig. 9-7.

The first and committed step in the pathway is catalyzed by **glutamine phosphoribosylpyrophosphate amidotransferase**. The reactants are PRPP, an activated form of the 5-phosphoribosyl group, glutamine, and water. The products include glutamate, PP_i, and 5-phosphoribosylamine. Note that inversion at C^1 of ribose occurs (see Fig. 9-7). C^1 changes from the α-configuration in PRPP to the β-configuration in 5-phosphoribosylamine. Next, the carboxylate group of glycine condenses with the amino group

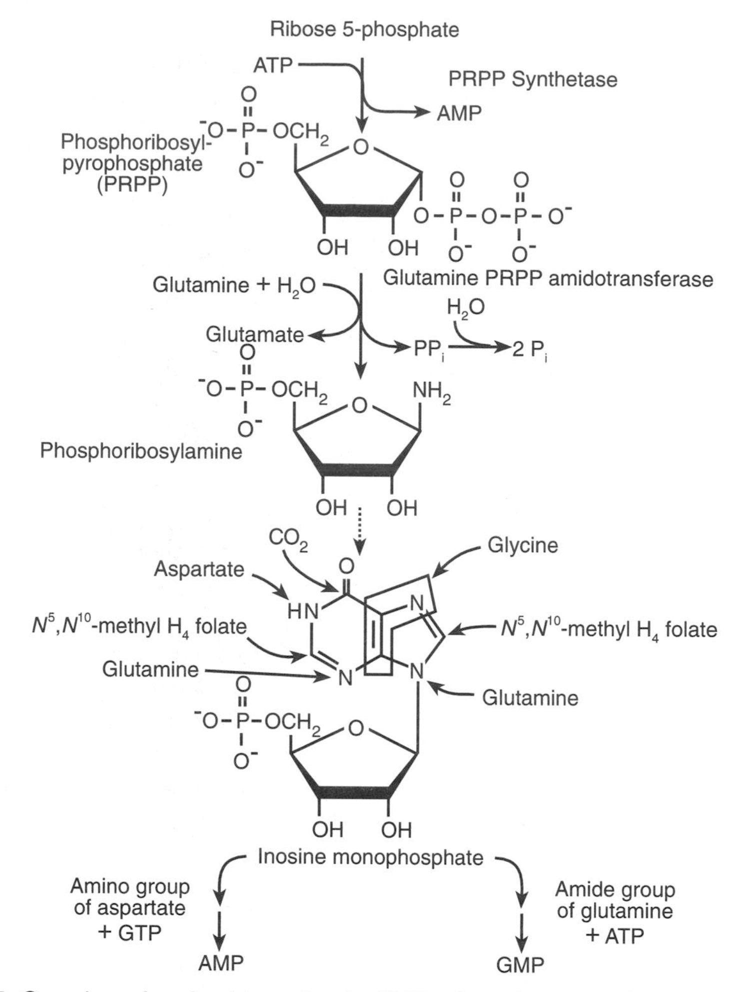

Figure 9-7. Overview of purine biosynthesis. AMP, adenosine monophosphate; ATP, adenosine triphosphate; GMP, guanosine monophosphate; GTP, guanosine triphosphate; P_i, inorganic phosphate; PP_i, inorganic pyrophosphate.

in an ATP-dependent reaction. Glycine contributes atoms 4, 5, and 7 of the final purine structure. *N*-formyl-THF then donates C^8. Glutamine contributes N^3 in an ATP-dependent reaction (ATP → ADP + P_i). The five-membered ring is formed in a subsequent ATP-dependent reaction (ATP → ADP + P_i). This constitutes a mechanism for removing the elements of water from the precursor to yield the five-membered imidazole ring. C^6 is next derived from CO_2 in an **ATP-independent** reaction. N^1 is derived from aspartate in a reaction analogous to that seen in urea biosynthesis. The amino group of aspartate condenses with a carboxylate to yield an *N*-succinylamide derivative in an ATP-dependent process. Unlike the reaction in urea biosynthesis (ATP → AMP + PP_i), in purine formation ATP is converted to ADP and P_i. As in urea formation, a lyase catalyzes the elimination of fumarate.

In summary, to add nitrogen, aspartate reacts with a compound in an ATP-dependent reaction and then fumarate is eliminated. Next, N^5,N^{10}-methenyl-THF donates C^2. Closure of the six-membered ring occurs with the elimination of water; no ATP is required for this reaction. The product of the reaction is 5'-IMP, as shown in Fig. 9-7.

IMP is the precursor of AMP and guanosine monophosphate (GMP) and occupies a branch point for these processes (see Fig. 9-7). A two-step reaction is required to convert IMP to **AMP**. This conversion involves the replacement of an oxygen (or hydroxyl group of the tautomer) by an amino group. The amino donor is again aspartate using

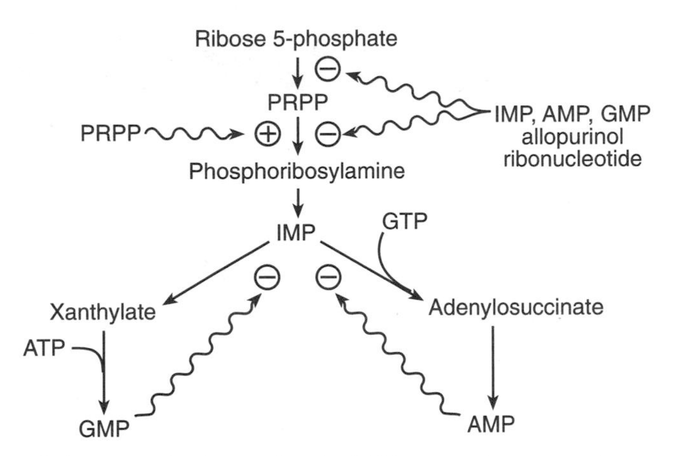

Figure 9-8. Regulation of purine biosynthesis. AMP, adenosine monophosphate; ATP, adenosine triphosphate; GMP, guanosine monophosphate; GTP, guanosine triphosphate; IMP, inosine monophosphate; PRPP, phosphoribosylpyrophosphate.

a familiar motif. IMP condenses with aspartate [as **guanosine triphosphate (GTP)** is converted to guanosine diphosphate (GDP) + P_i] to form *N*-succinyladenylate. A lyase catalyzes the elimination of fumarate to form AMP. Two steps are also required to convert IMP to GMP. First, IMP is oxidized in a reaction involving IMP, H_2O, and NAD^+. The products are NADH + H^+ and xanthosine monophosphate (XMP). Glutamine serves as the amido donor to yield the amino group found on C^2 of the purine ring. The reactants are XMP, glutamine, water, and ATP; the products are GMP, glutamate, AMP, and PP_i. Note that ATP is required for GMP formation and GTP is required for AMP formation. Kinases catalyze the phosphorylation of the purine nucleoside monophosphates to the corresponding diphosphates with ATP as donor. The nucleoside diphosphates are substrates for ribonucleotide reductase to yield the corresponding deoxyribonucleotides as necessary (see Fig. 9-5). GTP, deoxyguanosine triphosphate, and deoxyadenosine triphosphate are formed from the corresponding diphosphates and ATP. Several mechanisms exist for the formation of ATP from ADP by substrate-level and electron-transport phosphorylation.

In addition to the synthesis of purines from low–molecular-mass precursors (*de novo* synthesis), salvage pathways exist. The preformed bases (adenine, hypoxanthine, guanine) resulting from degradative reactions are reused for anabolic pathways. Two enzymes catalyze the salvage reactions. **Adenine phosphoribosyl transferase** catalyzes a reaction of substrate with PRPP to form AMP and PP_i. **Hypoxanthine-guanine phosphoribosyl transferase** (HGPRT) catalyzes a reaction of either base with PRPP to form IMP (from hypoxanthine) or GMP. The physiologic importance of salvage pathways was not fully appreciated until it was discovered that **Lesch-Nyhan syndrome** is due to a hereditary deficiency of HGPRT. This human genetic disorder is associated with extreme aggression and self-mutilation. It is X-linked, is recessive, occurs in males, and is rare.

The metabolism of PRPP is preeminent in both purine and pyrimidine metabolism and is important in metabolic regulation (Fig. 9-8). The committed step in purine formation is catalyzed by **glutamine PRPP amidotransferase**. This amidotransferase is also allosterically inhibited by AMP and GMP. Moreover, the amidotransferase is the most important regulatory enzyme of purine biosynthesis. Pyrimidine biosynthesis in humans is regulated at **carbamoyl-phosphate synthetase II**. Synthetase II is inhibited by UTP and is activated by ATP and PRPP. The regulatory step of pyrimidine biosynthesis in *Escherichia coli*, in contrast to humans, is at the level of the aspartate transcarbamoylase reaction, which is inhibited by CTP. Aspartate transcarbamoylase from *E. coli* is historically important in biochemistry because it was one of the first allosteric enzymes to be characterized. The conversion of IMP to AMP and GMP is differentially

TABLE 9-3. Selected Enzymes of Purine Catabolism

Enzyme	Reaction	Comments
5'-Nucleotidase	5'-Nucleotide + $H_2O \rightarrow$ nucleoside + P_i	Operates on all 5'-nucleotides and 5'-deoxynucleotides, including IMP and XMP
AMP deaminase	AMP + $H_2O \rightarrow NH_3$ + IMP	—
Adenosine deaminase	Adenosine + $H_2O \rightarrow NH_3$ + inosine	Hereditary deficiency associated with fatal immunodeficiency syndrome
Purine nucleoside phosphorylase	Purine nucleoside + $P_i \rightarrow$ purine + ribose 1-phosphate or deoxyribose 1-phosphate	—
Guanine deaminase	Guanine + $H_2O \rightarrow$ xanthine + NH_3	—
GMP deaminase (d)GMP + $H_2O \rightarrow$ (d)XMP + NH_3	NAD^+	—
Xanthine dehydrogenase	Hypoxanthine + O_2 + $H_2O \rightarrow$ xanthine + NADH + H^+ Xanthine + NAD^+ + $H_2O \rightarrow$ urate$^-$ + NADH + $2H^+$	—

AMP, adenosine monophosphate; (d)GMP, (deoxy)guanosine monophosphate; (d)XMP, (deoxy)xanthosine monophosphate; GMP, guanosine monophosphate; IMP, inosine monophosphate; NAD^+, nicotinamide-adenine dinucleotide; NADH, nicotinamide-adenine dinucleotide; P_i, inorganic; XMP, xanthosine monophosphate.

regulated. AMP inhibits its own synthesis, and GMP inhibits its own synthesis in the first step of the pathway from IMP.

Purine and Pyrimidine Catabolism

The end product of purine metabolism in humans is **uric acid** and **urate**, its salt. There are multiple pathways for converting AMP and GMP into xanthine and thence urate. Some of these enzyme activities are listed in Table 9-3. Through the action of various deaminases, a general 5'-nucleotidase, and purine nucleoside phosphorylase, purine bases result. These include adenine, guanine, and hypoxanthine. Hypoxanthine is derived from inosine, which is derived from adenosine in a hydrolytic reaction catalyzed by **adenosine deaminase**. A hereditary deficiency of adenosine deaminase results in combined immunodeficiency disease. Approximately 90% of these bases are reused by the salvage pathway in reactions with PRPP. The remaining 10% of these purines are converted to urate (Fig. 9-9). Hypoxanthine is oxidized to xanthine in a reaction catalyzed by **xanthine dehydrogenase**. Guanosine is deaminated (by hydrolysis) to yield xanthine. Xanthine dehydrogenase also catalyzes the oxidation of xanthine to urate.

A common derangement of purine metabolism is gout, with an incidence of approximately 3 in 1,000 persons. **Gout** is associated with **hyperuricemia**. Gout tophi in joints and urate calculi in the kidney occasionally result. Individuals are frequently treated with allopurinol, which is an inhibitor of xanthine dehydrogenase. The buildup of hypoxanthine resulting from inhibition of xanthine dehydrogenase provides substrate for the salvage pathway and the formation of IMP. IMP inhibits gluta-

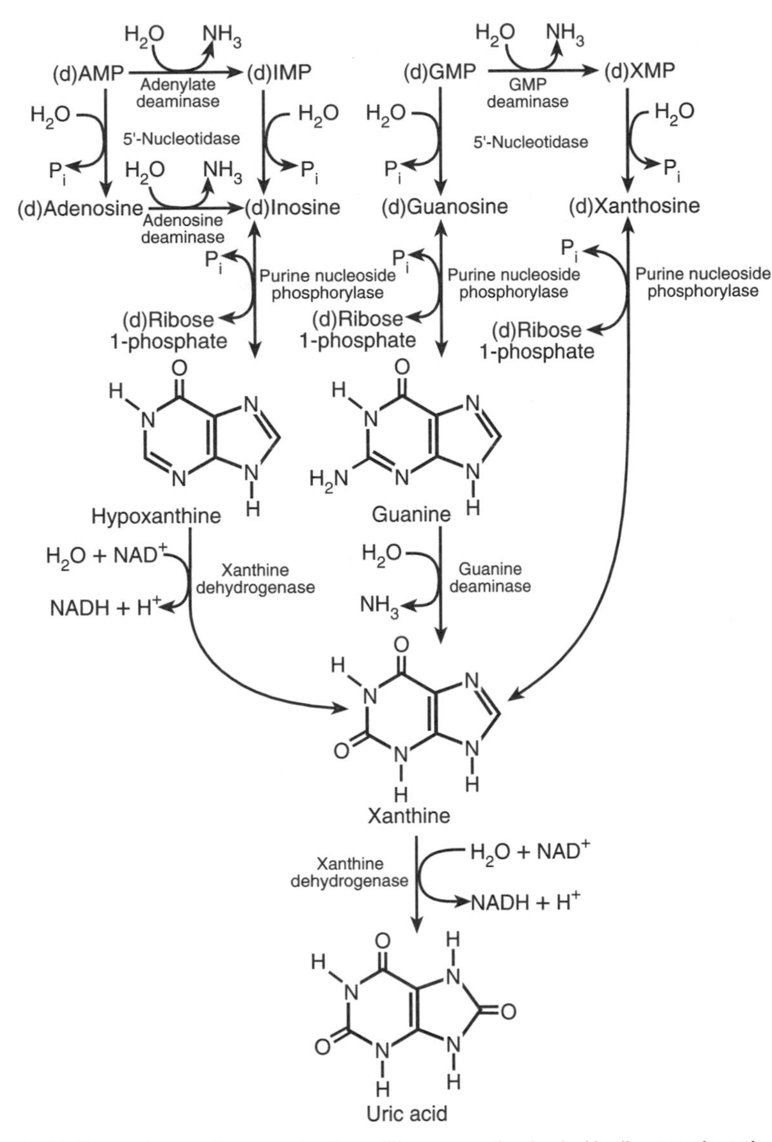

Figure 9-9. Pathway for purine catabolism. The parenthetical *d* indicates that the nucleotide in question can contain either deoxyribose or ribose. AMP, adenosine monophosphate; GMP, guanosine monophosphate; IMP, inosine monophosphate; NAD^+, oxidized form of nicotinamide adenine dinucleotide; NADH, reduced nicotinamide adenine dinucleotide; P_i, inorganic phosphate; XMP, xanthosine monophosphate.

mine PRPP amidotransferase (the rate-limiting step in purine biosynthesis). Inhibition of xanthine dehydrogenase and urate formation by allopurinol and secondary inhibition of glutamine PRPP amidotransferase by IMP and by allopurinol monophosphate (also formed by salvage enzymes) account for the therapeutic effects of allopurinol. The biochemical mechanisms responsible for the development of most cases of gout, however, have not been identified.

Pyrimidines are converted to **dihydrouracil** during their catabolism. Deaminases and nucleoside phosphorylase result in the conversion of the ribo- and deoxyribopyrimidines to uracil. Uracil is converted to dihydrouracil by NADPH and H^+ in a reaction catalyzed by dihydrouracil dehydrogenase. Dihydrouracil is hydrolyzed to β-ureido propionate, which is hydrolyzed to CO_2, ammonia, and β-alanine. Thymine catabolism yields β-aminoisobutyrate instead of β-alanine. β-Alanine is catabolized to acetyl-CoA, and β-aminoisobutyrate is catabolized to propionyl-CoA.

Chapter 10

Principles of Molecular Biology and Nucleic Acid Structure

OVERVIEW OF MOLECULAR BIOLOGY

Biology, the science of life, focuses on the nature of living organisms and how they reproduce, develop, function, adapt, and evolve. The goal of molecular biology is to understand biological phenomena at the molecular level. It has been known for millennia that progeny resemble their parents. During the twentieth century, the science of genetics has made prodigious advances; progress continues at an accelerating pace. The science of **molecular biology** concerns the structure, function, and expression of the gene. The **gene** is the unit of inheritance. In the first half of the twentieth century, the gene was a biological concept. Progress in science has occurred, and we now know that genes are made up of DNA. Genes consist of specific **sequences of bases** [adenine, thymine, guanine, and cytosine (A, T, G, and C)] along a sugar phosphate backbone, which code for specific RNAs and proteins or play a regulatory role in genetic expression.

The next section discusses the structure of DNA and how it is replicated. Studies indicate that the information present in parental DNA is used to direct the synthesis of two daughter molecules identical to the parent molecule in the process called **replication**. The pathway for the direction of information flow in biological systems is shown in Fig. 10-1. The information in the parental sequence of DNA serves as the source of information for progeny DNA in replication. It also dictates the sequence of nucleotides of RNA during gene **transcription** in RNA biosynthesis. RNA, in turn, directs the synthesis of proteins during **translation**. The four-letter alphabet of nucleic acids (corresponding to four bases) is translated into the 20-letter alphabet of proteins (corresponding to the 20 genetically encoded amino acids).

Messenger RNA (mRNA) dictates the sequence of amino acids found in proteins. mRNA in humans is derived from heterogeneous nuclear RNA (hnRNA) in a process involving splicing. Small nuclear RNA molecules (snRNA) play a role in the splicing process. **Transfer RNA** (tRNA) has two important functions: First, tRNA combines with its corresponding amino acid to produce a bioenergetically activated form of amino acid; and second, tRNA serves as an adapter in translating serial nucleic acid hydrogen-bonding patterns into an amino acid sequence of a protein. A third type of RNA, **ribosomal RNA** (rRNA), forms a scaffold for the ribosome and perhaps plays a functional role in ribosome action during protein synthesis. After ribosome-dependent protein synthesis (translation), proteins are subject to translocation to another part of the cell, exocytosis from the cell, and covalent modifications. The latter are called **posttranslational**, or **processing**, reactions and include, *inter alia*, proteolytic cleavage, glycosylation, phosphorylation, hydroxylation, methylation, carboxylation, farnesylation, and acetylation.

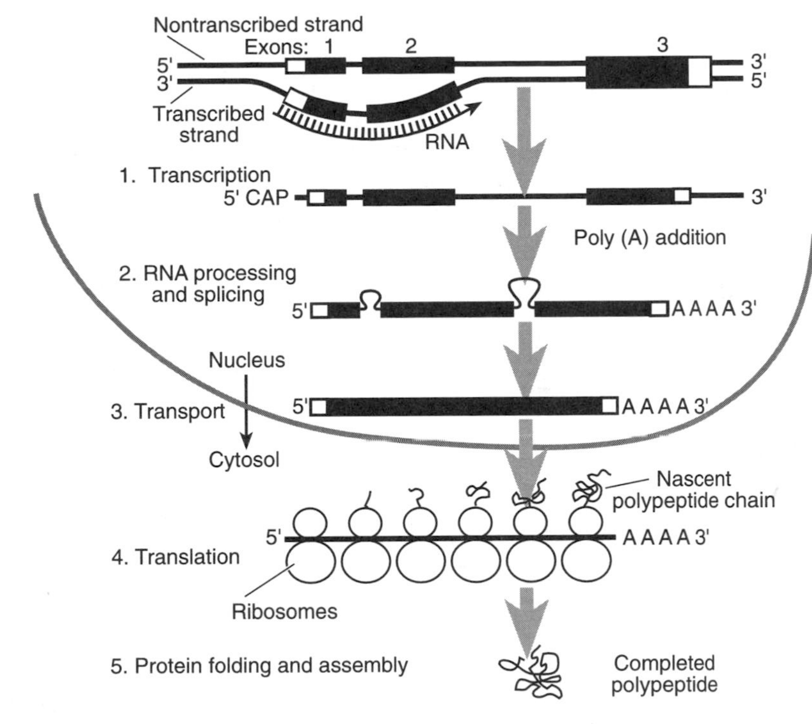

Figure 10-1. Information transfer from DNA to RNA to protein.

Before beginning a detailed consideration of nucleic acid structure and the mechanics of replication, transcription, and translation, the entire process by which the genetic material is copied and expressed in a cell is reviewed. To begin the review of gene expression, consider the flow of genetic information within a cell (see Fig. 10-1) and some of the means by which it is regulated. Evaluation of gene regulation involves more than just measuring the amounts of mRNA. There are a number of processes in both the synthesis of RNA (transcription and processing) and the generation of mature protein product (translation and posttranslational processing) that are subject to regulation.

Genomic DNA

Nearly every cell in an organism, exclusive of haploid reproductive germ cells, contains the same **genetic information** (**DNA**). For a cell to develop, only those genes within the DNA that are necessary for making that particular cell will be expressed. It is therefore important to think of genetic information as being simultaneously stable (a constant heritable legacy from cell to cell) and malleable (a changing pattern of expression of the inherited information depending on the cell and its physiologic state).

There are a number of powerful methods used to characterize genomic DNA and, therefore, the inheritance potential of humans. However, techniques such as Southern blots and the sequencing of genomic DNA are predicated on information about the specific gene(s) of interest; for instance, if a gene has been isolated (**cloned**) and it is known to be linked to a disease state, then individuals can be tested for the defect using a Southern blot, **polymerase chain reaction**, and/or **DNA sequencing**. The primary problem, therefore, is to identify the candidate genes involved in disease and later to identify the genetic polymorphism responsible for the defect.

A relatively recent addition to the study of replication is the characterization of **retroviruses** [especially human immunodeficiency virus (HIV)] and **reverse transcription**. As part of the life cycle of these devastating RNA viruses, the viral genome is copied from RNA into DNA (reverse transcription) by an *RNA-dependent* DNA polymerase. The resulting DNA is now compatible with the human genome and can be inserted into the chromosome, from where it directs viral replication and ultimately the death of the host. Much of current acquired immunodeficiency syndrome research is directed at developing specific inhibitors of the retroviral polymerase and the reverse transcription process. Azidothymidine is one agent that is used in the treatment of HIV, and its mechanism of action is the inhibition of **reverse transcriptase**.

Transcription

As stated earlier, whereas the information content within genomic DNA bears the potential of an individual, the subset of that information that is expressed (transcribed into RNA) determines the characteristics of an individual tissue or cell. The direct assessment of transcription and its regulation is accomplished using a number of different tools. These include **run-off transcription** (determining the number of active RNA polymerase molecules bound to a gene) as well as **gel mobility shift** assays and **DNA footprint analysis** (both of which assess specific DNA-binding protein activities under a given physiologic state).

RNA Processing

There is a large black box between the initial transcription of a gene, creation of the primary transcript, and the formation of the mature mRNA product that is collectively referred to as **RNA processing**. During this maturation, the primary RNA transcript is capped by adding a unique 7-methyl guanosine on the 5'-terminus. This cap serves as a pivotal signal for translation initiation and possibly prevents premature degradation of the message. Intron sequences, present within the gene, are removed from the transcript in a process termed **splicing**. In addition, a homopolymeric stretch of adenylate residues (the **polyA tail**) is added to the ends of nearly all mRNA molecules, a modification that increases the stability of mRNA and contributes to translation efficiency. Histones are unusual in that their mRNAs lack a polyA tail, and their genes lack introns. Although the mechanisms of these processing steps are known, their regulation remains obscure. It seems likely, however, that the regulation of processing will prove to be important for several areas of human pathophysiology.

The two primary means of assessing the combination of transcription and RNA processing are **Northern blot analysis** and ***in situ* hybridization**. In the former application, the total RNA from a tissue is analyzed for the presence of a specific message using nucleic acid hybridization. In the latter technique, hybridization is used to visualize the presence of mRNA in microscopic sections of tissues. The two approaches, in combination, provide information concerning the localization and magnitude of gene expression.

Translation

There is growing evidence that the regulation of translation plays a role in the expression of several genes. Most of the characterization of translational control has been limited to systems involving large homogeneous tissues (e.g., liver) or systems with well-characterized cell lines (e.g., anterior pituitary). Polysome analysis (i.e., analysis of mRNAs that are being actively translated by multiple ribosomes) and *in vitro* translation are the primary approaches to this issue.

Posttranslational Processing and Protein Stability

Finally, the modulation of posttranslational processing and the regulation of protein stability represent areas in which a great deal of work has occurred. In thinking about gene expression, it is important to keep in mind that a gene has not been successfully expressed until the functional protein has reached its physiologic destination and become active. Therefore, the myriad processesing events and second-messenger–mediated regulatory events have a major role in gene expression. These avenues of investigation make use of a variety of techniques that precede the advent of recombinant DNA technologies, yet remain very important in the characterization of functional gene expression. These techniques include enzyme assays, receptor binding protocols, and immunologic approaches. Subjects that should be explored in this regard include glycosylation, the regulation of protein transit through the Golgi apparatus, control of proteolytic processing, and lipid addition.

While reviewing the flow of genetic information, this section has illustrated the vast panoply of different experimental procedures that can be used. It is important to realize that changes in gene expression can occur anywhere in this genetic pathway. In characterizing the inherited and epigenetic phenomena associated with human health and disease, efforts must be made to view the entire process and not only one of the various steps.

STRUCTURE OF DNA

DNA is the **molecule of heredity**. DNA is a long, thin macromolecule that is made up of a large number of deoxynucleotides. It can be made up of millions of deoxynucleotides depending on the species or particular chromosome of a species. The specificity and uniqueness of a DNA molecule are determined by the **sequence of bases** constituting each deoxynucleotide (base-sugar-phosphate) unit. The information of a DNA molecule corresponds to the sequence of bases in the same way that the information in this sentence depends on the sequence of letters of the alphabet. The structure of a **tetranucleotide** is shown in Fig. 10-2. It illustrates the sugar-phosphate backbone with bases (A, T, G, and C) attached to the 1'-carbon of deoxyribose by a glycosidic bond. The bonds between the sugars are phosphodiesters linking a 3' group to the 5' group in an adjacent nucleotide unit. Linear molecules have a directionality or polarity from the 5' end to the 3' end. If an arrow is drawn from the 5'-carbon to the 3'-carbon of the same sugar, the direction of the arrow shows the 5' to 3' polarity (see Fig. 10-2).

Human DNA and almost all other DNA form a **double-stranded duplex** (the DNA of a few viruses exists as a single strand but forms a double strand during replication), described by James Watson and Francis Crick in 1953. The double-stranded DNA forms a **right-handed double helix**; the strands of the helix ascend as a right hand is turned clockwise (Fig. 10-3). The two strands of the double helix are composed of **complementary polydeoxynucleotides**. The sugar phosphate backbone of each strand is on the exterior and is represented by the ribbons; the bases face the interior and are represented by the horizontal lines. The hydrogen bonds between bases are represented by the vertical lines (see Fig. 10-3).

The crux of the Watson-Crick structure is the formation of **complementary base pairs**. The complementary base pairs are (1) **A** and **T** and (2) **G** and **C** (Fig. 10-4). The complementary nature is associated with specific base pairing involving hydrogen bonds (two hydrogen bonds between A and T and three hydrogen bonds between G and C). Each complementary pair is composed of a **purine** and **pyrimidine** (and not two purines or two pyrimidines). Note that this maintains a constant distance between the two sugar-phosphate backbones (essentially three rings and a set of hydrogen bonds). Other pair combinations would create pinches or bulges in the regular structure of the helix. Analysis of DNA molecules from a variety of species has shown that the mole fraction of adenine equals that of thymine; the mole fraction of guanine also equals that of cytosine. This is called **Chargaff's rule**. The G and C content, however, widely varies from approximately 30% to 70% in all species. (It is constant in a given species.)

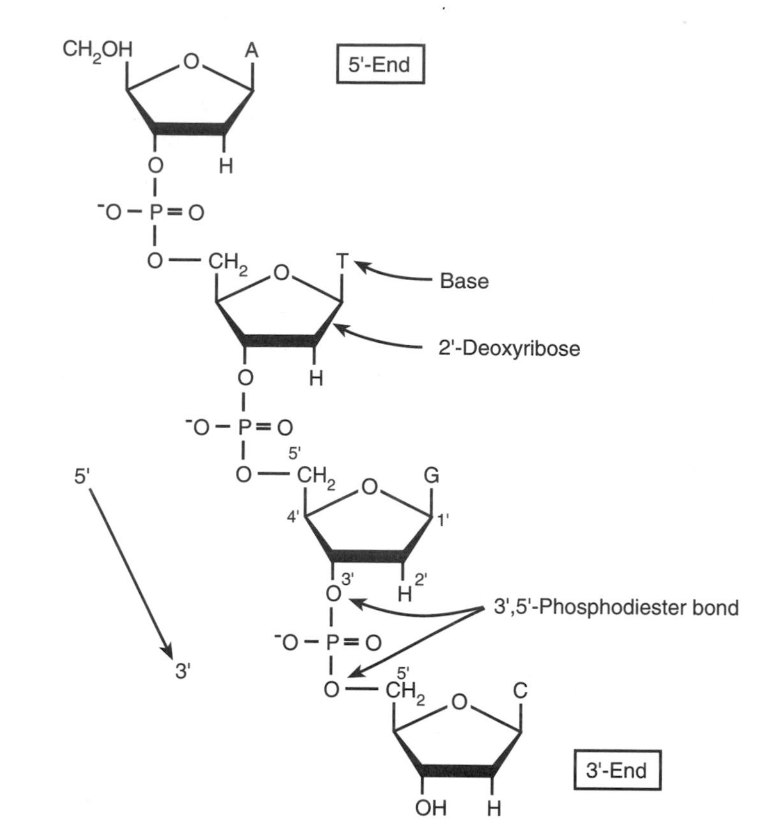

Figure 10-2. The 5' to 3' polarity of DNA.

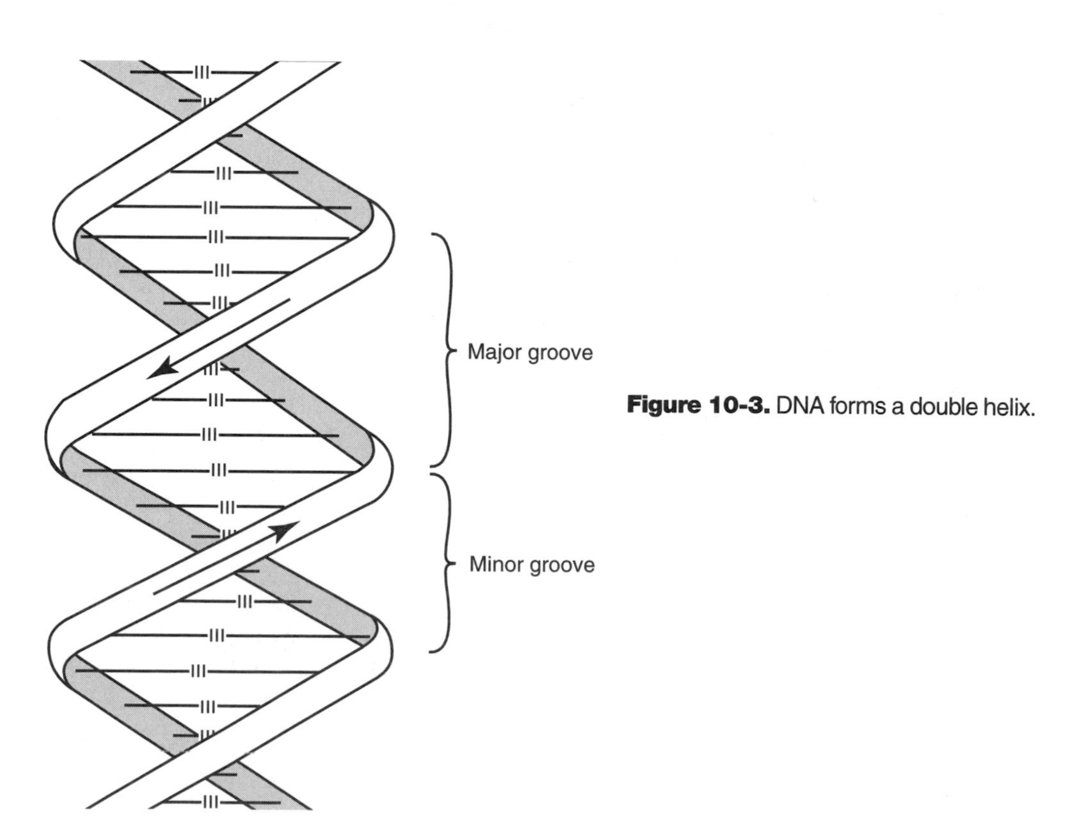

Figure 10-3. DNA forms a double helix.

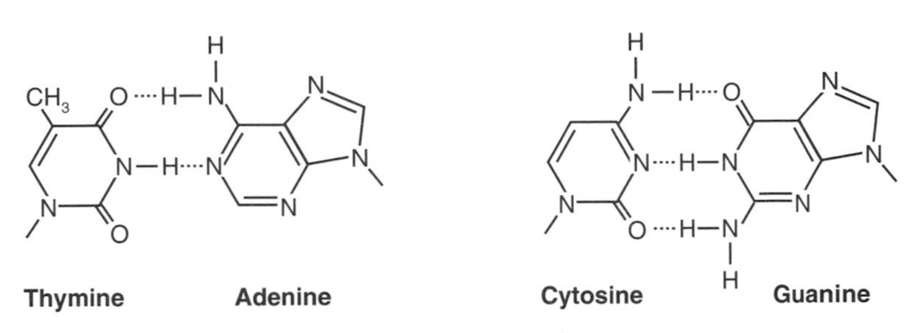

Figure 10-4. Watson-Crick complementary base pairing.

The two strands of the double helix exhibit opposite polarity. One courses in the 5'-to-3' direction, and the complementary strand extends in the opposite direction, as indicated by the arrows in Fig. 10-3. This is called the **antiparallel** property. Besides complementary hydrogen bonding between double strands in DNA, complementary base pairing occurs between DNA and RNA (during transcription) and between RNA and RNA (intramolecularly in the tRNA cloverleaf and intermolecularly between the anticodon of tRNA and the codon of mRNA). In all known instances, complementary base pairing is antiparallel in nature.

The existence of a double-stranded DNA structure with complementary base pairing has a number of theoretic and practical consequences. First, if we know the sequence of bases along one strand of DNA, then we can deduce the sequence of bases along the other because an A base pairs with T, and G with C. The sequence has the opposite polarity when compared with its complementary strand. Second, the duplex structure suggests a replication mechanism. If each strand serves as a template for the biosynthesis of a complementary strand, then two daughter DNA duplexes will result, and each will be identical with the parent duplex. Third, methods have been developed for studying the complementary interaction of natural or artificial segments of DNA or RNA with any DNA or RNA of interest. When nucleic acids form a complementary duplex, the polynucleotides are said to **anneal** or **hybridize**. The natural, functional DNA duplex exists in its native conformation. The native structure can be converted into two single-stranded polynucleotides by denaturation. The denatured form is produced by treatment with heat, alkali, or selected organic solvents such as formamide. Under appropriate conditions, the denatured DNA can reanneal to form a native duplex, and this hybridization is important in genetic diagnosis.

The familiar DNA double helix exists in the B form. The B form of DNA corresponds to a specific structure determined from its X-ray diffraction pattern. One complete turn of the double helix occurs every 3.4 nm. There are ten base pairs for each complete turn, and each pair extends 0.34 nm. The B form of DNA also exhibits a major and minor groove when viewed from the side (see Fig. 10-3).

X-ray diffraction studies have also revealed the existence of another form of DNA called **Z-DNA** (for zig zag). The major difference between the B form and Z form is that the Z form exists as a left-handed helix. The Z form still retains the Watson-Crick complementary base-pairing property. The sugar phosphates reside on the exterior, and the bases occur on the interior. The Z form is more elongated and slimmer than the B form. In Z-DNA, there are 12 base pairs per turn of the helix; one full turn is 4.6 nm in length. The physiologic role of Z-DNA is unclear.

Diploid nuclear DNA of human somatic cells contains approximately 6×10^9 base pairs. The DNA is distributed among 23 pairs of linear chromosomes. The haploid genome consists of 3×10^9 base pairs in the 23 individual chromosomes. The DNA in each chromosome consists of a single, linear DNA duplex. Each human mitochondrion contains approximately 100 small, circular DNA molecules. Mitochondrial DNA codes for 13 essential genes of oxidative phosphorylation, two rRNAs, and 22 tRNAs. The **genome** of a cell or organism refers to its total DNA. A gene is a portion of DNA that codes for a functional unit. The gene may code for a polypeptide chain, a tRNA, or an rRNA.

The manner in which the DNA of a cell is packaged or organized is formidable. The length of the DNA in a single human cell is approximately 2 m. The length of a typical human cell, however, is only 20×10^{-6} m, or 20 μm. DNA must be condensed into a compact structure. Because DNA is found in the nucleus and because the nucleus represents only a portion of the cell (see Table 1-1), the condensation required is even more formidable. **Chromatin** is the term applied to the condensed DNA-protein complex. **Euchromatin**, which is transcriptionally active, is decondensed and light-staining during metaphase. **Heterochromatin** is condensed and dark-staining during metaphase. Heterochromatin, which is not transcribed, is composed of the DNA that corresponds to centromeres, telomeres, acrocentric short arms of chromosomes, and nontranscribed DNA.

Chromatin is approximately 35% DNA, 5% RNA, and 60% protein by mass. Proteins in chromatin are divided into two classes: histones and nonhistones. The mass of DNA and protein in chromatin is nearly equal. In humans, **histones** are the most abundant proteins associated with DNA. Histones are a class of proteins that are rich in positively charged lysine and arginine residues. The positively charged residues interact with the negatively charged phosphates along the DNA backbone.

The lowest order of condensation of DNA and that which is best understood relates to the formation of nucleosomes. **Nucleosomes** resemble beads on a string when observed by electron microscopy and contain two loops of DNA containing approximately 150 base pairs wrapped around a protein core. The protein core consists of two molecules each of histone H2A, H2B, H3, and H4, which form an octomer. Adjacent nucleosomes are connected by a 50 base-pair region, associated with one molecule of histone H1, called the **linking region**. Higher order fibrils and chromatin fibers have been described. These structures account for only a small proportion of packaging necessary to delimit 2 m of DNA in an 8-μm diameter nucleus.

Nearly the entire **chromosome** (4×10^6 base pairs) of ***Escherichia coli*** contains information corresponding to proteins or functional RNA molecules. It was surprising to find that a significant fraction of human DNA consists of sequences that do not correspond to functional protein or RNA. The quantity of DNA in the human haploid genome is sufficient to encode 3×10^6 proteins. The actual number of proteins that humans produce in their lifetime, however, is estimated to range from 75,000 to 150,000. The function of the apparently excessive DNA, if any, is unknown. As we shall see, this excess information resides between genes (intergenic sequences) and within genes (intragenic sequences, better known as **intervening sequences** or **introns**).

Human DNA consists of various classes based on their copy number in a haploid genome. Approximately 75% of human DNA is unique and occurs only once. **Interspersed DNA** consists of short interspersed nuclear elements (SINEs), long interspersed nuclear elements (LINEs), a variable number of tandem repeats, and inverted repeats. SINEs consist of a few to only several hundred base pairs. The *Alu* family is one example of a short interspersed repeat. (*Alu* is a restriction enzyme used in characterizing this family.) There are 500,000 copies of the *Alu* family in humans, and *Alu* constitutes 5% of the genome. LINEs occur in unit lengths of 5,000 to 7,000 base pairs, and they also constitute approximately 5% of the genome. Inverted repeats are linear DNA sequences (from 100 to 1,000 bases) that can form stem loops and complementary structures with just one strand of DNA. The function of inverted repeats, if any, is an enigma.

STRUCTURE OF RNA

The chemical structure and function of **RNA** are important in understanding the nature of transfer of genetic information. RNA is a polyribonucleotide consisting of a sugar phosphate 3',5'-phosphodiester backbone to which either of two purine or two pyrimidine bases are attached. RNA shares many properties with DNA, but it also possesses some unique attributes. First, the pentose sugar in RNA is **ribose**. The presence of ribose confers alkaline lability to the molecule (0.1 N NaOH), a property that is advantageous for laboratory manipulations. Second, whereas both RNA and DNA contain adenine, guanine, and cytosine, RNA contains **uracil** in place of thymine. Third, RNA is **single stranded** and does not exist as a duplex. As a corollary, the content of guanine does not necessarily equal cytosine, nor does adenine equal uracil.

The **primary structure** of DNA and RNA refers to the sequence of bases along the molecule. By convention, sequences of each are given in the 5'-to-3' direction from left to right unless specified otherwise. The **secondary structure** of RNA refers to hydrogen-bonding properties (and also to specific hydrogen-bonding patterns). The single-stranded RNA molecule forms intramolecular loops when segments are self-complementary. These complementary regions are prominent in tRNA.

There are three major classes of RNA in all living organisms. Humans and other eukaryotes possess two additional RNA classes. The three universal classes are **tRNA**, **rRNA**, and **mRNA**. Some of their properties are noted in Table 10-1. Humans also contain **heterogeneous nuclear RNA**, which is the precursor of mRNA. Humans, moreover, contain **small nuclear RNA**, some of which plays a role in splicing reactions (removal of intervening sequences of RNA).

TABLE 10-1. General Classes of Eukaryotic and Prokaryotic RNA

Class	Size		Comments
Eukaryotic rRNA	18S	1,900 bases	18S, 28S, and 5.8S rRNA derived from common precursor; RNA polymerase I transcript
	28S	4,700 bases	—
	5.8S	160 bases	18S found in small ribosomal subunit; other three occur in large subunit
	5S	120 bases	RNA polymerase III transcript
Prokaryotic rRNA	16S	1,541 bases	Small subunit
	23S	2,904 bases	Large subunit
	5S	120 bases	Large subunit
Prokaryotic and eukaryotic tRNA	75–90 bases		Approximately 40 different tRNAs in cytosol of human cells; many bases modified posttranscriptionally; products include ribothymidine, dihydrouracil, pseudouridine, 4-thiouridine, inosine, and isopentenyladenosine, among others
Prokaryotic mRNA	600 bases and greater		5% of total cellular RNA; short half-life (minutes); may be polycistronic (translated into more than one protein)
Eukaryotic mRNA	600 bases and greater		5% of total cellular RNA; half-life from minutes to days; contains 5'-7 methyl G cap and polyA 3'-tail; derived from hnRNA; monocistronic
Eukaryotic hnRNA	May contain 100 kb of nucleotide or more		95% degraded in nucleus; precursor of mRNA; undergoes splicing reactions; RNA polymerase II transcript
Eukaryotic snRNA	100–300 bases		At least ten classes exist; each presents at 10^5–10^6 copies/cell; RNA polymerase III transcript

hnRNA, heterogeneous nuclear RNA; kb, kilobase; mRNA, messenger RNA; rRNA, ribosomal RNA; snRNA, small nuclear RNA; tRNA, transfer RNA.

Chapter 11

Replication—DNA Synthesis

REQUIRED COMPONENTS FOR REPLICATION

The **properties of the various enzymes** and proteins necessary for DNA replication (Table 11-1) are considered first, with emphasis on the replication process in *Escherichia coli*. The biochemistry and genetics of *E. coli* have been extensively characterized. Advances in identifying and understanding the properties of human replication enzymes have also been made. After the known components are described, the biochemistry and cell biology of DNA replication and repair are considered. DNA repair is necessary to eliminate adventitious changes that DNA has undergone as a result of the inherent instability of the bases or as a result of an environmental insult (e.g., irradiation, carcinogenic chemicals).

The **classes of enzyme activity** that are required for replication in *E. coli* include DNA polymerases and DNA ligase. Another enzyme, called **primase**, is required for the formation of a polynucleotide primer. A **helicase** is an adenosine triphosphate (ATP)-dependent enzyme that separates the bases of the double strand ahead of the site of polydeoxyribonucleotide biosynthesis. **DNA gyrase** is an enzyme that alters the supercoiling of DNA, which facilitates the polymerization reactions (see Table 11-1). Otherwise identical molecules of DNA with different degrees of supercoiling are called **topological isomers**. **Topoisomerases** catalyze the interconversion of these various forms. Topoisomerase I cleaves and reseals one strand of the DNA duplex to alter its supercoiling, and topoisomerase II cleaves and reseals two strands of the DNA duplex. Topoisomerase II is a target for drugs used in cancer chemotherapy. DNA gyrase is one type of topoisomerase.

DNA Polymerases

The elongation reactions of DNA biosynthesis are catalyzed by **DNA polymerases**. In *E. coli*, three polymerases have been described and designated I, II, and III in the order of their discovery. DNA polymerase III is responsible for the preponderance of *E. coli* DNA synthesis *in vivo*. DNA polymerase I is required for replacing the primer (a mixed ribodeoxyribonucleotide segment) and for repair synthesis. The function of DNA polymerase II is unknown. Five mammalian DNA polymerases have been described: α, β, γ, δ, and ε. Polymerase α mediates the synthesis of the lagging strand (see DNA Biosynthesis), and polymerase δ mediates the synthesis of the leading strand. Polymerase δ requires proliferating cell nuclear antigen (PCNA) for expression of its activity. Polymerases β and ε function in nuclear repair synthesis (Table 11-2).

All known **DNA polymerases** (both eukaryotic and prokaryotic) exhibit the following properties. They catalyze the elongation of an existing polynucleotide (designated as the **primer**) in the **5'-to-3' direction**. The enzyme requires all four deoxynucleoside triphosphates (as their Mg^{2+} complex) as substrates. The sequence of

TABLE 11-1. Replication Proteins in *Escherichia coli*

Protein	Role
DNA polymerase III	Synthesizes DNA
DNA polymerase I	Degrades primer and fills gaps; repair synthesis
DNA ligase	Eliminates nicks in phosphodiester backbone; NAD^+ serves as a source of phosphate bond energy
Primase	Initiates polymerization with hybrid ribodeoxyribonucleotides
Helicase	ATP-dependent separation of base pairs in the replication fork
DNA gyrase	A topoisomerase that introduces superhelical twists
Single-strand binding protein	Stabilizes single-stranded regions in replication fork

ATP, adenosine triphosphate; NAD^+, oxidized form of nicotinamide adenine dinucleotide.

deoxyribonucleotides in the growing chain is determined by a **template** strand of DNA by the principle of Watson-Crick base pairing. If a cytosine (C) base is present in the template strand, then guanine (G) is added to the growing chain (Fig. 11-1) and vice versa. If thymine (T) is present in the template strand, then adenine (A) is added to the growing chain and vice versa. The template strand is antiparallel to the growing polynucleotide chain (see Fig. 11-1).

The chemistry of the **elongation reaction** is shown in Fig. 11-2. (This diagram contains a deceptively large amount of information and should be understood by the reader.) It shows that the 3'-hydroxyl group of the growing chain attacks the α-phosphorus of the incoming deoxynucleoside triphosphate. A new phosphodiester bond forms, and inorganic pyrophosphate (PP_i) is displaced. From this diagram, it can be discerned that chain growth is in the 5'-to-3' direction. The polarity of the growing polynucleotide is such that the last residue added contains a free 3'-hydroxyl group.

TABLE 11-2. Replication Proteins in Humans

Protein	Function
DNA polymerase α	Synthesizes lagging strand of DNA
DNA polymerase β	Repair synthesis
DNA polymerase γ	Mitochondrial DNA synthesis
DNA polymerase δ	Synthesizes leading strand of DNA; PCNA cofactor
DNA polymerase ε	Repair synthesis
DNA ligase	Eliminates nicks in phosphodiester backbone; ATP serves as a source of phosphate bond energy
Topoisomerase II	ATP-dependent topoisomerase that introduces superhelical twists

ATP, adenosine triphosphate; PCNA, proliferating cell nuclear antigen.

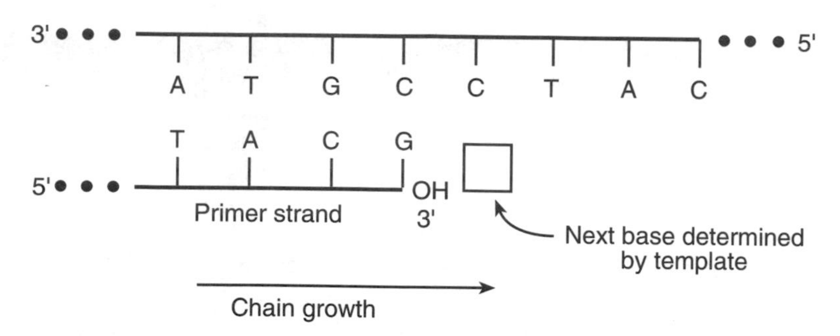

Figure 11-1. Role of template and primer in DNA biosynthesis.

DNA polymerases from bacteria exhibit **3'-exonuclease** activity. This seemingly paradoxical activity in a polymerase can catalyze the hydrolytic removal of the last nucleotide added to the growing polynucleotide chain. The product is a polynucleotide with one fewer residue and a free 3'-hydroxyl group. The best substrate for this 3'-exonuclease activity is a molecule that contains a mismatched 3'-deoxynucleotide (Fig. 11-3); for example, if G is added in a position complementary to T, then the deoxyguanosine monophosphate residue constitutes a very good substrate for hydrolytic removal. The 3'-exonuclease activity is called the **editing** or **proofreading** function of DNA polymerase. After this correction, A is incorporated. Editing increases the fidelity of enzymatic DNA replication. The polymerase selects the appropriate deoxynucleoside triphosphate by the base-pairing principle. The selection is monitored a second time by the proofreading function, and the occasional mistake is eliminated. The error frequency is reduced to 1 in 10^8 by this mechanism.

DNA polymerases from bacteria also exhibit **5'-exonuclease** activity. The enzyme can remove monomers and somewhat higher segments (perhaps up to ten) by a single hydrolytic cleavage. The enzyme is capable of degrading polynucleotide segments by 5'-exonuclease activity and filling in the resulting gaps by polymerase activity. These

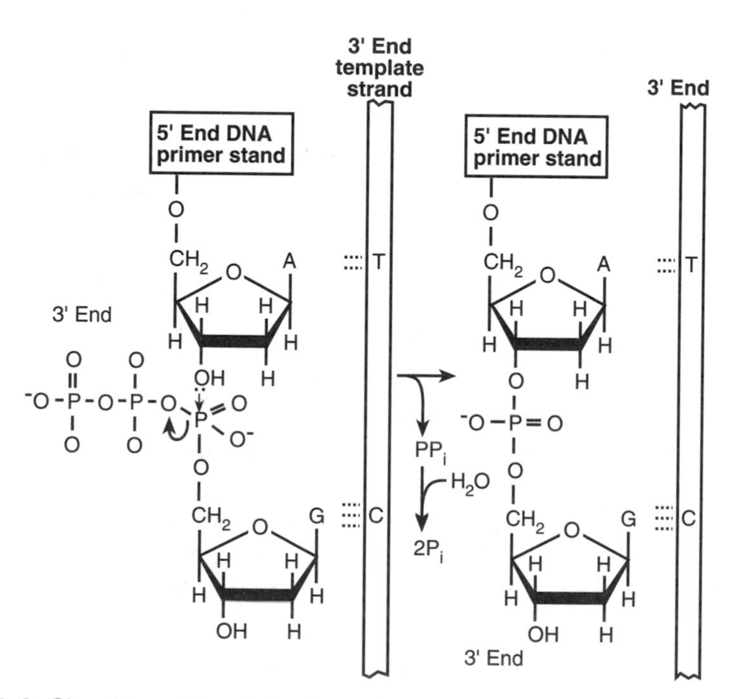

Figure 11-2. Chemistry of the chain elongation reaction of DNA biosynthesis.

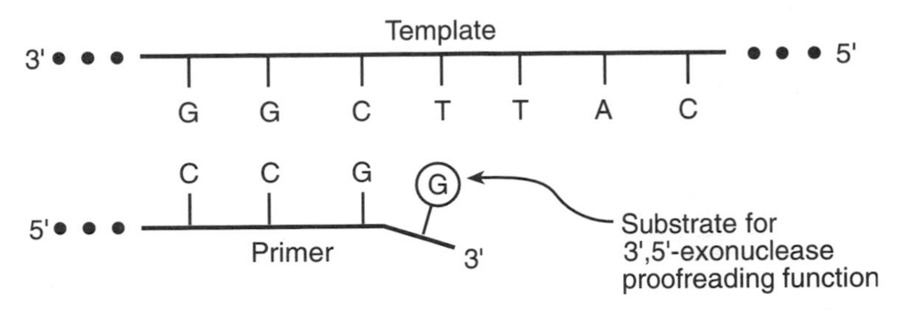

Figure 11-3. Substrate for the 3',5'-exonuclease proofreading activity of DNA polymerase.

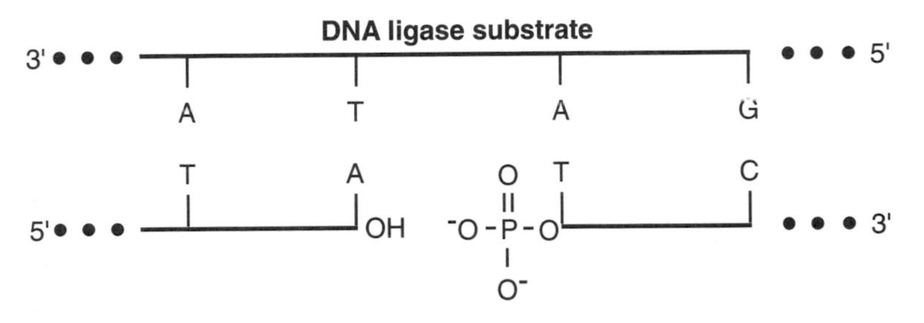

Figure 11-4. Substrate for DNA ligase.

dual characteristics are postulated to be important in removing the mixed **ribodeoxyribonucleotide primer** whose formation is mediated by primase, and in filling the gap resulting from the excision of the primer.

DNA ligase activity is responsible for linking a free 3'-hydroxyl with an adjacent 5'-phosphate occurring in a DNA duplex (Fig. 11-4). DNA ligase eliminates a nick from DNA. The elimination of nicks is important for both replication and DNA repair. In humans, DNA ligase catalyzes the adenylylation of the 5'-phosphate (Fig. 11-5). This results in activation of the 5'-phosphate; the molecule contains an acid anhydride,

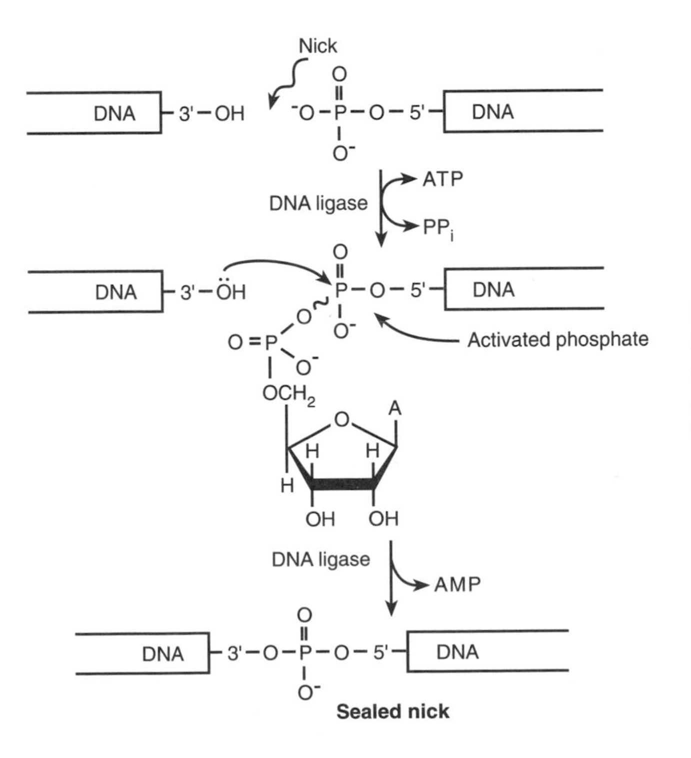

Figure 11-5. Bioenergetics of the DNA ligase reaction. AMP, adenosine monophosphate; ATP, adenosine triphosphate; PP_i, inorganic pyrophosphate.

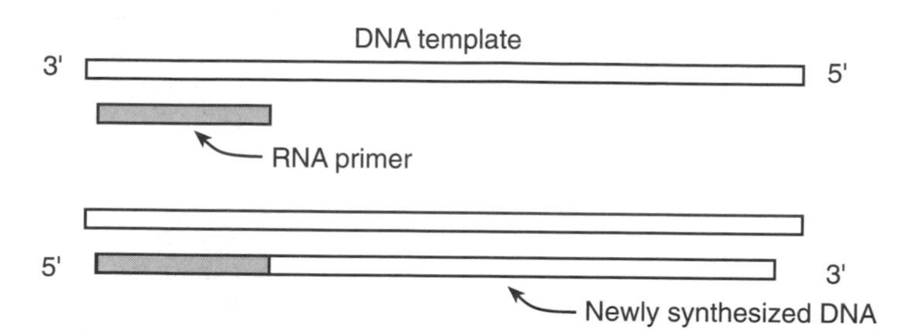

Figure 11-6. An RNA primer is required to initiate DNA biosynthesis.

high-energy bond. The enzyme catalyzes the reaction between the nucleophilic 3'-hydroxyl with the activated phosphate to produce a phosphodiester bond (the nick is eliminated), and adenosine monophosphate is released. PP_i is the other product, and PP_i is degraded by hydrolysis by pyrophosphatase. The reaction in *E. coli* is analogous except that nicotinamide-ribose-phosphate-phosphate-adenosine is the adenylyl (phosphoryl-adenosine) donor.

It was noted previously in this section that DNA polymerase requires a prefabricated polynucleotide (primer) to catalyze the formation of any phosphodiester bonds. **Primase** initiates polynucleotide formation. Primase uses both nucleoside and deoxynucleoside triphosphates as substrates and also requires a template. Chain growth is in the 5'-to-3' direction, and the reaction catalyzed is analogous to that shown in Fig. 11-2. After a primer of ten to 50 residues is produced, polymerase III uses the resulting primer and catalyzes the template-directed synthesis of polydeoxyribonucleotide (Fig. 11-6). The primer is recognized by cellular proteins as being distinct from the product of the elongation or polymerization reaction because the primer contains ribonucleotides. The primer is removed by the 5'-exonuclease activity of DNA polymerase I; DNA polymerase I also fills the resulting gap. DNA ligase completes the process by combining a free 3'-hydroxyl with a 5'-phosphate.

DNA BIOSYNTHESIS

Now consider a bird's-eye view of DNA replication. **Replication** involves the synthesis of DNA complementary to both strands of the DNA duplex, and replication forks move in both directions from a replication origin. In *E. coli,* there is a unique sequence of DNA, named **oriC**, that serves as replication origin. In human cells, there are hundreds of replication origins on each chromosome, occurring, on the average, approximately every 150 kilobases. Multiple origins of replication are illustrated in Fig. 11-7.

The following components play a role in the replication process in *E. coli.* A **helicase** separates the strands at the cost of one ATP per base pair. **DNA gyrase** (a topoisomerase) introduces supercoils to relieve the torsion. **Single-strand binding** proteins stabilize the single-stranded regions. **Primase** initiates synthesis of one strand toward the right. **Polymerase III** continues synthesis. Elongation toward the right is in the 5'-to-3' direction, and synthesis of the leading strand proceeds continuously (Fig. 11-8).

A major dilemma results in the synthesis of the opposite or lagging strand as the replication fork moves to the right. All DNA polymerases catalyze elongation in the 5'-to-3' direction. The solution to this quandary emerged when it was discovered that one strand of DNA is synthesized in short segments (**Okazaki fragments**) in a discontinuous fashion (Fig. 11-9; see Fig. 11-8). For this to occur, primase starts on the right, and polymerase III synthesizes the complementary strand toward the left. Primase then initiates synthesis of another segment farther to the right, and synthesis contin-

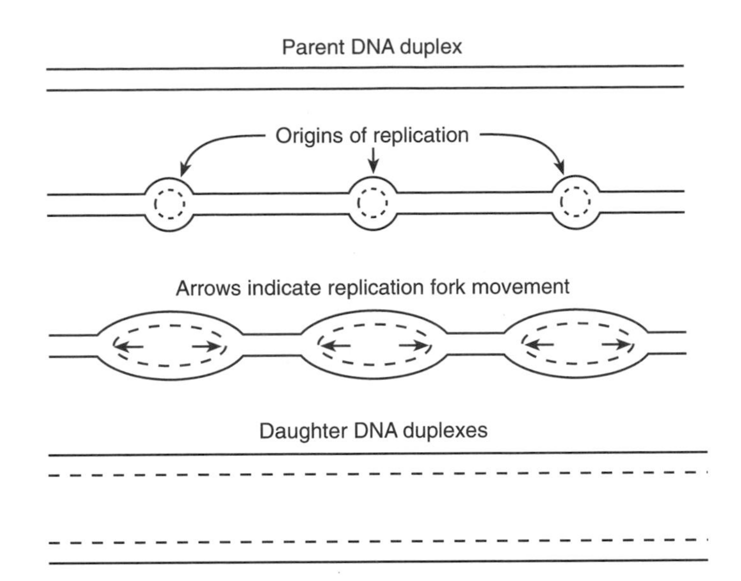

Figure 11-7. Eukaryotic DNA exhibits multiple origins of replication.

ues toward the left (see Fig. 11-9). In this fashion, both strands are elongated as the replication fork proceeds to the right. The trick used by nature is to synthesize the lagging strand in short (1,000-nucleotide) stretches to the left. This procedure emphasizes the importance of polymerase I in physiologic synthesis. At the many primer sites on the lagging strand, polymerase I 5'-exonuclease activity degrades the primer portion, and the enzyme's polymerase activity fills the gap. DNA ligase then seals the nick, and a continuous strand thereby results (see Fig. 11-9).

An analogous situation exists as the replication fork progresses to the left. Primase can initiate chain growth toward the left, and polymerase can keep it going in a continuous fashion on one of the two strands. Synthesis using the opposite strand as template begins on the left and continues to the right. A second strand begins farther to the left and continues to the first strand. Then, 5'-exonuclease activity of polymerase I removes the ribonucleotide-deoxyribonucleotide portion and fills the gap. DNA ligase seals the nick. As chain growth continues, the parental strands separate. The daughter DNA contains one parental strand and one newly synthesized strand. This property is referred to as **semiconservative replication** (**conservative replication** refers to a nonexistent situation in which both parent strands and both daughter strands are found

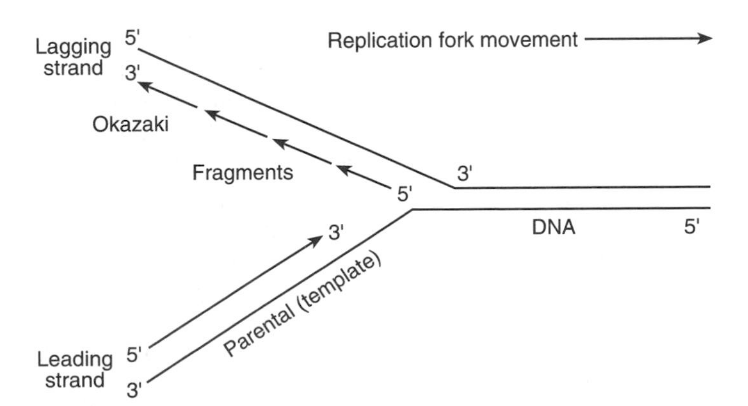

Figure 11-8. Leading and lagging strands in DNA replication.

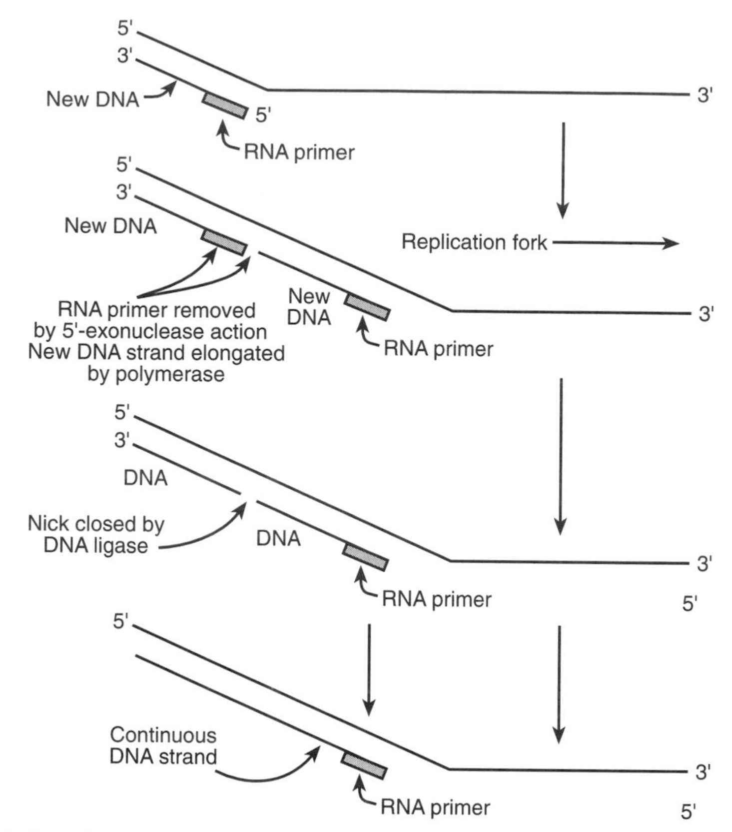

Figure 11-9. Lagging strand biosynthesis in DNA replication.

together). DNA gyrase aids in the process by converting the supercoiled DNA into more favorable topologic isomers. It removes twists that are produced by unwinding.

Human DNA replication is similar to that described for *E. coli*, although unlike *E. coli*, with a single origin of replication, thousands of replication origins exist. Replication is initiated at apparently random sites throughout the human genome. Human chromosomes also possess nucleosomes with their associated histones; bacterial chromosomes lack histones and nucleosomes. The negative charges of DNA phosphate in bacteria are neutralized by metals (e.g., Mg^{2+}, K^{+}) and organic cations such as spermine and spermidine. The histones associated with the parent human DNA duplex become associated with the leading strand and its template strand where reformation of nucleosomes occurs. Newly synthesized histones become associated with the lagging strand and its template strand. Human DNA polymerases progress along their templates with approximately one-tenth the velocity of that of *E. coli*. The discontinuous strands, or Okazaki fragments, are 10% to 20% of the length of those in *E. coli*. These differences may be related to the presence of nucleosomes and the greater degree of compactness of human DNA when compared with *E. coli* DNA.

DNA polymerase α is associated with primase activity. Primase activity is important in the initiation of DNA synthesis of the Okazaki fragments of the lagging strand. DNA polymerase α lacks 3',5'-exonuclease activity. DNA polymerase δ contains 3',5'-exonuclease activity but lacks primase activity. DNA polymerase α is dependent on **PCNA** for activity. PCNA is one type of cyclin. Cyclins are proteins necessary for replication or traversing the cell cycle, hence their name. A complex of DNA polymerase α and PCNA mediates the synthesis of the DNA leading strand. After chain initiation, the strand is synthesized continuously and there is no need for continued primase activity. Leading strand synthesis is initiated by DNA polymerase α.

DNA REPAIR

DNA, which is altered by physical and chemical agents in the environment, is inherently unstable. Altered DNA is repaired by mechanisms considered in this section. These repair mechanisms rely on information originally contained in the two strands. If only one strand is modified, information residing on the other can be used to effect the repair process. If both strands of the duplex are altered, then a **recombinational process** involving an **allele** on the other chromosome of the pair may provide the information necessary for repair. If neither type of repair occurs, then the damage is retained as a mutation (an inheritable change in DNA, which is passed on to progeny cells). Such insults accumulate in humans and have been postulated to be important factors in tumorigenesis, cell death, and aging.

Consider the types of chemical modifications that DNA undergoes and then the process of repair. The amino groups of cytosine and adenine in DNA undergo spontaneous (nonenzymatic) hydrolysis and form uracil and hypoxanthine, respectively. These structures in DNA are recognized by cellular proteins as abnormal. **Base excision repair** is the process used to correct such lesions. An **AP-glycosidase** catalyzes the hydrolytic removal of the abnormal base, leaving an intact sugar phosphate backbone. The result is an AP site where AP is an abbreviation for apurinic (lacking A or G) or apyrimidinic (lacking C or T). Endonucleases catalyze the hydrolysis of phosphodiester bonds on each side of the AP site removing deoxyribose. A repair polymerase, using a newly created 3′-hydroxyl group, replaces the deoxyribonucleotide. Then, DNA ligase seals the nick, and the original base sequence is restored. A similar sequence of reactions results when a purine is spontaneously hydrolyzed from one of the strands.

The **nucleotide excision repair** mechanism removes bulky lesions from DNA, including those resulting from tobacco smoke and ultraviolet light–induced thymine dimer formation. The four steps in this process include (1) hydrolysis on the affected strand of the 3' side of the lesion by **excinuclease** (excision endonuclease), (2) **hydrolysis** on the 5' side by excinuclease to give an oligonucleotide (approximately 30 residues) to contain the lesion, (3) **DNA synthesis** using the 3'-hydroxyl group produced by excinuclease as primer for repair polymerases (DNA polymerases δ and ε), and (4) **DNA ligation** to seal the nick (Figs. 11-4 and 11-5). The exonuclease complex requires ATP hydrolysis [adenosine diphosphate (ADP) and inorganic phosphate (P_i)] for the initial recognition step, and helicase activity to dissociate the oligonucleotide. A transcription factor is an integral component of nucleotide excision repair. **Xeroderma pigmentosum** is a hereditary disease that is due to a defect in nucleotide excision repair of thymine and other pyrimidine dimers. Affected individuals are prone to develop multiple skin cancers.

Bulky lesions such as pyrimidine dimers in the template strand block transcription. In contrast, lesions in the coding strand have no effect on progression of the transcription complex. **Transcribed DNA** is repaired faster than nontranscribed DNA, and the preferential repair is largely confined to the template strand of RNA polymerase II transcripts. **RNA polymerase II** stalls at a lesion, and the stalled complex is recognized by cellular proteins. **Transcription factor IIH** is recruited to the lesion and facilitates nucleotide excision repair. **Polymerase II** then continues the synthesis of the RNA strand.

REVERSE TRANSCRIPTION: RNA-DEPENDENT DNA SYNTHESIS

The direction of biological information flow was thought initially to extend from DNA to DNA (replication), DNA to RNA (transcription), and RNA to protein (translation). Later work with the avian Rous sarcoma virus (one of a large number of tumor viruses) showed that the viral genetic RNA possesses a DNA intermediate during its life cycle; furthermore, the DNA intermediate integrates into the host cell genome. The synthesis of DNA from RNA templates is termed **reverse transcription**; avian Rous sarcoma virus is one example of a retrovirus. The enzyme that catalyzes RNA-dependent DNA synthesis is **reverse transcriptase**; it requires the four deoxynucleoside triphosphates and an RNA template. Like DNA polymerases, reverse transcriptase cannot initiate DNA synthesis *de novo*; a primer is required. Reverse transcriptase, the RNA template (the genome), and the primer (a transfer RNA) are carried by the infectious virus. The elongation reactions proceed in the 5'-to-3' direction in a manner identical to the DNA-dependent DNA polymerases.

A large number of **retroviruses** produce cancer in a variety of species. Only one human neoplasm (**T-cell leukemia**), however, is caused by a retrovirus (human T-cell lymphotropic virus, or **HTLV-1**). Moreover, the human disorder **acquired immunodeficiency syndrome (AIDS)** is a result of infection by a retrovirus called **human immunodeficiency virus**. The study of retroviruses has been helpful in the study of tumorigenesis and AIDS. Moreover, reverse transcriptase is an extremely important tool in molecular biology and biotechnology.

Chapter 12

Transcription—RNA Synthesis

RNA BIOSYNTHESIS

RNA is transcribed from DNA. The enzymes that catalyze RNA biosynthesis are **DNA-dependent RNA polymerases**. The reaction catalyzed by prokaryotic and eukaryotic RNA polymerases is examined first; then, the properties of *Escherichia coli* and human RNA polymerases are described. After biosynthesis, the primary RNA transcript may be chemically modified in processing reactions. The signals in DNA that are important in dictating the start site for transcription and the sequences that play a regulatory role in determining the frequency of initiation are noted.

RNA polymerase, unlike DNA polymerase, is able to initiate polynucleotide synthesis; a primer is not required. This probably accounts for the role of an RNA polymerase (primase) in replication, as discussed in Chapter 11. The elongation reactions are analogous to those catalyzed by DNA polymerase. Chain growth proceeds in the 5'-to-3' direction. The sequence of nucleotides in the resulting RNA is determined by Watson-Crick base-pairing principles with the substitution of **uracil** (RNA) for **thymine** (DNA). The base-pairing rules are as follows: the adenine (A) template yields uracil (U), thymine (T) yields A, guanine (G) yields cytosine (C), and C yields G. The RNA strand is antiparallel to its template. During replication, both strands of DNA serve as a template to produce two daughter duplexes. In RNA synthesis, however, only one of the two strands of a particular genetic DNA functions as a template. One strand of the DNA duplex serves as a template in some genes, whereas the opposite strand is the template strand in other genes. The template strand, however, is antiparallel to the RNA in all cases. The **sense** strand of DNA is the strand with the sequence that corresponds to that of the RNA (except RNA contains U in place of T). The **antisense** strand is complementary to the sense strand, and the antisense strand serves as a template for RNA synthesis.

Besides the DNA template, the four **nucleoside triphosphates** (as their Mg^{2+} complexes) are required as substrates. The chemistry of the elongation reaction is illustrated in Fig. 12-1. From this diagram, you can see that chain growth occurs in the 5'-to-3' direction. RNA polymerases lack the 5'-to-3' proofreading exonuclease function that DNA polymerases exhibit. The physiologic consequence of an error in RNA synthesis is not as great as that for DNA synthesis. In the case of RNA synthesis, a mistake represents only a small part of the total expression of the gene. An error during replication, if undetected, becomes a part of the heritable legacy of the cell and so affects all expression of that gene.

The properties of **RNA polymerase** isolated from *E. coli* are intricate. The holoenzyme consists of four different protein subunits. They exhibit the stoichiometry $\alpha_2\beta\beta'\sigma$ (Table 12-1). The $\alpha_2\beta\beta'$ component constitutes the **core enzyme**, which possesses RNA polymerase activity. The **sigma subunit** (σ) confers the property of specific initiation to the core enzyme. After initiation at a physiologic start site, σ dissociates from the holoenzyme; σ can combine with another core enzyme to initiate synthesis of another

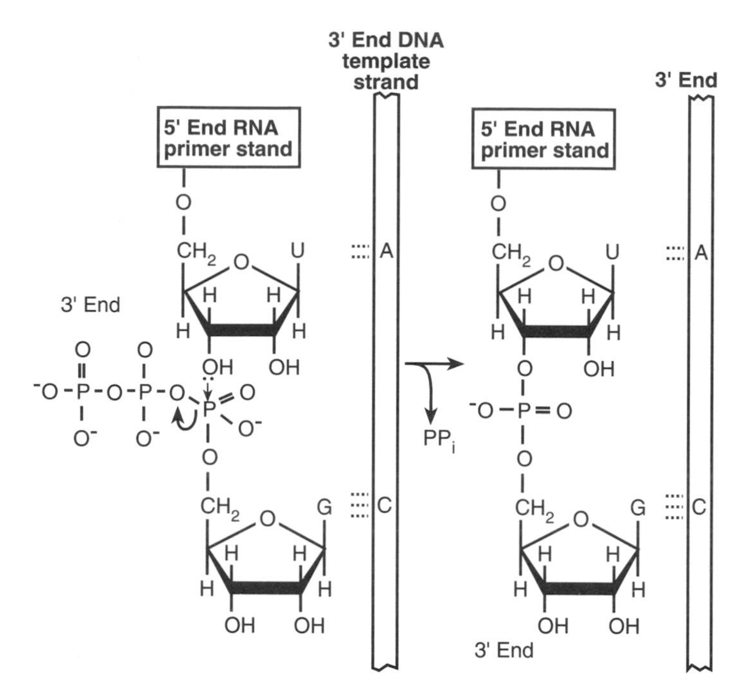

Figure 12-1. Chemistry of the RNA polymerase reaction.

chain of RNA. *E. coli* contains approximately five different sigma subunits that recognize different promoters and are responsible for differential gene expression. When the core enzyme approaches a transcriptional stop signal, a factor named **rho** (ρ) interacts with the core polymerase and results in appropriate chain termination.

In contrast to *E. coli* with its single RNA polymerase, humans have three RNA polymerases in the cell nucleus that catalyze the biosynthesis of different classes of RNA. Each of the three polymerases has more subunits (approximately ten) than the bacterial enzyme. The functions of the subunits, however, have not yet been determined. The three classes of human RNA polymerase are designated by Roman numerals I, II, and III. **RNA polymerase I** is responsible for ribosomal RNA (rRNA) synthesis in the nucleolus. **RNA polymerase II** mediates the formation of heterogeneous nuclear RNA (hnRNA), the precursor of messenger RNA (mRNA). **RNA polymerase III** catalyzes the formation of 5S RNA, transfer RNAs (tRNAs), and other small RNAs (see Table 10-1). RNA polymerase II has been studied extensively because it participates in the synthesis of mRNA, which code for proteins.

TABLE 12-1. RNA Polymerase Subunits and Transcription Factors in *Escherichia coli*

Subunit	Number in Enzyme	Function
β (Beta)	1	Catalytic site
β' (Beta prime)	1	DNA binding
α (Alpha)	2	Unknown
σ (Sigma)	1	Promotor recognition, initiation
ρ (Rho)	1	Termination

In addition to the polymerases, several auxiliary proteins called **transcription factors** are required for initiation of RNA synthesis at the correct sites on DNA. Such factors may play a role in the differential gene expression that leads to differentiation, development, and aberrant processes such as cancer. Preliminary work suggests that the number of transcription factors represents the most prevalent family of proteins encoded by the genome of eukaryotes. This indicates that the number of different transcription factors may exceed 10,000.

RNA PROCESSING

The primary transcripts of all three classes of RNA in humans undergo additional modifications called **processing reactions**. The conversion of hnRNA to mRNA by pro-

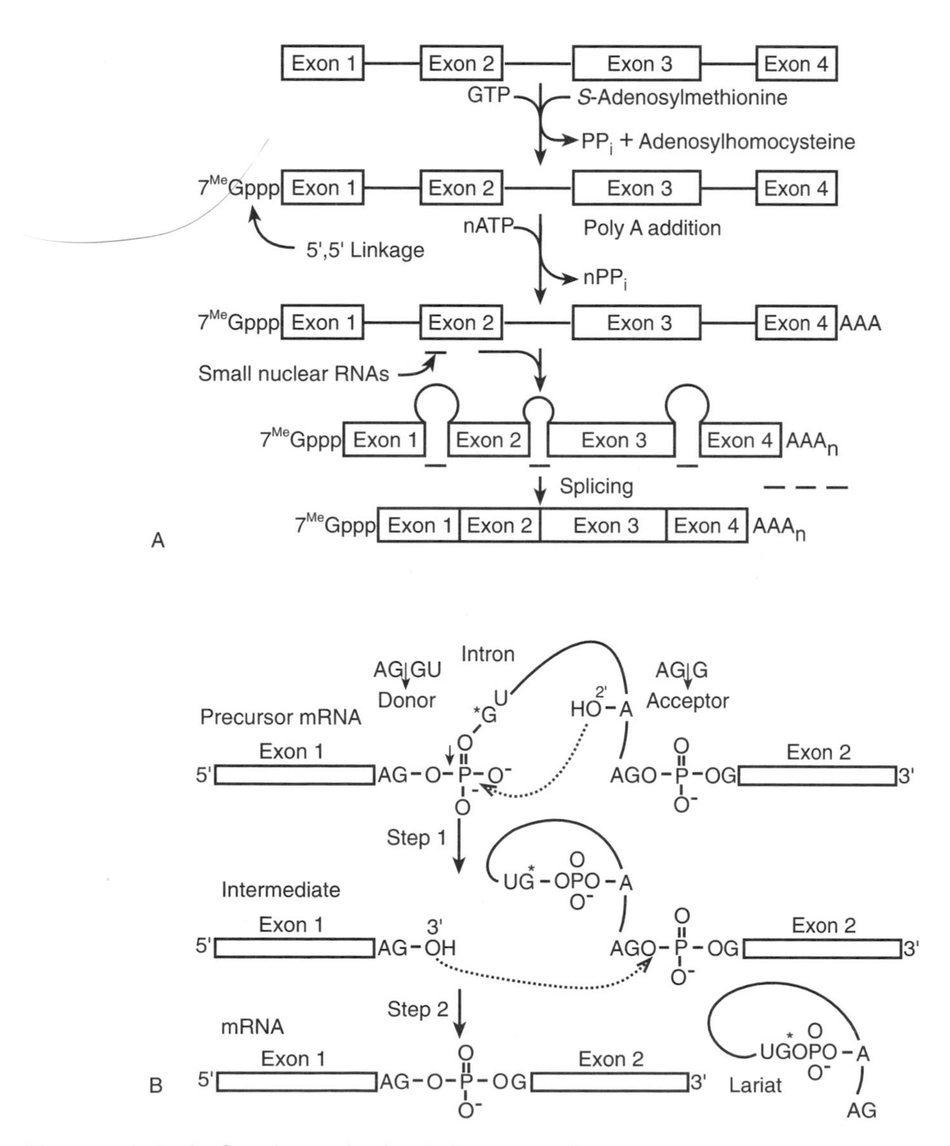

Figure 12-2. A: Capping, polyadenylation, and splicing reactions are involved in messenger RNA (*mRNA*) formation. **B:** Splicing involves the formation of a lariat form of RNA. GTP, guanosine triphosphate; nATP, variable number (n) of adenosine triphosphate precursors; nPP_i, variable number (n) of pyrophosphate molecules (corresponding to nATP); PP_i, inorganic pyrophosphate.

cessing is the most complex. Before completion of the primary transcript or RNA molecule, the 5' end of the growing nascent chain reacts with guanosine triphosphate (GTP) to form a 5' to 5' GTP terminus (Fig. 12-2A). The 5' terminus is methylated by ***S*-adenosylmethionine**, and the resulting structure is called the 5'-cap. After chain termination, the 3' end reacts with several (approximately 100) adenosine triphosphate (ATP) molecules to form a polyA tail structure (a series of covalently linked adenylate residues) and inorganic pyrophosphate (PP_i). Canonical sequences in the 3' region of hnRNA (AAUAAA) specify the site for adding the 3' polyA tail 10 to 30 nucleotides downstream from the AAUAAA signal. The cap is necessary for mRNA to interact with the ribosome, and the polyA tail stabilizes the message. Occasional mRNAs, such as those encoding histones, lack the polyA tail.

hnRNA undergoes splicing reactions. During this process, intervening sequences are removed by excision. It is imperative that splicing be performed accurately; otherwise, an aberrant protein would result. **Small nuclear RNAs** (snRNAs, pronounced "snurps") play a role in aligning hnRNA to ensure accurate splicing (see Fig. 12-2A). The signals for splicing out, or excising, intervening sequences within hnRNA consist of canonical sequences of approximately nine bases consisting of 5'-donor and 3'-acceptor sites. The first two bases of the excised RNA are GU, and the last two are AG (see Fig. 12-2B). hnRNA splicing occurs with the formation of an RNA lariat (see Fig. 12-2B). A 2'-hydroxyl group contributed by an adenosine residue in the RNA segment attacks the phosphorus atom at the 5' end of the splice junction, as shown in step 1 of Fig. 12-2B. A free 3'-hydroxyl group attacks the phosphorus atom at the 3' end of the splice junction. The two exons are joined, and the lariat-shaped RNA is displaced. These processing reactions occur in the cell nucleus before transport to cytoplasmic ribosomes.

Rifamycin is a drug that inhibits the initiation, but not the elongation or completion, of RNA biosynthesis in prokaryotes. Rifamycin binds specifically to the β-subunit of RNA polymerase. This suggests that the interaction of the β- and σ-subunits is important during the initiation process. Rifamycin is currently one of the three drugs used for the treatment of **tuberculosis**. It does not inhibit initiation of RNA synthesis in humans. **Actinomycin D** is a drug that binds to DNA and inhibits RNA elongation reactions catalyzed by RNA polymerases in both eukaryotes and prokaryotes. This substance is used in the treatment of several types of **malignant tumors**; however, it fails to inhibit DNA elongation reactions.

REGULATION OF GENE EXPRESSION

Gene regulation lies at the heart of differentiation, development, and cell maintenance. Major advances in the understanding of gene regulation have been made with the advent of gene isolation, gene sequence analysis, and recombinant DNA methodologies.

Before considering gene regulation, a few terms should be introduced. As previously stated, DNA-dependent RNA synthesis is called **transcription**. A bacterial **promoter** represents the RNA polymerase binding site on genetic DNA, and the **terminator** is the region of genetic DNA downstream from the promoter where RNA synthesis stops. The **transcription unit** extends from the promoter to the terminator, and the RNA product resulting from transcription is called the **primary transcript**.

A **cistron** is a unit of gene expression. In prokaryotes, the product of several contiguous genes may be transcribed to produce a **polycistronic message**. A single mRNA in *E. coli*, for example, codes for three proteins called β-galactosidase, permease, and acetylase. Initiation and termination of protein synthesis from a polycistronic message occurs independently for each of the components. In contrast to prokaryotes, messages for eukaryotes are generally **monocistronic**.

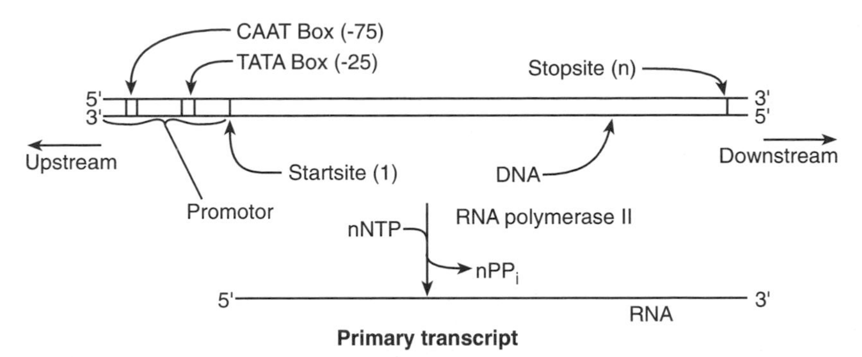

Figure 12-3. Organization of a eukaryotic gene encoding a messenger RNA. nNTP, variable number (n) of nucleoside triphosphate molecules; nPP$_i$, variable number (n) of pyrophosphate molecules (corresponding to the number of NTP molecules).

HUMAN TRANSCRIPTION SIGNALS

Through DNA sequence analysis and functional studies, **consensus sequences** for functional elements of DNA have been established. These constitute the predominant base at each position of the functional unit. Approximately 25 nucleotides upstream from the start sites of human hnRNA is an AT-rich region called the **TATA box** (Fig. 12-3). The TATA box corresponds to the following consensus sequence: TATAAAAG. Recall that there are two hydrogen bonds between A and T and three hydrogen bonds between G and C (see Fig. 10-4). The AT-rich region is therefore thought to participate in local strand separation, which allows for template-directed RNA biosynthesis. Approximately 75 nucleotides upstream from the transcription start site is the CAAT (pronounced "cat") box with its specific consensus sequence (see Fig. 12-3). Both CAAT and TATA sequences provide information on where hnRNA synthesis should originate. Note that for mRNA synthesis (i.e., RNA polymerase II transcription), the promoter itself is not transcribed into RNA.

Other classes of sequences called **enhancers** and **silencers** play a role in determining the frequency of transcription initiation. These sequences, which may be 50 to 100 nucleotides long, can be effective hundreds or thousands of nucleotides away from the promoter (either upstream or downstream). Moreover, they are effective in both orientations: left to right or vice versa with respect to the promoter. **Regulatory elements** represent DNA sequences that affect transcriptional regulation by hormones or second messengers such as cyclic adenosine monophosphate. Other genetic elements play a role in tissue-specific expression, so a given gene is transcribed only in liver, pancreas, or the hematopoietic system, and not in other cells. Proteins that bind to specific DNA sequences have been implicated in the overall process.

BACTERIAL TRANSCRIPTION SIGNALS AND THE *lac* OPERON OF *E. coli*

Bacteria such as *E. coli* have a TATA box approximately eight nucleotides upstream from an mRNA transcription start site. They also have a TTGACA consensus sequence 35 nucleotides upstream from start sites.

E. coli that are grown on glucose as a carbon source possess low levels of three proteins required to metabolize lactose. When the growth medium is changed to lactose as a carbon source, the levels of these three lactose-related proteins increase by several orders of magnitude. This system proved convenient for the study of the regulation of gene expression, and the concepts are outlined briefly.

The ***lac*** **operon** constitutes the regulatory and structural genes that play a role in lactose metabolism. The following genes are involved. A continuous DNA segment includes an **operator** sequence plus the *z*, ***y***, and ***a*** genes, which encode β-galactosidase, galactoside permease, and acetylase, respectively. An ***I*** gene, which is separated from these four genes, is also involved. The operator is a regulatory sequence and does not code for a protein, but the ***z***, ***y***, ***a*** and ***I*** genes are structural genes and code for proteins. A group of contiguous genes controlled by a single operator is called an **operon**.

The *I* gene codes for a **repressor**. The repressor is a protein that binds to the operator (a specific 17-nucleotide DNA sequence) and inhibits transcription of the *z*, *y*, and *a* genes. When lactose serves as the carbon source, lactose or a metabolite (allolactose) binds to the repressor. The sugar-repressor complex is inactive and can no longer bind to the operator. Transcription and translation of the proteins of the *z*, *y*, and *a* genes then transpires. The corresponding proteins are synthesized, and their levels increase by several orders of magnitude. The three proteins participate in the use of lactose as a fuel to support cellular metabolism. The operator lies between the promoter and start site of the *z* gene. The binding of the repressor inhibits gene expression and is a negative regulator. Moreover, examples of positive regulators, or **inducers**, are known for both bacterial and animal cells.

CARCINOGENESIS

Great advances in understanding the pathogenesis of cancer have been made, and progress continues at an accelerating rate. Epidemiologic studies have indicated that many chemicals are carcinogenic, such as asbestos (to lung), vinyl chloride (to liver), benzo[*a*]pyrene (to skin and lung), and β-naphthylamine (to the bladder). More recent work suggests that many agents undergo oxidation reactions catalyzed by the hepatic cytochrome P450 system to yield the active carcinogen.

Most carcinogens are mutagens. The **Ames test** is one method for testing the mutagenicity and possible carcinogenicity of compounds. The test substance is first incubated with a liver extract containing the cytochrome P450 system, followed by incubation with a histidine-requiring *Salmonella* strain. Mutagens produce revertants not requiring histidine; the mutagens then can be further characterized in animal systems. The Ames test is approximately 90% accurate in identifying carcinogens.

Certain DNA and RNA viruses produce tumors in animals. A study of the mechanisms of tumorigenesis by viruses has been enlightening in understanding human cancer; for example, Rous sarcoma virus contains an oncogene—a nonreceptor protein-tyrosine kinase—that is responsible for tumorigenesis. Each viral oncogene was originally derived from the host organism. The host gene is called a **proto-oncogene**. The action of **proto-oncogenes** is thought to be required for normal growth control or development.

Proto-oncogenes are carefully conserved evolutionarily and closely related genes occur in yeast, *Drosophila* (fruit flies), and humans. Oncogenes and their gene products are classified into a few families; one large family, for example, possesses protein-tyrosine kinase activity (Table 12-2). This is reminiscent of plasma membrane receptors for insulin, platelet-derived growth factor, and epidermal growth factor. The *src* gene of the Rous sarcoma virus is an example of a nonreceptor protein-tyrosine kinase. Myc and Fos are nuclear proteins that function as transcription factors. The

TABLE 12-2. Oncogenes

Designation	Virus	Gene Product Activity or Location
abl	Abelson murine leukemia	Protein-tyrosine kinase
fes	Feline sarcoma	Protein-tyrosine kinase
fps	Fujinami sarcoma	Protein-tyrosine kinase
*src**	Rous sarcoma*	Protein-tyrosine kinase
erb B*	Avian erythroblastosis	Truncated epidermal growth factor receptor and tyrosine kinase*
*sis**	Simian sarcoma	Platelet-derived growth factor B chain*
H-*ras**	Harvey murine sarcoma	Binds GTP*
K-*ras**	Kirstin murine sarcoma	Binds GTP*
*myc**	Myelocytomatosis	Nucleus*
fos	Murine osteosarcoma	Nucleus
erb-A	Avian erythroblastosis	Cytoplasm
ets	Avian E26 myeloblastosis	Cytoplasm

**Noteworthy.*

GTP, guanosine triphosphate.

production of the neoplastic state might result from either abnormal expression of proto-oncogenes or from a mutation producing an abnormal product. Evidence for both processes contributing to neoplasia has been uncovered. Mutant *ras* genes are found in approximately one-fourth of all human cancers. Proto-oncogenes can be dominant, and the mutation of a single allele can sometimes result in tumorigenesis.

Tumor suppressor genes block abnormal growth and malignant transformation. These genes are recessive, and both copies of normal diploid suppressor genes must undergo mutation to allow for malignant transformation. One tumor suppressor gene is the retinoblastoma tumor suppressor gene, called *RB1.* The gene product, RB1, appears to function at the G_1-S transition in the cell cycle. A second tumor suppressor gene, *p53,* has a protein gene product with a molecular mass of 53 kd. Alteration of this gene occurs in approximately half of all human neoplasms. p53 is usually located in the nucleus, and it normally binds to DNA. p53 can put the brakes on cell growth and division, and it prevents the unruly amplification and mutation of DNA. Normal p53 can turn on the synthesis of p21, a protein that inhibits cyclin-dependent protein kinases. As a result of this inhibition, the cell is unable to pass a checkpoint in the cell division cycle. A normal rhythm of suppressor/enhancer expression is therefore responsible for the normal regulation of the cell cycle. Perturbation of this complex interplay probably underlies most neoplastic events.

Chapter 13

Translation—Protein Synthesis

INFORMATION TRANSFER IN LIVING ORGANISMS

Crick's law of molecular biology states that information flows from DNA to RNA to protein. The alphabet of nucleic acids consists of four letters: A (adenine), T (thymine), G (guanine), and C (cytosine) in DNA and A, uracil (U), G, and C in RNA. The alphabet of proteins consists of 20 letters corresponding to the 20 amino acids that participate in ribosome-dependent protein synthesis. The conversion of the four-letter nucleic acid alphabet to the 20-letter protein alphabet is called **translation**. Translation, or protein synthesis, requires very elaborate biochemical machinery consisting of more than 150 molecular components.

Three classes of RNA are involved in translation: ribosomal RNA (rRNA), transfer RNA (tRNA), and messenger RNA (mRNA). The **rRNA** constitutes approximately half the mass of ribosomes, and protein constitutes the remainder. Ribosomes are made up of two subunits: a small subunit and a large subunit. **Ribosomes** are the subcellular machines where peptide bond formation and protein synthesis occur. **tRNA** has two functions: to serve as an adapter that recognizes the nucleic acid code, and to carry activated amino acids bound through a high-energy bond. **mRNA** carries information as a specific sequence of bases (codons), which specify the sequence of amino acids found in the corresponding protein.

Amino acyl–tRNA synthetases are a family of enzymes that catalyze the attachment of the amino acids to their corresponding or cognate tRNAs (one amino acyl–tRNA synthetase per amino acid). Several nonribosomal proteins participate in protein biosynthesis. These proteins include initiation factors; elongation factors; and termination, or release, factors. Before considering the steps involved in protein synthesis, the properties of the genetic code are enumerated.

THE GENETIC CODE

A two-letter code constructed from any four letters (A, T, G, or C in DNA or A, U, G, or C in RNA) yields 4^2, or 16, different code words, or **codons**. This is insufficient to uniquely specify the 20 different amino acids. A three-letter, or triplet, code yields 4^3, or 64, different codons, which is more than adequate to specify the 20 different amino acids. Experiments show that the genetic code is triplet in nature. The codons and their corresponding amino acids are tabulated in Table 13-1.

Based on the methodology used to decipher the code, the codons are expressed as a sequence of RNA beginning from the 5' end of each triplet. During protein synthesis, mRNA is translated triplet by triplet in the 5'-to-3' direction.

TABLE 13-1. Genetic Code

First Position (5' End)	Second Position: U	C	A	G	Third Position (3' End)
U	Phe	Ser	Tyr	Cys	U
	Phe	Ser	Tyr	Cys	C
	Leu	Ser	Stop	Stop	A
	Leu	Ser	Stop	Trp	G
C	Lou	Pro	His	Arg	U
	Leu	Pro	His	Arg	C
	Leu	Pro	Gln	Arg	A
	Leu	Pro	Gln	Arg	G
A	Ile	Thr	Asn	Ser	U
	Ile	Thr	Asn	Ser	C
	Ile	Thr	Lys	Arg	A
	Met	Thr	Lys	Arg	G
G	Val	Ala	Asp	Gly	U
	Val	Ala	Asp	Gly	C
	Val	Ala	Glu	Gly	A
	Val	Ala	Glu	Gly	G

Ala, alanine; Arg, arginine; Asn, asparagine; Asp, aspartic acid; Cys, cysteine; Gln, glutamine; Glu, glutamic acid; Gly, glycine; His, histidine; Ile, isoleucine; Leu, leucine; Lys, lysine; Met, methionine; Phe, phenylalanine; Pro, proline; Ser, serine; Stop, termination codon; Thr, threonine; Trp, tryptophan; Tyr, tyrosine; Val, valine.

Of the **64 possible codons**, all but three (i.e., 61) correspond to an amino acid. The three exceptions (UAG, UAA, UGA) are stop codons and code for chain termination. Methionine (AUG) and tryptophan (UGG) have a single codon. **AUG** is the initiating codon for protein biosynthesis, and **methionine** is the initiating amino acid in eukaryotes and prokaryotes. AUG also codes for the methionine residues that occur in the interior of proteins. The 18 other amino acids are represented by more than one codon, and this property is called **degeneracy**. Nine amino acids are represented by two codons. In this group of nine, the first two bases are the same, and the third position is either a pyrimidine (Py) or a purine (Pu). The codons are XYPy or XYPu. Five amino acids are represented by four codons. In each case, the first two bases are the same, and the third can be any of the four bases (A, U, G, or C). Three amino acids are represented by six codons. These constitute a combination of four codons and two codons with the aforementioned properties. Isoleucine is the only amino acid represented by three codons. Again, the first two bases are the same (see Table 13-1).

Because of the variation in the third position of the codon and because it participates in some nonstandard Watson-Crick base pairing with the tRNA anticodon, the third codon position was designated the **wobble** position by Crick. As a result of this property, some tRNA molecules are able to interact with (or adapt to) two or even three different codons. Approximately 40 tRNA molecules will interact with the 61 codons. (The precise nature of wobble base pairing is not considered here.) The interaction of tRNA with mRNA occurs in an antiparallel orientation. The wobble position is on the 3' end of the mRNA codon and corresponds to the 5' end of the anticodon triplet.

As mentioned previously, AUG codes for the first and initiating amino acid, which is methionine. The genetic code is read triplet by triplet in the 5'-to-3' direction until a termination or stop codon is reached. No punctuation signal is required to indicate the end of one codon and the beginning of the next (the code is **commaless**). Each base of the triplet is used only once per polypeptide synthesized (the code is **nonoverlapping**). The genetic code is (almost) **universal** and is the same in both prokaryotes (e.g., *Escherichia coli*) and the cytosol of eukaryotes (e.g., humans). This genetic code is called the **standard genetic code**, which for mitochondrial protein synthesis is exceptional and is somewhat different from the one shown in Table 13-1.

THE RIBOSOME

The ribosome is the biochemical machine on which protein biosynthesis occurs. The ribosome is where mRNA and aminoacyl-tRNA interact and where peptide bond formation occurs. The size of ribosomes and their subunits is expressed by their sedimentation coefficients [Svedberg (S) values]. The *E. coli* ribosome is a 70S ribosome. It is composed of a large subunit (50S) and a small subunit (30S). The composition is shown in Table 13-2. The human ribosome is 80S in nature and consists of a large (60S) and small (40S) subunit (Table 13-3). The primary structures of the rRNAs and most of the ribosomal proteins have been determined.

Ribosomes contain two functional sites termed the **A site** and the **P site**. Aminoacyl-tRNA binds at the A site; peptidyl-tRNA and the initiating methionine-$tRNA_I$ bind at the P site. The small and large subunits make contributions to both sites. **Peptidyltransferase** activity, which catalyzes peptide bond formation, resides on the large subunit and is intrinsic to the ribosome; this activity, moreover, may be associated with rRNA and not protein (Fig. 13-1).

TABLE 13-2. Composition of the Ribosomes of *Escherichia coli*

Property	Ribosome	Small Subunit	Large Subunit
Sedimentation coefficient	70S	30S	50S
RNA	—	16S	23S 5S
Protein	—	21 polypeptides	35 polypeptides

TABLE 13-3. Composition of Mammalian Ribosomes

Property	Ribosome	Small Subunit	Large Subunit
Sedimentation coefficient	80S	40S	60S
RNA	—	18S	28S 5.8S 5S
Protein	—	33 polypeptides	49 polypeptides

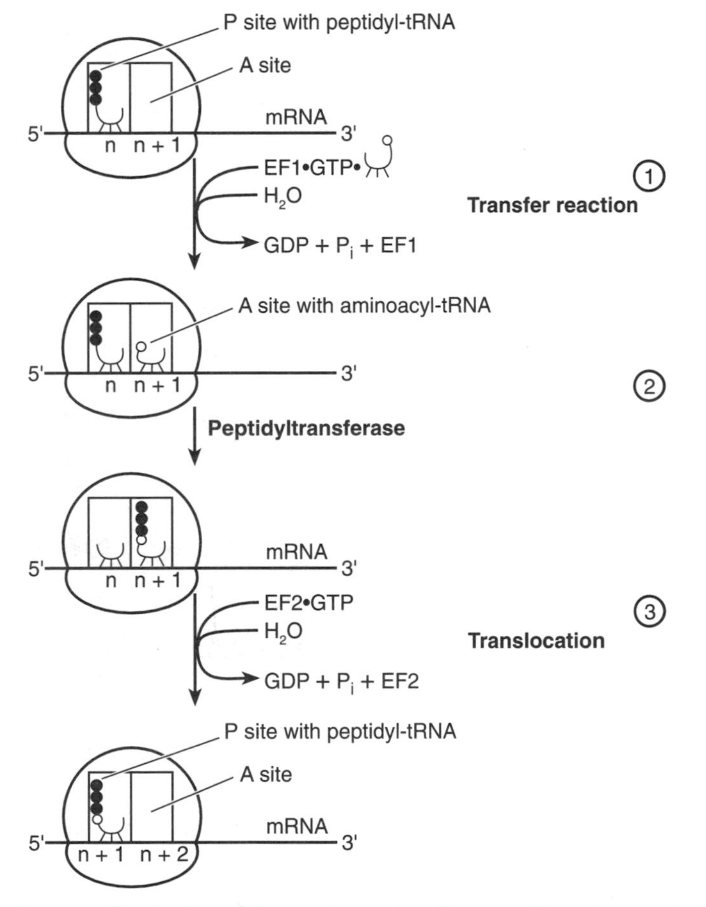

Figure 13-1. Topography of elongation and translocation reactions in protein biosynthesis: the A and P sites. The circled numbers denote the three key steps involved in introducing a single amino acid residue. EF, elongation factor; GDP, guanosine diphosphate; GTP, guanosine triphosphate; mRNA, messenger RNA; P_i, inorganic phosphate; tRNA, transfer RNA.

AMINO ACID ACTIVATION

The enzymes that catalyze the formation of aminoacyl-tRNA are termed **aminoacyl-tRNA synthetases** (or ligases). There is one enzyme in *E. coli* and one cytoplasmic enzyme in humans corresponding to each of the 20 genetically encoded amino acids. These enzymes attach the specific amino acid to each of the tRNA molecules that correspond to that amino acid. The degeneracy of the genetic code necessitates the use of more than one tRNA for several amino acids. The tRNA molecules that correspond to a given amino acid are called **isoacceptors**. The mechanism of amino acid activation is analogous to that for fatty acid activation. It involves a pyrophosphate split from adenosine triphosphate (ATP) to yield an aminoacyl–adenosine monophosphate (AMP) intermediate. Aminoacyl-AMP reacts with its corresponding tRNA to form aminoacyl-tRNA and AMP. The reaction can be outlined as follows:

$$\text{Amino acid} + \text{ATP} \rightleftharpoons \text{aminoacyl-AMP} + \text{PP}_i$$
$$\text{Aminoacyl-AMP} + \text{tRNA} \rightleftharpoons \text{aminoacyl-tRNA} + \text{AMP}$$

Both steps of this reaction are catalyzed by a single enzyme. There is no loss of energy-rich bonds in this process. The hydrolysis of inorganic pyrophosphate (PP_i) catalyzed by a separate pyrophosphatase, however, is exergonic and serves to pull the reaction in the forward direction. The 3' sequence of eukaryotic and prokaryotic tRNA molecules ends with CCA. The amino acid is covalently linked through an energy-rich bond to the 2'- or 3'-hydroxyl of ribose on the 3'-terminal adenosine-containing nucleotide of tRNA.

PROTEIN SYNTHESIS FACTORS

Protein synthesis is divided into initiation, elongation, and termination. There are protein factors that transiently associate with the ribosome to perform specific func-

TABLE 13-4. Factors Involved in Protein Synthesis in *E. coli*

Factor	Molecular Weight	Function
Initiation factors (IFs)		
IF1	9,000	Keeps ribosomes dissociated
IF2	100,000	Binds GTP and formyl-Met-$tRNA_1$
IF3	23,000	Keeps ribosomes dissociated
Elongation factors (EFs)		
EF-T is made of		Transfer
EF-Tu	43,000	Unstable; binds aminoacyl-tRNA/GTP
EF-Ts	74,000	Stable; displaces GDP
EF-G	77,000	GTPase; translocates mRNA along ribosome
Release factors (RFs)		
RF1	—	Recognizes UAA, UAG
RF2	—	Recognizes UAA, UGA

GDP, guanosine diphosphate; GTP, guanosine triphosphate; GTPase, guanosine triphosphatase; mRNA, messenger RNA; tRNA, transfer RNA.

TABLE 13-5. Eukaryotic Protein Synthesis Factors

Factor	Function
Initiation	
eIF1	Assists mRNA binding
eIF2	Binds initiator Met-tRNA$^{Met}{}_{I}$ and GTP
eIF2A	Binds Met-tRNA$^{Met}{}_{I}$ to 40S ribosome via AUG in mRNA
eIF2B	Exchanges GTP/GDP
eIF2C	Stabilizes ternary complex
eIF3	Binds to 40S subunit before mRNA binding
eIF4A	Unwinds secondary structure of mRNA via ATP-dependent helicase activity
eIF4B	Assists mRNA binding
eIF4C	Assists mRNA binding
eIF4D	Plays role in formation of the first peptide bond
eIF4E	Recognizes mRNA cap
eIF4F	Is a complex made of eIF4A, eIF4E, and p220
eIF5	Promotes GTP hydrolysis and release of other initiation factors
eIF6	Dissociates subunits
Elongation	
eEF1	—
EF1α	Binds aminoacyl tRNA and GTP
eEF1$\beta\gamma$	Assists in the exchange of GTP and GDP in EF1α
eEF2	Translocates mRNA along ribosome; hydrolyzes GTP; is inhibited by ADP-ribosylation catalyzed by diphtheria toxin
Release	
eRF	Promotes the hydrolysis of peptidyl-tRNA to form peptide and tRNA; the factor also binds and hydrolyzes GTP (GDP + P_i)

ADP, adenosine diphosphate; eEF, eukaryotic elongation factor; EF, elongation factor; eIF, eukaryotic initiation factor; eRF, eukaryotic release factor; GDP, guanosine diphosphate; GTP, guanosine triphosphate; mRNA, messenger RNA; Met-tRNA$^{Met}{}_{I}$, methionyl-transfer RNA with special initiator methionine; P_i, inorganic phosphate; tRNA, transfer RNA.

tions. In prokaryotes, the **initiation factors** are designated **IF**, the **elongation factors** are designated **EF**, and the **termination or release factors** are designated **RF**. The factors in humans are designated similarly but with an *e* prefix for eukaryotic (e.g., **eIF** for **eukaryotic initiation factor**). The various factors are given in Tables 13-4 and 13-5.

Three **initiation factors** are required in prokaryotes, and several more occur in humans. The functions of the first three initiation factors for prokaryotes and eukaryotes are similar. **IF2** binds guanosine triphosphate (GTP) and formyl-Met-tRNA$_I$ (where *I* refers to a special initiator molecule) in prokaryotes; **eIF2** binds GTP and Met-tRNA$_I$ in humans. These factors place the initiator methionine-tRNA$_I$ into the P site of the ribosome during the initiation process. *N*-Formyl-methionine-tRNA$_I$ (fMet-tRNA$_I$) is the initiating residue in prokaryotes, and Met-tRNA$_I$ (unformylated) is the initiator in humans.

The formyl group in prokaryotes is added after methionine has been linked to its cognate tRNA; *N*-formyltetrahydrofolate is the formyl donor. The initiation factors do not form a complex with methionine-$tRNA^{Met}$; the latter is the source of the methionine placed into internal positions of nascent polypeptide chains. **Methionine-$tRNA^{Met}$** is not formylated. **EF-Tu** (where *T* refers to transfer and *u* refers to unstable) in prokaryotes and **eEF-1** in humans form a complex with GTP and methionine-$tRNA^{Met}$ and all other aminoacyl-tRNAs (one at a time) required for protein biosynthesis. The complex consisting of the elongation factor, GTP, and aminoacyl-tRNA is called a **ternary complex**. These factors place the aminoacyl-tRNA into the A site during protein synthesis.

STEPS IN PROTEIN SYNTHESIS

Initiation

To initiate biosynthesis, mRNA forms a complex with the ribosome and fMet-$tRNA_I$ (prokaryotes) or Met-$tRNA_I$ (eukaryotes). In prokaryotes, or bacteria, the small ribosomal subunit is prevented from associating with the large subunit by the action of IF1 and IF3, which are bound to the small subunit. The initiating AUG codon is not at the 5' end of mRNA but is 50 to 100 nucleotides or more from the 5' end. To initiate synthesis in bacteria, mRNA binds to the small ribosomal subunit at five bases (CCUCC) near the 3' end of 16S RNA called the **Shine-Dalgarno** sequence in a process that is negotiated by IF3; this places the initiating AUG in the correct position relative to the ribosome. Next, IF2•GTP•fMet-$tRNA_I$ binds to the mRNA–small subunit complex, IF3 is released, and then the large subunit binds to the small subunit. After this association, GTP is hydrolyzed to guanosine diphosphate (GDP) and inorganic phosphate (P_i); IF2 dissociates from the complex along with IF1. At the end of this process, fMet-$tRNA_I$ is bound to the P site of the ribosome with its anticodon bound in an antiparallel fashion with the initiating AUG of mRNA.

Initiation of protein synthesis in eukaryotes, which involves an **AUG scanning** mechanism, is similar in outline but differs in detail. Initiation differs in that Met-$tRNA_I$ is unformylated, more initiation factors participate, and there is no comparable interaction of mRNA with a Shine-Dalgarno sequence. In contrast to bacterial protein synthesis, the ternary complex formed from eIF2•GTP•Met-$tRNA_I$ with the small subunit occurs before binding to mRNA. eIF4E recognizes the guanine cap of mRNA. The 40S subunit binds to the 5' end of monocistronic mRNA and scans along the message until the first AUG is encountered. Several other initiation factors participate in binding the small subunit to mRNA. The anticodon of Met-$tRNA_I$ plays a role in scanning. After the first AUG is encountered, the large subunit binds to form the initiation complex. eIF5 catalyzes GTP hydrolysis and the dissociation of initiation factors from the ribosomes; ATP hydrolysis also accompanies the formation of the human initiation complex. At the end of the initiation process in both bacterial and human protein synthesis, the **initiator methionine-tRNA** is found in the **P site**. Initiator methionine tRNA is the only aminoacyl-tRNA that is delivered physiologically to the P site; the other aminoacyl-tRNAs are delivered to the A site.

Elongation

After the formation of the initiation complex, a series of repetitive elongation reactions occur. For a protein containing 300 amino acids, there is a unique **initiation** event followed by **299 elongation events** and a unique **termination** event. A series of ternary complexes of EF-Tu•GTP•aminoacyl-tRNAs (eEF-1α•GTP•aminoacyl-tRNAs in humans) interact with the A (aminoacyl-tRNA) site of the ribosome. When a match occurs between the triplet codon immediately after the AUG (on the 3' side) and the

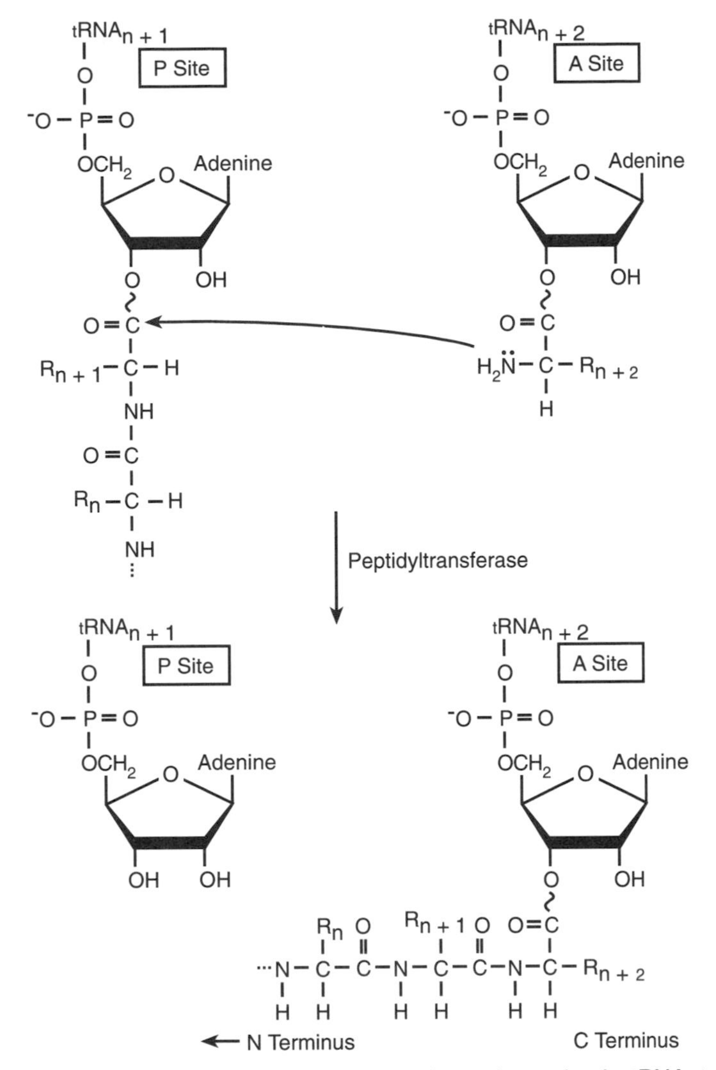

Figure 13-2. Chemistry of the elongation reaction of protein synthesis. tRNA, transfer RNA.

corresponding aminoacyl-tRNA anticodon, then GTP is hydrolyzed, and EF-Tu dissociates from the ribosome. The appropriate aminoacyl-tRNA is implanted in the A site. Next, the endogenous **peptidyltransferase** activity of the large ribosomal subunit catalyzes a reaction between the amino group in the A site with the activated carboxyl group in the P site to form the first peptide bond. After this reaction, the **dipeptidyl-tRNA** is bound to the A site and free tRNA is found in the P site. Before the next elongation reaction, **EFG•GTP** interacts with the complex and mediates the translocation of mRNA and peptidyl-tRNA from the A site to the P site. In the process, GTP is hydrolyzed to GDP and P_i, and the free tRNA is ejected from the P site. This completes the first elongation cycle.

The **dipeptidyl-tRNA** now occupies the P site. The second and subsequent cycles of elongation occur as described for the first. Aminoacyl-tRNA is brought to the A site as a ternary complex with EF-Tu and GTP. After interaction of the matching codon with anticodon, **GTP hydrolysis** occurs. Peptide bond formation and translocation follow. The process is shown in Fig. 13-1.

The reaction catalyzed by **peptidyltransferase** is the same in prokaryotes and eukaryotes and is shown in Fig. 13-2. From the diagram, you can determine that chain growth occurs from the amino to the carboxyl terminus. You can also see that **peptidyl-**

TABLE 13-6. Antibiotics That Inhibit Protein Synthesis

Antibiotic	Sensitive Organisms	Process Inhibited
Streptomycin	Prokaryotes	Initiation; produces mistakes in translation
Tetracycline	Prokaryotes	Aminoacyl-tRNA attachment to the ribosome
Chloramphenicol	Prokaryotes	Ribosomal peptidyltransferase
Puromycin	Prokaryotes and eukaryotes	Causes premature chain termination
Cycloheximide	Eukaryotes	Ribosomal peptidyltransferase

tRNA, transfer RNA.

tRNA occupies the A site after the peptidyltransferase reaction. This necessitates a translocation reaction, as illustrated in Fig. 13-1.

Termination

The elongation cycles continue until a **stop codon** (UAA, UAG, or UGA; see Table 13-3) occupies the A site. An RF interacts with the complex and discharges the polypeptide from tRNA by hydrolysis. The ribosomal subunits dissociate, and the tRNA is liberated. The process in prokaryotes and eukaryotes is analogous.

Several **antibiotics inhibit specific steps of translation**; the actions of several of these are given in Table 13-6. In addition, diphtheria toxin catalyzes a reaction between eEF-2 and the oxidized form of nicotinamide adenine dinucleotide to give adenosine diphosphate (ADP) ribosyl–eEF-2 and nicotinamide; the process called **ADP-ribosylation**. The covalently bound ADP ribosyl group inactivates the eukaryotic elongation factor. The potent toxin, made in *Corynebacterium diphtheriae*, has no effect on bacterial protein synthesis.

Bioenergetics of Protein Synthesis

The bioenergetic cost of **peptide bond formation** is appreciable. Two high-energy bonds of ATP are expended to form aminoacyl-tRNA (ATP → AMP + 2 P_i), and one GTP is hydrolyzed (to GDP and P_i) in the transfer reaction, in which aminoacyl-tRNA is placed in the A site. Peptide bond formation per se does not require any additional energy expenditure. The high-energy bond of aminoacyl-tRNA is converted to the low-energy peptide bond in an exergonic reaction. A fourth high-energy bond is expended as GTP is hydrolyzed (to GDP and P_i) in the translocation reaction. A total of four high-energy bonds are thus expended per peptide bond formed. The cost of peptide bond formation therefore is considerable. The chemical energy and the complex ribosomal machinery are required to convert the language of the four-letter nucleic acid alphabet to the 20-letter alphabet of proteins.

POSTTRANSLATIONAL PROTEIN MODIFICATION

Two common posttranslational modifications that occur either during or shortly after synthesis of polypeptides include **proteolytic cleavage** and **glycosylation**. Although **methionine** is the universal initiating amino acid, it is found on the amino terminus of only a small proportion of proteins. The initiating methionine is hydrolytically removed from those proteins lacking methionine at the amino terminus by the action of an aminopeptidase. Acetylation of the resulting amino terminus by acetylcoenzyme A to give an ***N*-acetylprotein** is a common posttranslational modification in humans. Additional posttranslational modification occurs in proteins destined for secretion from the cell, insertion into the plasma membrane, or translocation into the lysosome. Proteins with these properties are synthesized on ribosomes found with the rough endoplasmic reticulum. The properties of proteins that lead to their synthesis at the rough endoplasmic reticulum are described by the signal peptide hypothesis.

Signal Peptide Hypothesis

Proteins destined for insertion into membranes or secretion exhibit a leader sequence of 15 to 40 amino acids at their amino terminus. The signal contains at least one positively charged residue and a hydrophobic stretch of 10 to 15 residues. Translation begins on free ribosomes and stops shortly after the leader or signal sequence has been synthesized. A **signal recognition particle** (SRP) is responsible for arresting biosynthesis as SRP recognizes the hydrophobic leader sequence and binds to the nascent polypeptide-ribosome complex.

SRP, which is a G-protein, consists of several proteins and 7SL RNA, a 300-nucleotide RNA. The arrested complex binds to an SRP receptor and two ribosome receptors in the rough endoplasmic reticulum. This interaction allows biosynthesis to resume as the nascent polypeptide chain with its signal sequence directed into the lumen of the endoplasmic reticulum. GTP is hydrolyzed after the formation of the **SRP-SRP receptor complex**, and SRP is recycled. Before translation is complete, the signal sequence is hydrolyzed from the precursor (also known as the **preprotein**) to produce the protein within the lumen of the rough endoplasmic reticulum. **Signal peptidase** catalyzes the hydrolytic removal of the leader sequence (Fig. 13-3).

The mechanism of **insulin biosynthesis** in the β-cells of the pancreas illustrates many facets of posttranslational modification. Insulin consists of an A chain and a B chain linked by two disulfide bonds. Insulin is synthesized from a precursor called **pre-**

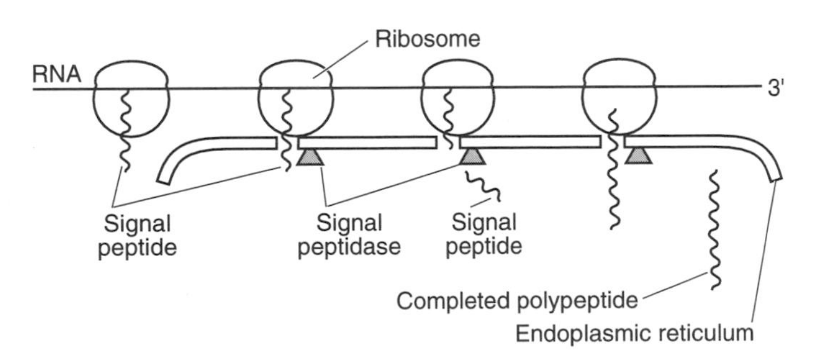

Figure 13-3. The signal peptide directs nascent proteins to the rough endoplasmic reticulum.

proinsulin. The order of synthesis of preproinsulin is leader sequence, B chain, connecting peptide, and finally A chain. All of these reside initially on a single polypeptide, preproinsulin; the discrete chains form as a result of proteolytic processing. After synthesis is initiated, preproinsulin is channeled into the lumen of the rough endoplasmic reticulum in a process involving a signal peptide sequence (the presequence of preproinsulin), the SRP, and the SRP receptor and two ribosome receptors on the rough endoplasmic reticulum membrane. A signal peptidase catalyzes the hydrolytic removal of the signal sequence, and synthesis continues to yield **proinsulin**.

The three disulfide bonds found in native insulin (two between the A and B chain, and one within the A chain) form within the single polypeptide chain constituting proinsulin as catalyzed by **protein disulfide isomerase**. The polypeptide is cleaved at paired basic residues at two sites by enzymes with trypsin-like specificity called **prohormone convertases** to yield a C, or connecting, peptide and insulin composed of its two chains (A and B). Carboxypeptidase H removes extra carboxyterminal basic residues on the B chain to yield insulin.

Insulin is packaged into secretory vesicles, stored, and released on demand. It undergoes no further posttranslational modifications. Insulin is released in response to elevated glucose within the β-cell as sensed by glucokinase, the high Michaelis constant (10 mmol/L) isoform of hexokinase that is found in liver, and the β-cell that is not inhibited by glucose 6-phosphate.

Protein Glycosylation

Many proteins found in the plasma membrane and many circulating proteins, such as antibodies and hormones (e.g., thyroid-stimulating hormone, luteinizing hormone, follicle-stimulating hormone), are glycoproteins. Glycosylation of proteins occurs within the **Golgi complex** and before their transport into secretory vesicles. Albumin, the most abundant plasma protein, is synthesized in and secreted by the liver. Albumin, however, is not glycosylated. This finding indicates that glycosylation is not a requirement for secretion.

There are two major classes of glycoprotein. A single protein, however, can be a member of both classes. The two classes of glycoprotein are ***O*-linked** (involving protein-serine or threonine) and ***N*-linked** (involving protein-asparagine residues). The *O*-linked class, which is the simpler of the two, is considered first. There are approximately 80 types of sugar linkages in glycoproteins. Each type of linkage is determined by enzyme specificity. The donor substrates are listed in Table 13-7. Formation of these oligomeric sugar derivatives is unlike that of nucleic acids because the sequence is not determined by a template mechanism.

The ABO blood group pentasaccharide is an example of a carbohydrate attached to protein by an *O*-linkage. A specific transferase catalyzes the addition of each residue to its acceptor substrate. Biosynthesis is specified and determined by enzymes acting sequentially. For the **ABO oligosaccharide**, *N*-acetylgalactosamine is attached to an acceptor protein-serine or threonine. Then galactose, *N*-acetylglucose, galactose, and fucose are added sequentially to yield the H antigen, which corresponds to blood type O. The addition of *N*-acetylgalactose yields A antigen, or the addition of galactose yields B antigen.

There are four nucleotide differences in the genes encoding the ***N*-acetylgalactosyltransferase** and the **galactosyltransferase** that distinguish between the A and B blood types. The corresponding H antigen and the O allele result from a defect in the gene that encodes the terminal transferase (*N*-acetylgalactosyltransferase or galactosyltransferase). There is a deletion in the gene for the O allele that produces a nonfunctional protein; as a consequence, H antigen is not further metabolized to produce the A or B antigen.

The **oligosaccharides** attached at *N*-linkages are larger than those attached at *O*-linkages, and the mechanism of biosynthesis is more formidable. There are three

TABLE 13-7. Carbohydrate Donors Required for *O*-Linked and *N*-Linked Glycoprotein Synthesis

Donor	Both *O*- and *N*-Linked or *N*-Linked	Comments
UDP-galactose	Both	—
CMP-sialic acid	Both	A nine-carbon sugar acid; same as *N*-acetylneuraminic acid
GDP-fucose	Both	—
UDP-*N*-acetylgalactosamine	Both	—
UDP-glucose	*N*	Glucose present as an intermediate but eliminated in final product
GDP-mannose	*N*	—
UDP-*N*-acetylglucosamine	*N*	—

CMP, cytidine monophosphate; GDP, guanosine diphosphate; UDP, uridine diphosphate.

subclasses of *N*-linked oligosaccharides: **high mannose**, **complex**, and **hybrid**. The complex and hybrid subclasses are derived from the high mannose variety by additional reactions. The pathway for biosynthesis of high mannose chains involves the following three operations: (1) formation of a 14-member oligosaccharide linked via a high-energy bond to dolichol by a pyrophosphate group; (2) transfer of the oligosaccharide to the acceptor protein; and (3) hydrolytic removal (trimming) of specific sugars. The subsequent addition of several sugars from nucleotides to the high mannose subclass results in the formation of the hybrid and complex classes of oligosaccharide. These processes begin in the endoplasmic reticulum and are completed in the Golgi bodies.

The first phase involves **dolichol**, which is a polyisoprenoid compound (17 to 20 five-carbon, or isoprenoid, units) with an alcohol at one terminus. Dolichol is found within the membranes of the endoplasmic reticulum. The alcohol group is phosphorylated by cytosine triphosphate (CTP) to form cytosine diphosphate (CDP) and dolichol phosphate. Dolichol phosphate reacts with uridine diphosphate-*N*-acetylglucosamine to form dolichol diphospho-*N*-acetylglucosamine and uridine monophosphate. Six additional sugars are transferred from nucleotide donors, and then seven sugars are transferred from dolichol phosphate derivatives. The 14-member core is transferred *en bloc* to the acceptor protein at a specific asparagine residue. Transfer occurs to sites in the acceptor protein in the sequence . . . AsnXSer (Thr) . . . , where X is nearly any amino acid. Terminal glucosyl residues (four of them) are removed by hydrolysis reactions catalyzed by glycosidases. A single mannose is removed. When the process stops here, the product is a **high-mannose oligosaccharide**.

To form the **complex** and **hybrid chain oligosaccharide**, *N*-acetylglucosamine, fucose, galactose, and sialic acid are added from the derivatives indicated in Table 13-7. These activated monomers serve as precursors or donors for condensation or polymerization reactions. The provision of a terminal mannose 6-phosphate on an *N*-linked oligosaccharide of a glycoprotein targets that protein to the lysosome. The inability to produce the terminal mannose 6-phosphate results in **I-cell disease** (where I denotes inclusion body), a rare condition that affects many organ systems.

Chapter 14

Recombinant DNA Technology

GENETIC CHEMISTRY

The previous chapters have covered the general principles of genetic chemistry. Use of recombinant DNA technology and genetic engineering began in the mid-1970s and, in addition to advancing the understanding of physiologic and pathologic processes, has led to advances in the diagnoses and understanding of many genetic disorders. This chapter gives a general overview of progress in this rapidly developing field. With the Human Genome Project scheduled to give the complete DNA sequence of the 3×10^9 nucleotide found in human chromosomes by 2003, understanding of human diseases (and not just so-called genetic diseases) will improve significantly. The effects that this information will have on the practice of medicine will be profound.

Since the 1970s, important advances in manipulating DNA and RNA in the laboratory have been made, making it possible to obtain considerable amounts of purified DNA for analysis and study. The techniques involve the production of **recombinant DNA** molecules. Recombinant DNA is constructed from any DNA of interest (**target DNA**) and a **vehicle DNA**, which can be combined with the target DNA. The vehicle serves as a molecular handle and provides a means for amplifying DNA by cloning procedures to produce adequate amounts of DNA for study. Vehicles are generally bacterial **plasmids** (extrachromosomal DNAs that replicate autonomously) or bacterial **viruses**. Sometimes animal virus sequences are added to recombinant DNA so that the DNA in either bacterial or animal cells in culture can be propagated. Recombinant DNA can also be used to produce **chimeric genes**, which encode products that are derived from two different organisms. The design of DNA molecules with specific properties is called **genetic engineering**.

DNA can be sequenced, which permits the deduction of amino acid sequences of proteins (both normal and mutant) through the use of the **genetic code**. **Restriction endonucleases** are the class of enzymes that revolutionized the study of DNA and permitted the production, manipulation, and analysis of recombinant DNA molecules. These molecules cleave DNA only at specific nucleotide sequences. It is also possible to prepare radioactive or fluorescent DNA for use as probes. The probes interact with related sequences of DNA obtained from the genome by complementary base pairing or **annealing**. This practice has permitted the development of genetic fingerprinting techniques and is playing a role in the diagnosis of a variety of diseases through the use of **restriction fragment length polymorphisms** (RFLPs). The **polymerase chain reaction** (PCR) is a technique for amplifying target DNA molecules in an exponential fashion during multiple cycles of DNA synthesis *in vitro*. This procedure permits the amplification and subsequent study of minuscule amounts of DNA for cloning, genetic analysis, and diagnosis of infectious diseases.

The process of introducing human DNA into *Escherichia coli* and other microorganisms is called **transformation**. By appropriate genetic engineering, cells can be induced to produce large amounts of protein corresponding to the human DNA. Human insulin and human growth hormone produced by recombinant DNA technology are currently available for treatment of **diabetes mellitus** and **dwarfism**, respectively. Prokaryotic cells are unable to mediate certain posttranslational modifications such as glycosylation. When such modifications are important for stability or activity of the gene product, expression of human genes is performed in animal cells in culture. Tissue plasminogen activator, interferons, interleukins, blood-clotting factors, and erythropoietin are among the glycoprotein therapeutic agents produced by this technology. **Erythropoietin** is available for the treatment of the **anemia** associated with **end-stage renal disease** (erythropoietin is produced in the kidney); this treatment has markedly enhanced the quality of life of the recipients. Besides having the protein with the human sequence (as opposed to that of another species), the shortcoming that only minuscule amounts can be obtained from human sources is obviated by these expression systems.

Before considering some of the experimental detail of the molecular biology explosion in biomedical science, it is interesting to note that the entire revolution has been made possible by five essential major technical advances: (1) **characterization and purification of restriction endonucleases**, (2) **design of prokaryotic genetic vectors**, (3) various **blotting and hybridization methodologies**, (4) **DNA sequencing techniques**, and (5) the **PCR**. Taken together, these technologies have enabled biomedical scientists to create recombinant nucleic acids, to isolate and propagate genetic sequences in a time- and cost-effective manner, and to amplify minute quantities of DNA.

Restriction Enzymes

Scientists have isolated more than 100 enzymes, called **restriction endonucleases**, from bacteria. **Restriction endonucleases** catalyze the hydrolysis of a phosphodiester bond in each of the complementary strands of DNA containing a specific sequence of nucleotide bases. The great advantage of these enzymes is that cleavage is sequence specific and not random. The sequence recognized by restriction endonucleases constitutes a palindrome. A **palindrome** is a word, sentence, or number that reads the same forward and backward. Examples include Otto, able was I ere I saw elba, and 2002. In the case of nucleic acids, a different definition of palindrome is used. A **DNA palindrome** occurs when the sequence on one strand of nucleic acid (from 5' to 3') is identical to its complementary strand (from 5' to 3'):

5' GGCC 3'
3' CCGG 5'

5' GAATTC 3'
3' CTTAAG 5'

In each case, the bottom strand is identical to that of the top strand (in an antiparallel fashion). Although it is easy to identify palindromic sequences presented in the absence of extraneous nucleotides, it is much more difficult to do so within long stretches of DNA. In practice, computers are used to analyze sequences of many kilobases (kb) for palindromes and other sequence characteristics.

The sequences and positions of cleavage of several restriction endonucleases are provided in Table 14-1. Some enzymes nick or hydrolyze the DNA at staggered positions (e.g., *Pst*I). Other enzymes produce blunt ends (*Bal*I). The ends produced by *Pst*I are self-complementary and are termed **cohesive ends**, or sticky ends. The product of the *Pst*I cleavage has an extended 3'-end, termed a 3'-overhang (Fig. 14-1). If two

TABLE 14-1. Restriction Endonuclease Cleavage Sites

Source	Enzyme Designation	Sequence 5' → 3' 3' ← 5'
Bacillus amyloliquefaciens H	*Bam*HI	G↓GATCC CCTAG↑G
Brevibacterium albidum	*Bal*I	TGG↓CCA ACC↑GGT
Escherichia coli RY13	*Eco*RI	G↓AATTC CTTAA↑G
Haemophilus aegyptius	*Hae*III	GG↓CC CC↑GG
Providencia stuartii 164	*Pst*I	CTGCA↓G G↑ACGTC

different DNAs (e.g., human and bacterial) treated with the same restriction enzyme are mixed (producing sticky ends), some of the strands of human DNA will combine at their cohesive ends with the complementary ends of *E. coli* DNA. This is one of the important strategies used in producing **recombinant DNA molecules**. The restriction endonucleases yield a strand with a 5'-phosphate, and the other contains a free 3'-hydroxyl group. Annealed cohesive ends are good substrates for DNA ligase so that the recombinant strands can be covalently attached to each other.

The probability of having a **tetranucleotide sequence** of **GGCC** in a DNA molecule is $1{:}(4 \times 4 \times 4 \times 4)$, or 1:256. Therefore, this sequence would occur on a random basis once every 256 nucleotides. In contrast, the probability for restriction endonuclease sites for **CTGCAG** is $1{:}4^6$, or 1:4,096. The number and size of the DNA fragments produced by restriction endonucleases depend on the actual DNA sequence and are not totally predictable. SV40 DNA (from an animal virus) is 5,226 nucleotides long and contains a single *Eco*RI site. Bacteriophage T7 DNA is 40,000 nucleotides long but lacks a single *Eco*RI site.

Plasmids and DNA Cloning

Plasmids are small, autonomously replicating circular DNA molecules characteristic of bacteria. Many copies (up to 50) of small plasmids (4 kb or fewer) can be produced per bacterium. Plasmids replicate independently of the main chromosome. Many naturally occurring plasmids carry genes that confer antibiotic resistance; antibiotic resistance markers have been used in the production of genetically engineered plasmids for recombinant DNA research. Plasmids are used as vectors for cloning foreign, or

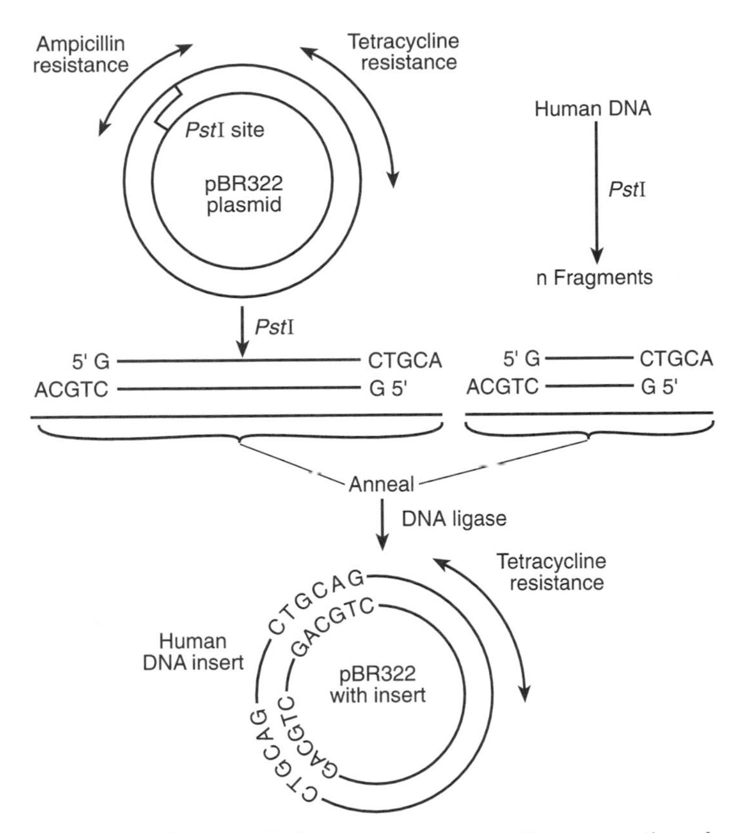

Figure 14-1. Use of the *Pst*I restriction endonuclease in the preparation of a recombinant DNA molecule.

nonbacterial, DNA. Cloning the DNA of a gene is the process of preparing a large number of identical DNA molecules. A **clone** is an exact copy of an original form (DNA, cell, or organism). Plasmids can be used to clone DNA up to 4 or 5 kb in length; other vehicles are used to produce DNAs of longer length. Bacteriophage lambda, for example, can be used to clone DNAs of 10 to 20 kb; cosmid vectors can be used to prepare DNAs up to 50 kb in length; and yeast artificial chromosomes (YACs) are used to prepare even larger DNA fragments.

Because human genes are interrupted and contain sequences that are removed by splicing from the primary transcript, the DNA corresponding to messenger RNA (mRNA) is contained in a much longer nucleotide segment; for example, the 2-kb mRNA of phenylalanine hydroxylase is derived from a DNA sequence of nearly 100 kb. The 1.8-kb mRNA of tyrosine hydroxylase is derived from a DNA sequence of 10 kb. One reason that scientists are interested in cloning large DNAs is to obtain segments that contain the entire gene, which may extend for many kilobases.

A plasmid genetic element termed **pBR322** served as the progenitor for most plasmids in use today. Typically, a plasmid will contain 2,500 to 5,000 base pairs (bp) and genes that code for resistance to antibiotics such as ampicillin, tetracycline, or kanamycin. They will also contain a stretch of unique restriction sites in a contiguous region termed the **multiple cloning site** (or polylinker). This region has been genetically engineered to possess convenient choices for constructing recombinant DNA molecules (Fig. 14-2).

If a sample of target DNA previously treated with *Pst*I is mixed with plasmid, also treated with this enzyme, a certain proportion of the plasmid will anneal with the target DNA. After treatment with (bacteriophage) T4 DNA ligase and adenosine triphosphate (ATP), a covalently closed circle results. Some plasmid will reanneal (not bind

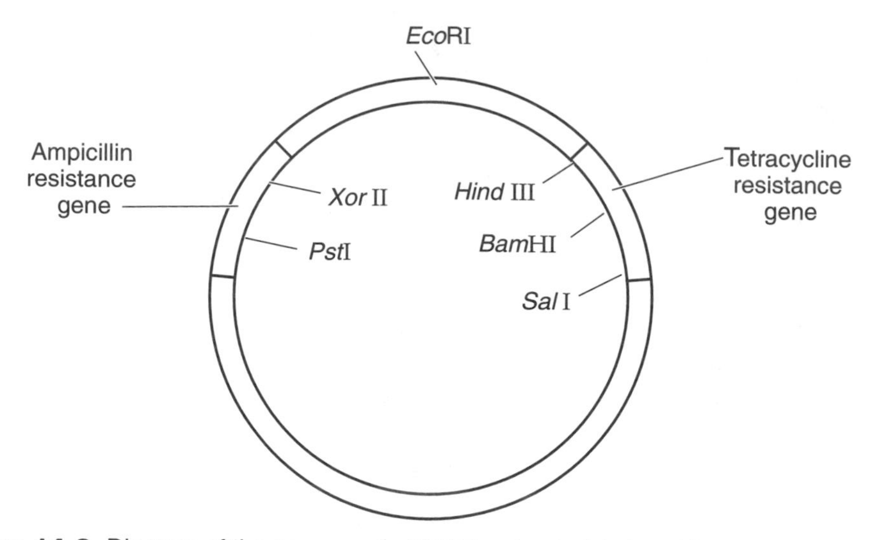

Figure 14-2. Diagram of the genome of pBR322, a bacterial plasmid.

to target DNA). To minimize the yield of plasmid without inserts, the plasmid is treated with bacterial alkaline phosphatase to remove its 5' phosphate, which was generated during restriction enzyme treatment, so that reannealed plasmid lacking an insert is not a substrate for DNA ligase. Some target DNA will reanneal (not bind to plasmid DNA); such DNA is a substrate for DNA ligase, but it is not capable of autonomous replication and will not direct its own amplification in bacteria. Some complexes consisting of target DNA and plasmid DNA will result. These complexes are substrates for DNA ligase, and such recombinant molecules are capable of replication. Bacteria (e.g., *E. coli*) lacking antibiotic resistance genes are exposed to the recombinant plasmids (in the presence of $CaCl_2$), and some of them take up the DNA (and are transformed). The cells are then grown in the presence of antibiotic. Untransformed bacteria do not grow under such conditions because they lack the antibiotic resistance gene. Individual cells with recombinant plasmids form colonies of cells containing identical copies (clones) of the original recombinant molecule.

A **gene library** refers to a collection of **plasmids**, **phage**, **cosmids**, or **YACs** containing the DNA corresponding to the entire genome of an organism. A bacteriophage lambda library of 250,000 different recombinant molecules is sufficient to contain the entire human genome. This library can be contained in a single drop of medium. To identify the DNA sequence or gene of interest requires ingenuity, and there are a number of successful strategies. One common procedure is to screen thousands of plasmid-containing bacterial colonies grown on Petri dishes. The few positive colonies of the library that hybridize with a radioactive DNA probe of interest are taken for further study.

Nucleic Acid Resolution by Electrophoresis

Negatively charged (anionic) DNA and RNA migrate in an electric field toward the positive electrode, or anodally. The distance of **migration** is inversely related to the molecular mass or number of monomeric residues in the polynucleotide; small molecules migrate more rapidly than large molecules. The **matrix**, or medium, in which the macromolecule is electrophoresed is synthetic acrylamide or a carbohydrate polymer called **agarose**. Basically, the matrix serves as a sieve that retards the migration of increasingly large nucleic acid molecules. The concentration and composition of the matrix is varied depending on the size of the nucleic acid of interest. In some cases, nucleic acids of 2 to 10 kb or larger are under study, and in others the size ranges from

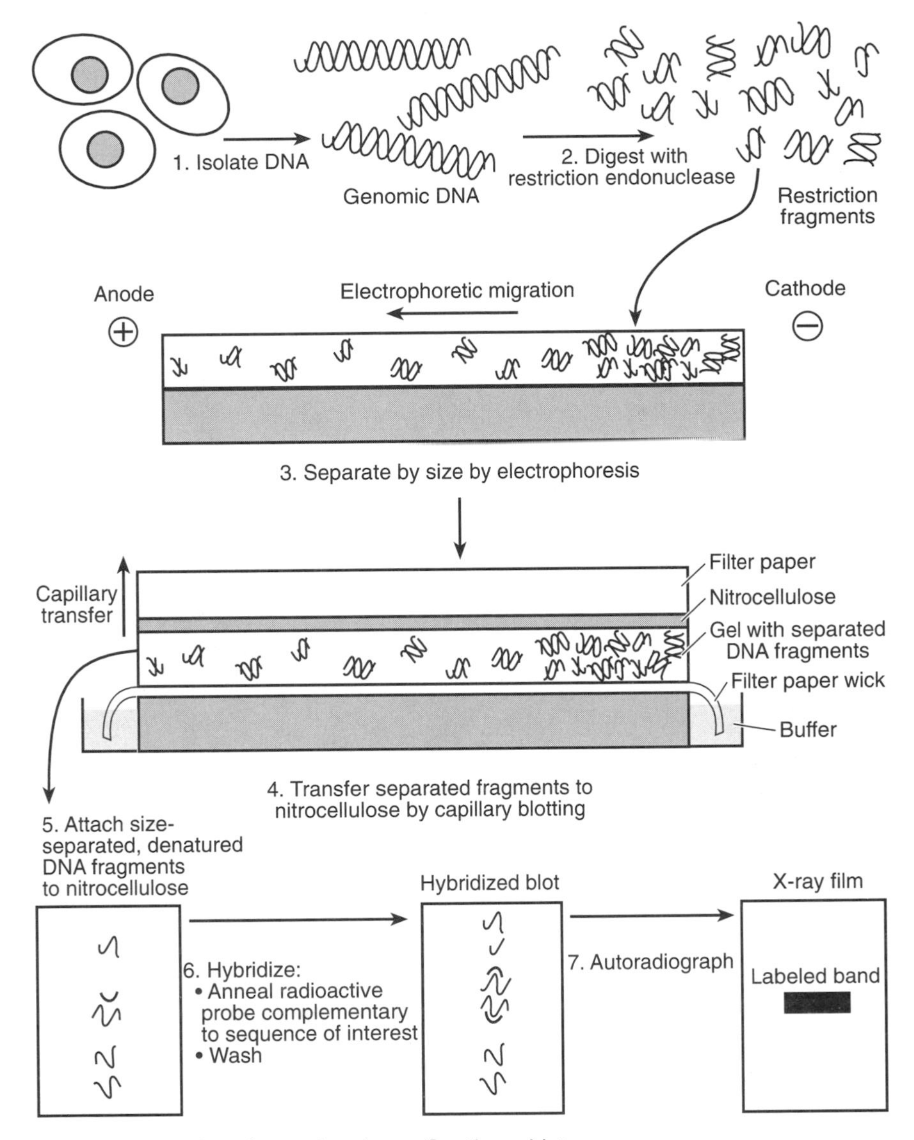

Figure 14-3. Procedure for performing a Southern blot.

30 to 300 nucleotides. Systems have been developed for resolving DNA molecules of one million nucleotides and greater. Handling such large pieces of DNA is important to the proposed sequencing of the human genome.

DNA samples can be characterized by **Southern blotting** (Fig. 14-3). A sample of DNA (e.g., human DNA) is treated with a restriction endonuclease and then subjected to agarose gel electrophoresis. The resulting fragments are resolved on the basis of size. The DNA on the gel is transferred to a thin support system, such as nitrocellulose paper, by capillary action or pressurized buffer flow. Buffer moves through the slab gel and nitrocellulose; in the process, DNA (or RNA) is transferred from the gel and is retained by nitrocellulose. The DNA (or RNA) is bonded to the nitrocellulose by treatment with heat or ultraviolet light. The mechanics are such that the relative positions of DNA (or RNA) on the gel are maintained on the nitrocellulose paper.

The resolution of DNA by electrophoresis and transfer to nitrocellulose is called a **Southern blot** (named after Edwin Southern, the originator). A similar analysis of RNA was dubbed a **Northern blot**. Electrophoresis and transfer of proteins is called a **Western blot**. The principles involve electrophoretic resolution of macromolecules based on size and a transfer to a bonding agent by a method that ensures maintenance of resolution. In the next section, methods for identifying specific DNA or RNA segments by annealing or hybridization are described.

Preparation of Radioactive DNA Probes and Polynucleotide Hybridization

Southern blotting of an *Eco*RI restriction endonuclease enzyme digest of a sample of human DNA might yield 1,000,000 different fragments. To identify the fragment of interest requires the use of a probe. This is generally accomplished by preparing a radioactive (^{32}P) DNA sample. Several strategies are used. If the primary structure of the DNA can be retrieved from the literature, a complementary DNA of 30 to 50 nucleotides in length can be synthesized by automated organic chemical procedures. This resulting oligonucleotide can be labeled at the free 5'-hydroxyl group in a reaction between radioactive ATP and the synthetic oligonucleotide catalyzed by **polynucleotide kinase**:

$$\text{Oligonucleotide} + \left[\gamma^{32}\text{P}\right]\text{ATP} \rightarrow \left[^{32}\text{P}\right]\text{oligonucleotide} + \text{ADP}$$

If the primary structure of a segment of protein of interest is known, corresponding **oligonucleotides** deduced from the genetic code can be constructed. Because of codon degeneracy, a family of oligomers corresponding to each possible codon has to be synthesized. Although these descriptions represent the simplest approach, they have shortcomings. Small probes are not very sensitive, and end-labeling (the polynucleotide kinase reaction) does not produce probes with the required radioactivity.

With a sample of a larger, double-stranded or duplex DNA corresponding to a gene, labeled probes can be produced by using random sequence oligonucleotides (typically six bases in length) as primers with heat-denatured DNA as a template (**random priming**). Random sequence oligonucleotides are mixtures that contain each of the four bases at each of the six positions. In essence, they coat the probe DNA and provide efficient priming sites for DNA polymerase I (*E. coli*) to bind and commence DNA synthesis. If one of the four deoxynucleotides is radiolabeled in the α-position with ^{32}P, the resulting double-stranded molecules will be intensely labeled with radioactive phosphorus and will serve as excellent probes in subsequent hybridization experiments.

Next, the Southern or Northern blots are **incubated** with the radioactive probe under specific temperature and salt concentrations (see Fig. 14-3). During this procedure, the probe hybridizes or anneals to any immobilized complementary DNA (a Southern blot) or RNA (a Northern blot) present on the filter. After hybridization, the blot is washed with buffer to remove nonannealed probe. The samples are autoradiographed to locate the position of the probe and the polynucleotide(s) to which it is bound. Standard nucleic acids are run in parallel so that the length or size of the DNA or RNA bands can be ascertained.

When a restriction enzyme digest of DNA from several humans is performed, the resulting Southern blots may show different patterns in response to a specific DNA probe. This might reflect **RFLPs** or different forms of a gene (**alleles**). Some individuals may show a 10-kb band, and others may show a 6-kb fragment. It may be that one form is closely associated with a specific disease. For example, RFLPs associated with Huntington's disease, cystic fibrosis, and Duchenne's muscular dystrophy have been described. It is not necessary that the restriction enzyme cleave the defective gene; often, the altered site is near the defective gene and serves as a marker.

Similar tests are feasible in diagnosing hereditary diseases in cells isolated from amniotic fluid after amniocentesis. Extensive experimentation and painstaking efforts are required to establish and validate such diagnostic tests. This type of analysis is also

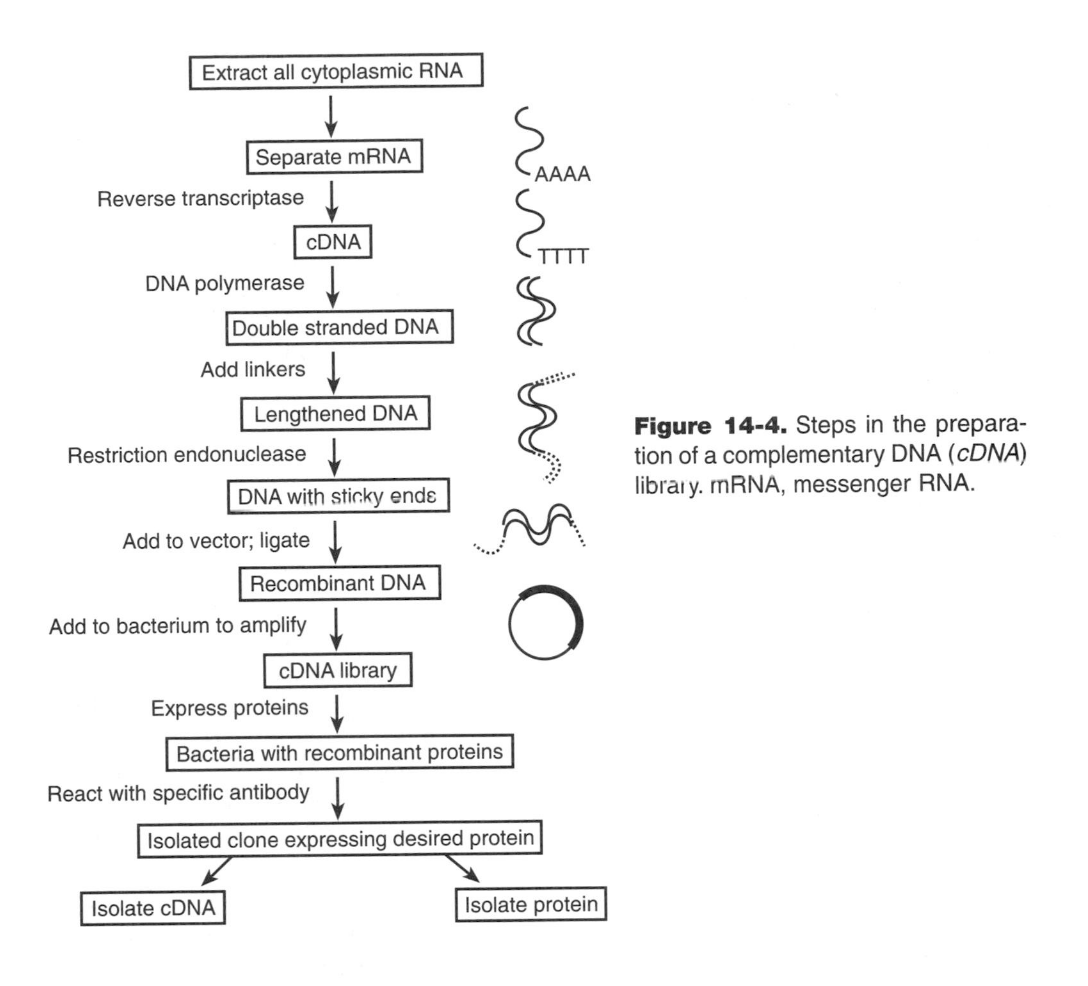

Figure 14-4. Steps in the preparation of a complementary DNA (*cDNA*) library. mRNA, messenger RNA.

the basis for the **DNA fingerprinting** that has become common in forensic analysis. Each human has a unique pattern of these polymorphisms, and if enough different polymorphic states are analyzed, biological samples can be uniquely associated with one individual.

Complementary DNA Preparation

Complementary DNA (**cDNA**) generally denotes DNA that is complementary to RNA. It is easier to clone, sequence, and manipulate DNA in the laboratory than it is to perform these functions with RNA. A cDNA library corresponds to many clones of DNA complementary to an assortment of isolated sequences of RNA. In some instances, it is possible to enrich or purify RNA so that the cDNA produced is limited. When cDNA corresponding to mRNA is the goal, mRNA can be resolved from ribosomal RNA (rRNA) and transfer RNA (tRNA) through the use of oligo-dT cellulose chromatography. Most mRNAs bind to oligo-dT by their 3'-polyA tail. Ribosomal RNA and tRNA, however, fail to bind. Very few mRNAs lack the 3'-polyA tail and fail to bind. Histone mRNA, however, is a prominent example of an mRNA lacking the 3'-polyA tail. In addition to enriching the mRNA, genes can be enriched through selection of an appropriate tissue; for instance, if searching for a liver gene, isolation of mRNA from this tissue automatically eliminates brain-specific genes, muscle-specific genes, and so on.

The first step in preparing cDNA (Fig. 14-4) is to incubate RNA with the four deoxynucleoside triphosphates, reverse transcriptase (RNA-dependent DNA polymerase derived from animal retroviruses), and a primer. **Oligo-dT**, which is complementary to the polyA tail at the 3' end of mRNA, is often used as a primer in the preparation of an mRNA library. After synthesis of the first strand (an RNA-DNA hybrid results), the sample is treated with an enzyme termed RNAse H. This enzyme is an endonuclease with specificity for RNA in an RNA/DNA heteroduplex. The resulting nicked RNA serves as an efficient primer for second strand DNA synthesis by DNA polymerase I and deoxynucleoside triphosphates. Note the analogy between this reaction and the synthesis of radioactive probes by the random priming reaction.

There are a number of procedures for combining the cDNA with plasmid or bacteriophage DNA for molecular cloning. One procedure involves the attachment of synthetic oligonucleotides to the duplex DNA in an ATP-dependent process catalyzed by bacteriophage T4 DNA ligase. The sequence of the linkers can be designed to produce cohesive ends complementary to those of a cloning vector that is treated with the appropriate restriction enzyme.

DNA Sequence Analysis

One of the great technical advances in biochemistry is the development of methods for rapidly determining the sequence of large molecules of DNA. It is possible to determine the sequence of tens of thousands of residues in a day. From such sequences, the sequence of corresponding proteins can be determined by using the genetic code. What might require several years' work in protein sequencing can be accomplished in months or days by DNA sequence analysis (DNA sequence analysis may be performed in a few days, but considerable time may be required to obtain the desired DNA for analysis). Computers are essential for analyzing the sequence of DNA and the corresponding protein. Present-day cloning and DNA sequencing techniques provide vastly more protein sequence information than direct protein sequencing by chemical methods.

The **dideoxynucleotide** or **chain termination** method of DNA sequence analysis is described in abbreviated form (Fig. 14-5). DNA of interest is denatured and incubated with a sequence-specific oligonucleotide primer. This annealed complex is divided into four identical reaction mixtures. To each mixture is added a solution of the four **deoxynucleoside triphosphates** (one or more of which contains ^{32}P in the α-position or ^{35}S in place of oxygen on the α-phosphate) and DNA polymerase I. A different chain terminator is included in each of the four mixtures. These terminators include ddATP, ddTTP (thymidine triphosphate), ddGTP (guanosine triphosphate), and ddCTP (cytidine triphosphate), where dd indicates dideoxy. Dideoxynucleotides lack hydroxyl groups at both the 2'- and 3'-carbons of ribose. When these dideoxynucleotides are incorporated into the nascent chain at positions complementary to the cloned insert, elongation is impossible (there is no free 3'-hydroxyl group) and chain termination results. In the case in which ddATP is included, chain termination occurs randomly at every position complementary to a thymine residue, and a family of polydeoxynucleotides results.

Chain extension to synthesize a family of **polynucleotides** ending at specific residues is possible because of the presence of deoxy (d)ATP in the reaction mixture. The incorporation of the other dideoxynucleotides indicates the positions corresponding to their complementary base. The four mixtures are subjected to gel electrophoresis and autoradiography. The gels resolve every polydeoxynucleotide by size. It is possible, for example, to resolve a polynucleotide of 197 from one of 198 bases. You can read from the shortest to longest and identify the residues terminating in A, T, G, or C by the order of their appearance on the autoradiograph. The information can be entered into a computer data bank for full analysis.

The steps in preparing the four samples corresponding to a given DNA sequence are shown in Fig. 14-4. The figure illustrates many fundamental aspects of DNA biosynthesis and therefore is worth all the time it takes to understand the principles outlined. These principles are more important to the reader than the DNA sequencing procedure itself. Although this approach to DNA sequencing is still widely used, it is rapidly being supplanted by automated techniques that provide higher throughput and longer

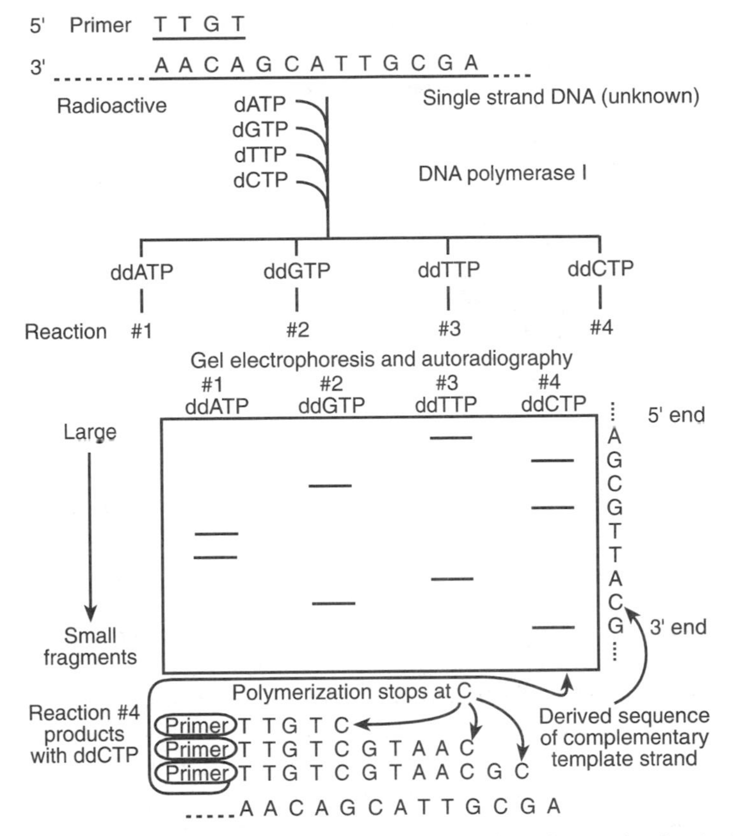

Figure 14-5. DNA sequence analysis by the dideoxy (*dd*) chain termination technique. ATP, adenosine triphosphate; CTP, cytidine triphosphate; d, deoxynucleoside; GTP, guanosine triphosphate; TTP, thymidine triphosphate.

"reads." The main technological advance has been the development of **fluorescently labeled dideoxynucleotides** that can be differentiated spectrophotometrically. They are all mixed in a single reaction and the resulting products resolved on a gel mounted on an instrument that automatically "samples" the bands as they migrate past a detector. The four colors are entered directly into a computer, thereby bypassing the need for a manual reading of the sequence. It is possible to automate the DNA sequencing process, and robots have been developed that perform all of the manipulations so that numerous samples of DNA can be sequenced without human intervention.

Polymerase Chain Reaction

PCR is a technique that is revolutionizing genetic biochemistry and genetic engineering. PCR is used to amplify DNA for study; for example, minute quantities of DNA from a few hair follicle cells can be characterized. This technique is used in forensic medicine, genetic and microbiological diagnostics, and gene manipulation and engineering. PCR technology is being used to study Duchenne's muscular dystrophy, hemophilia A, the detection of human T-cell lymphoma, human immunodeficiency virus, hepatitis B virus, and retinoblastoma. PCR technology can also be used for preparing mutant DNAs with site-specific changes.

Chain reaction (Fig. 14-6) refers to the use of multiple cycles (25 to 35) of primer-directed DNA elongation. The requirements include a target DNA for amplification, two DNA oligonucleotide primers complementary to the ends of each strand of the target DNA and extending toward each other, deoxynucleoside triphosphates, and a heat-stable DNA polymerase. The heat-stable polymerase (called ***taq* DNA polymerase**)

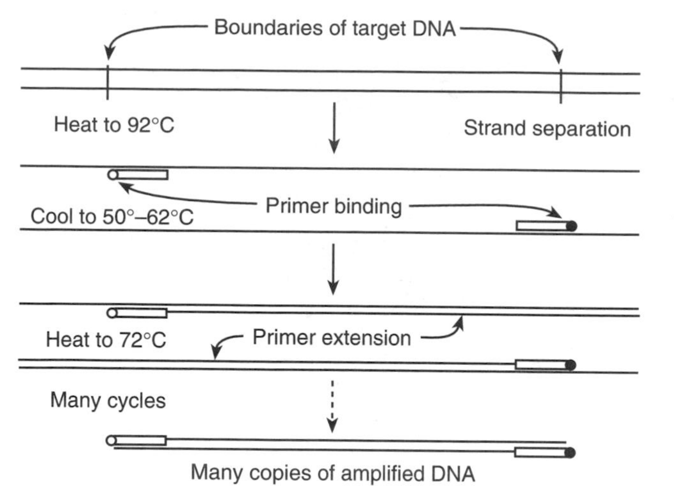

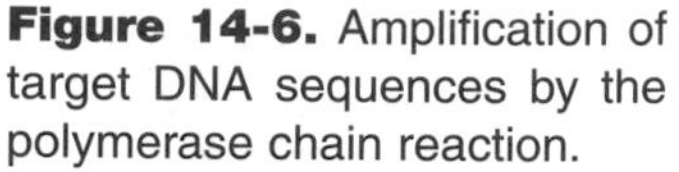
Figure 14-6. Amplification of target DNA sequences by the polymerase chain reaction.

is currently obtained from a bacterium, *Thermus aquaticus,* which grows in hot springs at 100°C or higher.

The **components of a PCR reaction** are **incubated** in a thermal cycler. First, the temperature is elevated to approximately 92°C for 1 minute, which causes the two strands of a DNA duplex to dissociate. The reaction components are **cooled to 50° to 62°C** for 1 minute so that the oligonucleotide primers can anneal with DNA. The reaction components are then **heated to 72°C** for 1 minute so that the primers can be elongated over the length of the target DNA. The process is repeated 25 to 35 times. Each DNA duplex present initially results in the formation two daughter strands for a total of four strands.

During the second cycle, the four strands will serve as a template and eight strands will result. At the end of the second cycle, **polynucleotides** that extend from only one end to the other end of the target DNA result. This produces a discrete fragment termed an **amplicon**. Subsequent cycles will result in the doubling of the amount of target DNA.

This procedure can amplify the amount of DNA by several million-fold. Scientists predict that PCR technology will affect genetic biochemistry in the same fundamental way that the use of restriction enzymes has altered molecular biology.

Chapter 15

Cell Signaling

NEUROCHEMISTRY

The brain and spinal cord constitute the **central nervous system** (CNS). The **peripheral nervous system** (PNS) lies outside the skull and vertebral column. The PNS is made up of the voluntary nervous system (nerves to skeletal muscle), the autonomic nervous system (sympathetic and parasympathetic nerves), and the sensory nervous system. The neurotransmitter of the voluntary, or motor, nervous system is acetylcholine. **Acetylcholine** (ACh) is also the neurotransmitter of the preganglionic sympathetic and pre- and postganglionic parasympathetic nervous system. **Norepinephrine** is the neurotransmitter of the postganglionic sympathetic nervous system. The identity of the neurotransmitters of the sensory nervous system has not yet been definitively established.

Considerable work has been performed to ascertain the identity of the neurotransmitters of the CNS. These include ACh, the catecholamines (dopamine, norepinephrine, and epinephrine), serotonin, glutamate, γ-aminobutyric acid (GABA), glycine, and several neuropeptides.

Acetylcholine Metabolism

For practical purposes, ACh synthesis is limited to nerves that are **cholinergic** in nature. Mature noradrenergic nerves, for example, do not synthesize ACh. The enzyme that catalyzes ACh formation is **choline acetyltransferase**, which catalyzes the reaction between acetylcoenzyme A (acetyl-CoA) and choline as indicated:

$$\text{Acetyl-CoA} + \text{choline} \rightleftharpoons \text{ACh} + \text{CoA}$$

Acetyl-CoA is an energy-rich donor of the acetyl group. ACh is packaged into synaptic vesicles. When a nerve action potential invades a cholinergic nerve terminal, ACh is released in discrete quanta or packets by calcium-dependent exocytosis. ACh diffuses across a synapse or a junctional region, interacts with an effector cell, and brings about the biological response.

To effect their physiologic responses, ACh and other neurotransmitters interact with their cognate receptor (Table 15-1). There are two major classes of **ACh receptors**: nicotinic and muscarinic. These receptors differ in their molecular structure, anatomic location, and action. Moreover, several subclasses of nicotinic and muscarinic receptors exist. The **nicotinic receptor** occurs at the neuromuscular junction, on postganglionic autonomic cell bodies, and in the CNS.

The receptor at the **neuromuscular junction** has been extensively studied. The form of the receptor in adults consists of four different polypeptide chains with the following composition: $\alpha_2\beta\delta\varepsilon$; the embryonic form is $\alpha_2\beta\gamma\delta$. The nicotinic receptor acts as an ion channel or ion gate. After binding ACh, Na^+ flows into a cell through the receptor.

TABLE 15-1. Classification of Neurotransmitter Receptors

Receptor Type	Function
Acetylcholine	
Muscarinic	
M_1 and M_4	Inhibit adenylyl cyclase; activate potassium channels
M_2 and M_3	Activate phospholipase C
M_5	Inhibits adenylyl cyclase and may couple to phospholipase C
Nicotinic	
N_1	Forms an ion channel at the neuromuscular junction
N_2	Forms an ion channel at autonomic ganglia and in the central nervous system
Adrenoceptor	
α_1	Activates phospholipase C
α_2	Inhibits adenylyl cyclase
β_1, β_2, and β_3	Activate adenylyl cyclase
Dopamine	
D_{1A} and D_{1B}	Activate adenylyl cyclase
D_{2S} and D_{2L}	Inhibit adenylyl cyclase; open potassium channels
D_3 and D_4	Unknown
γ-Aminobutyric acid (GABA)	
$GABA_A$	Increases chloride conductance
$GABA_B$	Inhibits adenylyl cyclase and affects calcium and potassium channels via G-protein action
Glutamate	
Glu_1 [*N*-methyl-D-aspartate (NMDA)]	Activated by NMDA, a synthetic drug; functions as a cation channel and may play a role in memory
Glu_2 [α-amino-3-hydroxy-5-methyl-4-isoxazole propionate (AMPA)]	Activated by AMPA, a synthetic drug; functions as a cation channel and is responsible for most fast excitatory neurotransmission in brain and spinal cord
Glu_3 (kainate)	Activated by kainate, a synthetic drug; functions as a cation channel
Glu_4 [1-amino-cyclopentane-1,3-decarboxylate (APCD)]	Activated by APCD; activates phospholipase C
Glu_5 [phosphonylglutamate (AP4)]	Activated by AP4; activates cyclic nucleotide phosphodiesterase in retina and hyperpolarizes cells
Histamine	
H_1	Linked to G-protein and phospholipase C activation
H_2	Linked to G-protein and adenylyl cyclase activation; blocked by cimetidine
H_3	Linked to G-protein and calcium and potassium ion fluxes
Opioid	
μ (mu)	Morphine (and β-endorphin) receptor; linked to G_i
δ (delta)	Enkephalin receptor; linked to G_i
κ (kappa)	Dynorphin receptor; linked to G_i

(continued)

TABLE 15-1. *(continued)*

Receptor Type	Function
Serotonin	
5-HT_{1A}	Inhibits adenylyl cyclase and opens potassium channels; activates adenylyl cyclase in some cells
5-HT_{1B}	Inhibits adenylyl cyclase
5-HT_{1C}	Activates phospholipase C
5-HT_{1D}	Inhibits adenylyl cyclase
5-HT_2	Activates phospholipase C and closes potassium channels
5-HT_3	Ligand-gated cation channel depolarizes cells
5-HT_4	Activates adenylyl cyclase and closes potassium channels

Myasthenia gravis, an autoimmune disease, is due to the adventitious production of antibodies against the nicotinic receptor at the neuromuscular junction. The **muscarinic receptor** is responsible for the actions of ACh liberated from the postganglionic parasympathetic system; ACh is the postganglionic parasympathetic effector. The muscarinic receptor is specifically blocked or antagonized by atropine.

There is experimental evidence that different subclasses of nicotinic and muscarinic receptors exist in different brain regions and in different regions innervated by the PNS. Results from cloning the genes corresponding to various receptors indicate that the actual number of receptor subtypes is greater than the number of receptor subtypes based on pharmacologic specificity; for example, although there is a generalized class of muscarinic ACh receptors, there are five different muscarinic receptor genes with distinct expression profiles.

Neurotransmitters such as ACh can be **excitatory** or **inhibitory**; the physiologic response depends on the receptor and its action. Activation of the nicotinic receptor at the neuromuscular junction and activation of the muscarinic receptor in the pancreas is excitatory. Activation of the muscarinic receptor in the heart, on the other hand, is inhibitory.

Inactivation of ACh is mediated by metabolic degradation. **Acetylcholinesterase**, localized on the exterior of the cell surface of many neural and nonneuronal cells, catalyzes the hydrolysis of ACh:

$$\text{ACh} + H_2O \rightarrow \text{acetate} + \text{choline}$$

After its liberation in the acetylcholinesterase reaction, choline is transported into nerve cells by a sodium-dependent high-affinity transport system and can be reused. A circulating form of acetylcholinesterase called **pseudocholinesterase** also exists in humans. Its physiologic function is unclear.

The hereditary **deficiency of pseudocholinesterase** results in increased sensitivity to succinylcholine, used as a muscle relaxant during surgery. **Diisopropylfluorophosphate** is an irreversible inhibitor of acetylcholinesterase. It forms a covalent adduct with an active-site serine hydroxyl group of the esterase. Diisopropylfluorophosphate (and the related Sarin) has been used in chemical warfare as a nerve gas. Related inhibitors of acetylcholinesterase (e.g., Parathion) are used as insecticides.

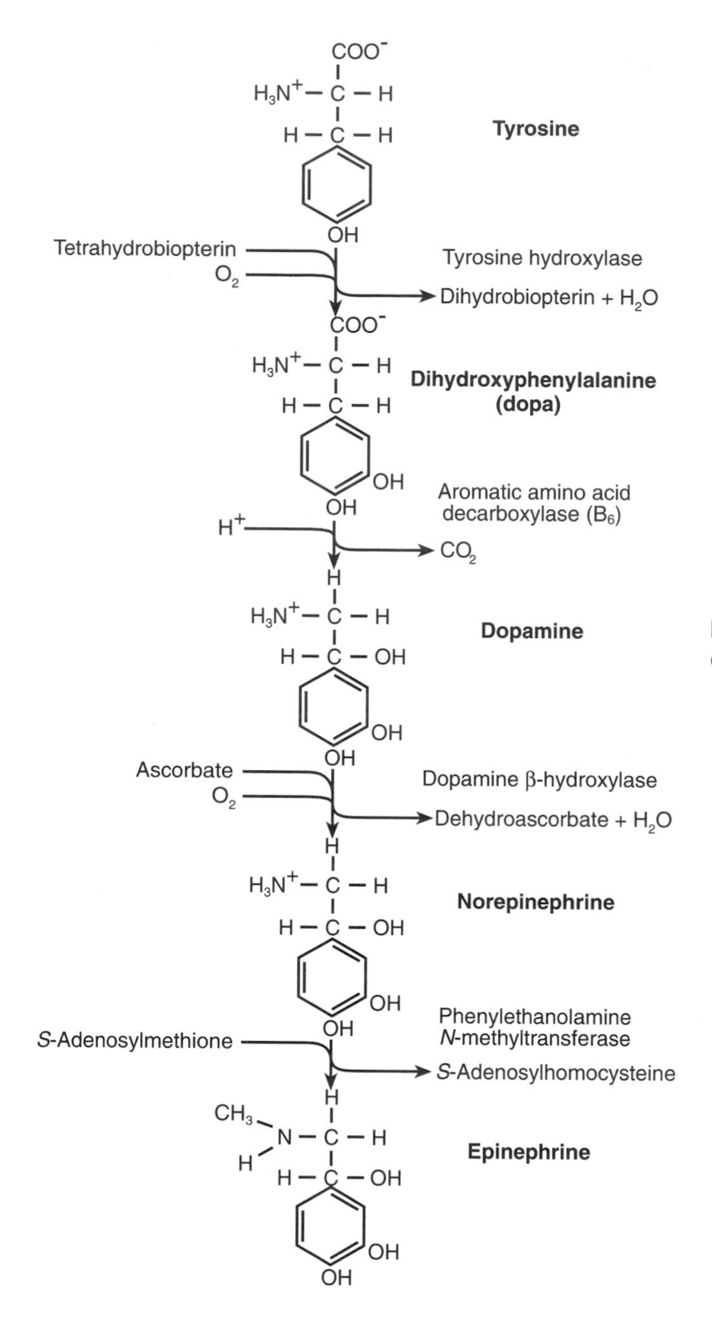

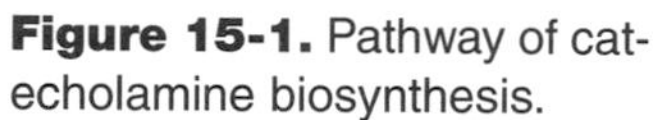
Figure 15-1. Pathway of catecholamine biosynthesis.

Catecholamine Metabolism

Catecholamines are derived from tyrosine. The first, rate-limiting, and regulatory step in catecholamine biosynthesis is catalyzed by **tyrosine hydroxylase**. Tetrahydrobiopterin and oxygen are the other reactants. Dihydroxyphenylalanine (**dopa**), water, and dihydrobiopterin are the products. Tyrosine hydroxylase is activated by the action of protein kinase A and other protein kinases and is inhibited by the catecholamine pathway end products. **Aromatic amino acid decarboxylase** (a pyridoxal phosphate–dependent enzyme) catalyzes the conversion of dopa to dopamine and carbon dioxide (Fig. 15-1). This is the end product in dopaminergic cells of the brain, including those of the nigrostriatal pathway. In other cells, **dopamine** reacts with oxygen and ascorbate to yield norepinephrine, water, and ascorbate radical in a reaction catalyzed by **dopamine β-hydroxylase**. This is the end product in noradrenergic neurons.

In other brain cells and in the adrenal medulla, norepinephrine is methylated by *S*-adenosylmethionine to yield epinephrine and *S*-adenosylhomocysteine (see Fig. 15-1). The corresponding enzyme is **phenylethanolamine *N*-methyltransferase** (PNMT). Epinephrine and norepinephrine are the hormones of the adrenal medulla.

Catecholamines are released from cells and interact with their specific receptor and bring about their physiologic responses (see Table 15-1). The β-adrenergic receptor activates adenylyl cyclase at many locations. The action of catecholamines, in contrast to ACh, is not terminated by metabolic degradation within the synaptic cleft or junctional region. Catecholamines are inactivated by sodium-dependent transport systems. The most active systems reside in the plasma membrane of the nerve cells that release the catecholamines by exocytosis.

Transport from the exterior to the interior of cells is the major mechanism for most neurotransmitter inactivation. The two exceptions are ACh and the neuropeptides. Inactivation of these two classes of neurotransmitter is mediated by hydrolytic cleavage catalyzed by degradative enzymes. The metabolism (not inactivation) of intracellular catecholamines is mediated by catechol-*O*-methyltransferase, monoamine oxidase, aldehyde reductase, and aldehyde dehydrogenase. The structures of the many intermediates are not considered here.

The chief metabolite of norepinephrine is **vanillylmandelate** (VMA). VMA is often measured in the urine of patients with suspected **pheochromocytoma** [a tumor of the adrenal medulla or abdominal paraganglia (organs of Zuckerkandl)] to aid in diagnosis.

The nigrostriatal pathway of the brain uses dopamine as its neurotransmitter. **Parkinson's disease** is associated with a degeneration of these neurons and a decrease in dopamine content in these brain regions. The mechanism of degeneration is unknown. The oral administration of **L-dopa** constitutes one treatment. L-dopa is transported across the blood–brain barrier and is taken up by the surviving cells and is converted into dopamine. **Dopamine** does not cross the blood–brain barrier and is ineffective in the treatment of Parkinson's disease. In addition to dopa, an aromatic decarboxylase inhibitor that does not cross the blood–brain barrier is administered. This is performed to reduce the metabolism of the dopa in the periphery, thus enhancing the therapeutic levels that reach the brain.

Serotonin Metabolism

Serotonin is found in the brain and gut. In fact, many neurotransmitters and neuropeptides are found in both locations, perhaps accounting for the adage of "having a gut feeling" about something or someone. Serotonin is derived from **tryptophan** in a two-step pathway.

First, **tryptophan hydroxylase** catalyzes a reaction with substrate, oxygen, and tetrahydrobiopterin to yield 5-hydroxytryptophan, water, and dihydrobiopterin. This reaction is analogous to those catalyzed by tyrosine hydroxylase and phenylalanine hydroxylase. The three enzymes exhibit homologous primary structures and constitute a family of aromatic amino acid hydroxylases.

Next, **5-hydroxytryptophan** is converted to serotonin and carbon dioxide by aromatic amino acid decarboxylase. This is the same pyridoxal phosphate–dependent enzyme that converts dopa to dopamine. Serotonin is packaged in synaptic vesicles and released by exocytosis. After diffusion to its receptor to bring about its effect, serotonin is inactivated by sodium-dependent uptake. Serotonin is metabolized intracellularly to 5-hydroxyindoleacetic acid (**5-HIAA**) by the action of monoamine oxidase and aldehyde dehydrogenase. Quantification of 5-HIAA in urine is undertaken in patients suspected of having the **carcinoid syndrome** associated with the production of excessive serotonin.

Excitatory Amino Acids

Considerable physiologic and pharmacologic evidence suggests that glutamate is the main excitatory neurotransmitter in the human brain. Glutamate is packaged into synaptic vesicles, released into the synaptic region to interact with its receptors to bring about its effect, and undergoes inactivation by uptake. In contrast to the restricted cellular distribution of ACh, the catecholamines, and serotonin, glutamate is present in all cells. The **glutamate α-amino-3-hydroxy-5-methyl-4-isoxazole propionate receptor** is responsible for most "fast" excitatory neurotransmission in brain and spinal cord. The **glutamate *N*-methyl-D-aspartate receptor** functions as a cation channel and may play a role in memory.

GABA Metabolism

GABA is the chief inhibitory neurotransmitter of the human brain. It is formed from glutamate in a one-step reaction catalyzed by **glutamate decarboxylase**; GABA and CO_2 are the products. As in many decarboxylase reactions, **pyridoxal phosphate** serves as a cofactor.

During the 1950s, infants inadvertently fed formulas deficient in vitamin B_6 developed seizures. Seizures were apparently produced by a paucity of inhibitory GABA. GABA is released by exocytosis, acts on its receptor to bring about a response, and is inactivated by its specific high-affinity sodium-dependent uptake system. Its degradation is mediated by a pathway called the **GABA shunt.** GABA (C_4) undergoes transamination with α-ketoglutarate to form succinate semialdehyde (C_4) and glutamate, as catalyzed by **GABA transaminase**. Succinate semialdehyde dehydrogenase catalyzes a reaction of substrate with the oxidized form of nicotinamide adenine dinucleotide to form succinate, reduced nicotinamide adenine dinucleotide (NADH), and H^+. The latter donates electrons to the electron-transport chain, which results in adenosine triphosphate (ATP) formation.

The content of the biogenic amines (ACh, serotonin, and the catecholamines) in the brain is in the μmol/kg range. On the other hand, the content of glutamate and GABA is in the range of mmol/kg.

Nitric Oxide Metabolism

Nitric oxide is a gas with the formula NO, and was known for many years as **endothelium-derived relaxation factor**. NO is derived from a reaction of arginine, molecular oxygen, and reduced nicotinamide adenine dinucleotide phosphate; the products include NO, citrulline, water, and the oxidized form of nicotinamide adenine dinucleotide phosphate. NO is an unstable free radical (it contains an unpaired electron) that self-inactivates. It diffuses from its site of synthesis and acts on adjacent cells; NO is not packaged in vesicles and is not released by exocytosis. The action of NO is to bind to the heme cofactor of soluble guanylyl cyclase and to stimulate its activity.

Neuroactive Peptides

More than 50 peptides are known to exist in the brain, as demonstrated by immunochemical and complementary DNA techniques. A selected list of **neuropeptides** is shown in Table 15-2. Most of these peptides are small (3 to 50 amino acids). They are synthesized as polypeptide precursors in a messenger RNA (mRNA)-dependent fashion and converted to the active agent by specific proteolysis or processing. The peptides are packaged, released by exocytosis, interact with their receptor to bring about their physiologic effect, and inactivated by proteolytic degradation. Besides their

TABLE 15-2. Selected Neuroactive Peptides

Angiotensin II
Cholecystokinin
Dynorphin*
Endorphin*
Leucine enkephalin*
Methionine enkephalin*
Neuropeptide Y
Neurotensin
Oxytocin
Secretin
Somatostatin
Substance P
Vasoactive intestinal peptide
Vasopressin

**Opioid peptide.*

exclusive release from their own specific neurons, they may be co-released with other low–molecular-mass neurotransmitters.

The list of possible coexisting neuropeptides and low–molecular-mass transmitters is long; for example, some cortical neurons contain and co-release GABA and somatostatin. **Thyrotropin-releasing hormone** (TRH) is a tripeptide that is synthesized in the hypothalamus and other neuronal cells. It is secreted from hypothalamic cells into the hypophyseal portal venous system and stimulates cells in the anterior pituitary to synthesize and release thyrotropin. TRH has a short half-life (approximately 15 minutes) and is not therapeutically efficacious. It can be used diagnostically in the evaluation of **thyroid diseases**. TRH is notable because, as a tripeptide, it is the smallest of the neuropeptides.

General Aspects of Nervous System Metabolism

In physiologic conditions, the brain uses glucose as a metabolic fuel. The brain is unable to use fatty acids as a source of chemical energy; however, it may be able to use ketone bodies after fasting for a few days or during starvation to account for nearly half of its ATP production. The brain is dependent on a continual supply of blood for both glucose and oxygen. A decrease of blood glucose to less than 40 mg/dL produces coma. Occlusion of the cerebral blood supply produces a stroke because of impaired delivery of oxygen and fuels.

The brain contains a small amount of **glycogen**, which may play an important function during hypoglycemia or during other forms of stress. Glycogen is clearly inadequate to sustain metabolism under conditions of severe hypoglycemia. It is stored in a hydrated state. Substantive changes of the glycogen content of the brain are prohibited because increases in volume are limited in the rigid cranium. Regardless of the underlying mechanisms, the brain is dependent on the liver and kidney to maintain adequate blood glucose levels.

Nearly all cells maintain a high intracellular potassium and a low intracellular sodium concentration. The diffusion of potassium through the plasma membrane and out of the cell accounts, in part, for the negative intracellular electrochemical potential ranging from –50 mV to –70 mV. Sodium is impermanent under resting conditions. During the propagation of a nerve action potential, sodium courses down its electrochemical gradient (from outside to inside) through sodium channels, and the polarity is briefly reversed.

The **sodium/potassium adenosine triphosphatase (ATPase)**, an integral membrane protein, is responsible for generating these ion gradients in nerve, heart, muscle, and other cells. A considerable proportion of ATP generated in nerve cells is required to maintain these ion gradients. The ATPase is an integral membrane glycoprotein with a subunit composition of $\alpha_2\beta_2$. The enzyme reacts with ATP to form a phosphoaspartyl-enzyme intermediate and transports three sodium ions to the cell exterior. The ATPase transports two potassium ions into the cell and undergoes a hydrolytic dephosphorylation reaction. The sodium-potassium/ATPase is a receptor for cardiac glycosides such as digitalis, which interact with the α-subunit. **Cardiac glycosides** exert their pharmacologic effects by altering sodium and secondarily calcium levels in the failing heart.

ENDOCRINE BIOCHEMISTRY

Hormones are substances produced by **endocrine cells**, released into and transported by the circulatory system to their target organs where they exert their effects. **Paracrine** function, in contrast to endocrine action, involves the release and action of substances on neighboring cells. **Autocrine** function involves the release of an effector, which then acts on the cell that released the substance. The paracrine and autocrine functions do not involve transport by the circulatory system.

Steroid Hormones

Hormones are conveniently classified as **lipophilic** (lipid soluble) or **hydrophilic** (water soluble). The **lipid-soluble hormones** listed in Table 15-3 include steroid hormones, thyroid hormones (triiodothyronine; and tetraiodothyronine), retinoic acid, and 1,25-dihydroxycholecalciferol (derived from vitamin D). Although these lipid-soluble hormones enter cells by diffusion, their sites of action are restricted to those cells that possess specific intracellular protein receptors that function in the nucleus. The hormone-receptor complex interacts with target sequences in the genome, and enhances or diminishes the transcription of hormone-responsive genes. Current studies suggest that the hormone-receptor complex recognizes a specific sequence of nucleotides upstream from the target genes called a **hormone-responsive element**.

The structures of the **steroid hormones** are shown in Fig. 15-2. Steroid hormones are synthesized from cholesterol in the appropriate endocrine cell. In contrast to the water-soluble hormones, steroid hormones are not stored to a significant extent but are synthesized and released on demand. Steroid hormones and other hormones are effective at nanomolar (1×10^{-9} mol/L) concentrations.

Hydrophilic Hormones

The **water-soluble hormones** include catecholamines, small peptides, polypeptides, proteins, and glycoproteins. The biosynthesis of norepinephrine and epinephrine (catecholamines) from tyrosine was considered earlier in this chapter (see Fig. 15-1). The peptides and polypeptides are synthesized by ribosomes in an mRNA-dependent

TABLE 15-3. Lipid-Soluble Hormones

Hormone	Structural Properties	Transport Protein
Steroid hormones		
Pregnane group (C21)		
Progesterone	Acetyl group attached to C17	Transcortin
Aldosterone	C18 is an aldehyde group (Ald)	None
Cortisol	11,17-Dihydroxyl groups	Transcortin
Corticosterone	11 Hydroxyl group	Transcortin
Androstane group (C19)		
Testosterone	17-Hydroxyl group; no aromatic ring	Sex hormone-binding globulin
Estrane group (C18)		
Estradiol	Aromatic A ring; 3,17-dihydroxyl groups	Sex hormone-binding globulin
Thyroid hormones		
Triiodothyronine	Three iodines	Thyroxine-binding globulin
Tetraiodothyronine	Four iodines	Thyroxine-binding globulin
Cholecalciferol		
1,25-Dihydroxycholecalciferol	Vitamin D ring system	α_1-Globulin transport protein

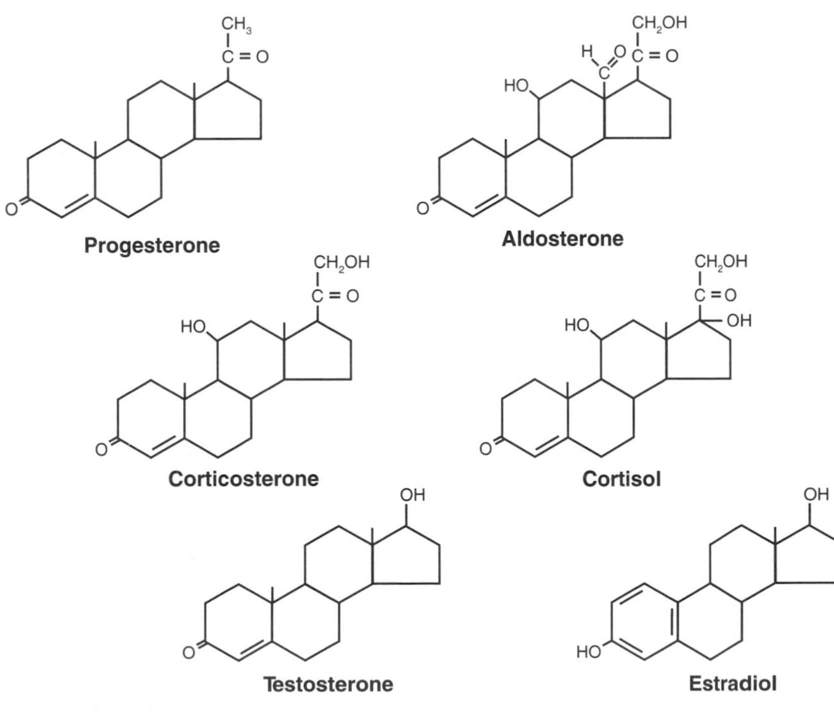

Figure 15-2. Steroid hormone structures.

fashion. Removal of signal peptides and processing by prohormone convertases are required to produce the active hormone. **Glycosylation** may also occur. These modifications occur in the Golgi. Hydrophilic hormones are stored in secretory granules and released from the endocrine cell by exocytosis.

Many aspects of hormonal action have important clinical applications. The most common endocrinopathy in the United States is **diabetes mellitus**; possibly 2% to 3% of the population will eventually develop the disorder. Diabetes mellitus is due to a lack of **insulin** or to the lack of adequate insulin action, but much of the pathophysiology of diabetes is related to the unopposed action of **glucagon**. **Addison's disease** is an uncommon disorder that is due to decreased adrenocorticotropic steroid production. In response to low circulating adrenocorticotropic hormones, there is a feedback increase in circulating **adrenocorticotropin** (ACTH). ACTH is produced by the anterior pituitary and is a polypeptide that contains the sequence that corresponds to melanocyte-stimulating hormone. As a result of increased levels of ACTH, people with Addison's disease commonly exhibit increased skin pigmentation.

MECHANISMS OF SIGNAL TRANSDUCTION

The Cyclic AMP Second-Messenger System

After the interaction of norepinephrine and epinephrine with the β-adrenergic receptor on target cells, the intracellular concentration of cyclic adenosine monophosphate (cAMP) increases. The activated receptor does not stimulate directly the activity of **adenylyl cyclase** (the enzyme that catalyzes the formation of cAMP from ATP). The activity of adenylyl cyclase is regulated by intermediary guanine nucleotide–binding proteins or G-proteins. The **β-adrenergic receptor** interacts with stimulatory guanine nucleotide–binding protein (**G_s**); the muscarinic ACh receptor in many, but not all, instances interacts with inhibitory guanine nucleotide–binding protein (**G_i**).

G_s and **G_i** consist of three different **polypeptides** (heterotrimers): α, β, and γ (see Fig. 6-19). The α-subunits differ; the β- and γ-subunits are similar. The epinephrine/β-receptor complex activates G_s; activation involves the exchange of guanosine triphosphate (GTP) for guanosine diphosphate (GDP) and the dissociation of α_s/GTP from the βγ-dimer. The α_s/GTP complex activates adenylyl cyclase. After hydrolysis of GTP to GDP and inorganic phosphate (P_i) catalyzed by an intrinsic GTpase activity of the α-subunit, inorganic phosphate is released and GDP remains bound to the subunit. α_s/GDP binds to βγ to form inactive α/GDP/βγ. The G-protein trimer can be reactivated by the epinephrine/β-receptor or by other activated stimulatory receptors present in the plasma membrane of the cell.

The interaction of an inhibitory hormone receptor complex with G_i results in the formation of **α_i/GTP** and a dissociated βγ-dimer; α_i/GTP inhibits adenylyl cyclase activity. Hydrolysis of GTP to GDP + P_i is followed by formation of the inactive α/GDP/βγ protein trimer of G_i. The hydrolysis reactions are catalyzed by intrinsic guanosine triphosphatase activity of α_s or α_i.

Another prominent G-protein is **G_q**, a GTP-binding protein that activates phospholipase C, leading to the hydrolysis of phosphatidylinositol 4,5-bisphosphate and thus yielding diglyceride (an activator of protein kinase C) and **inositol 1,4,5-trisphosphate** (which leads to the release of calcium from the endoplasmic reticulum and an increase in cytosolic calcium). G_q is also a heterotrimer made of α-, β-, and γ-subunits, which function in a manner analogous to that of G_s and G_i.

As noted in the discussion of glycogen metabolism, **adenylyl cyclase** mediates the formation of cAMP, which in turn activates **protein kinase A** (cAMP-dependent protein kinase). Activated protein kinase A catalyzes the phosphorylation of acceptor proteins, which then have altered activity (either increased or decreased, depending on the acceptor protein). The predominant residue phosphorylated by protein kinase A is protein-serine; in a few cases, protein-threonine is phosphorylated. To reverse the activation process, **cAMP phosphodiesterase** catalyzes the hydrolysis of cAMP to inactive 5'-AMP. cAMP dissociates from the protein kinase, and protein kinase A returns to the inactive form. **Phosphoprotein phosphatases** catalyze the hydrolytic dephosphorylation of the phosphorylated substrate proteins; the activity of the unphosphorylated form is again expressed.

Renin-Angiotensin System

Essential hypertension is a common malady in the United States. **Angiotensin-converting enzyme inhibitors** are commonly prescribed in its treatment. The renin-angiotensin system plays a pivotal role in the regulation of arterial pressure.

Angiotensinogen, the prohormone, is synthesized by the liver and released into the circulation. Angiotensinogen contains more than 400 amino acids. **Renin**, a proteolytic enzyme with an active-site aspartate similar to the protease pepsin, is secreted into the circulation by the juxtaglomerular cells of the renal afferent arterioles. Norepinephrine released from the postganglionic sympathetic fibers interacts with the juxtaglomerular cells of the kidney via a β_1-adrenergic receptor that activates adenylyl cyclase. cAMP promotes both renin transcription and release. Renin catalyzes the hydrolytic cleavage of **angiotensin I**, a decapeptide, from angiotensinogen. Angiotensin I is converted to **angiotensin II**, an active octapeptide hormone, by an exergonic hydrolysis reaction that removes a dipeptide from the carboxyl end. The reaction is catalyzed by **angiotensin-converting enzyme**, the target for a widely prescribed class of antihypertensive drugs. Angiotensin II increases blood pressure, and it stimulates the production of aldosterone. **Aldosterone** is a steroid hormone produced by the adrenal cortex, and this hormone promotes sodium retention.

A partial list of hormones that increase or decrease cAMP levels in target cells is given in Table 15-4.

TABLE 15-4. Hormones That Affect Cyclic AMP Levels in Appropriate Target Cells by Altering Adenylyl Cyclase Activity

Stimulatory
Adrenocorticotropic hormone
Catecholamines (at β-adrenergic receptors)
Follicle-stimulating hormone
Glucagon
Luteinizing hormone
Thyroid-stimulating hormone
Vasopressin (antidiuretic hormone)
Inhibitory
Acetylcholine (muscarinic)
Catecholamine (at α_2-adrenergic receptors)
Angiotensin II
Somatostatin

The Cyclic Guanosine Monophosphate Second-Messenger System

Most human cells contain cyclic guanosine monophosphate (**cGMP**). The components of the cGMP system parallel those of cAMP, and include guanylyl cyclase, phosphodiesterase, cGMP-dependent protein kinase (protein kinase G), acceptor substrates, and phosphoprotein phosphatases. Adenylyl cyclase is localized exclusively in the plasma membrane. In contrast, guanylyl cyclase occurs in both a plasma membrane particulate form and a cytosolic form. The particulate guanylyl cyclase is regulated by cardionatrin I, or **atrial natriuretic factor**. The soluble form of **guanylyl cyclase** is stimulated by NO, which functions as a messenger in the CNS and PNS and other cells. NO is formed from the amino acid arginine by an oxidation process.

Cells also possess **protein kinase G**. It is a dimer of identical subunits. This protein kinase is activated in an allosteric manner by cGMP. In contrast to protein kinase A, protein kinase G does not dissociate into separate catalytic and regulatory subunits. Activation can be expressed as follows:

$$E_2 + 4\,cGMP \rightleftharpoons E_2 - cGMP_4$$

The physiologic substrates of protein kinase G are largely unknown. In some cases, this enzyme phosphorylates the same substrates as protein kinase A; however, other protein kinase A substrates are phosphorylated poorly if at all by protein kinase G. Once hypothesized that cAMP and cGMP have antagonistic functions, the notion now has fallen out of favor.

The Calcium Second-Messenger System

Calcium serves as a very important intracellular regulator: It is required for **exocytosis** and **muscle contraction**. Calcium also plays an important role as a second messenger. A number of bioactive substances that exert their action, at least in part, by affecting intracellular calcium levels are shown in Table 15-5.

TABLE 15-5. Hormones That Affect Calcium Levels in Appropriate Target Cells

Adrenocorticotropic hormone
Acetylcholine (muscarinic)[a]
α_2-Adrenergic catecholamines
Angiotensin II
Cholecystokinin
Gastrin
Histamine (H_1)[b]
Luteinizing hormone
Thyrotropin-releasing hormone
Vasopressin

[a] *Neurotransmitter.*

[b] *Autocrine agent.*

There are at least two mechanisms by which a first messenger acting through its receptor can increase intracellular calcium. The **hormone-receptor complex** may enhance calcium influx through the plasma membrane in exchange for intracellular sodium. Another mechanism involves the **phosphoinositide system**. The initial substrate for this process is **phosphatidylinositol 4,5-bisphosphate** (PIP_2); PIP_2 is present on the inner aspect of the plasma membrane. The hormone-receptor complex activates **phospholipase C** (see Fig. 7-11), which in turn catalyzes the hydrolysis of phosphatidylinositol 4,5-bisphosphate to form 1,2-diglyceride and inositol 1,4,5-trisphosphate (IP_3). Inositol trisphosphate, liberated into the cytosol, interacts with the endoplasmic reticulum and promotes the release of stored calcium. The endoplasmic reticulum possesses a protein receptor that binds IP_3.

To effect a response, calcium interacts in many cases with a calcium-binding protein called **calmodulin**. Calmodulin is a low–molecular-mass protein (17,000 daltons) that binds up to four calcium ions with high affinity (binding constants are in the micromolar range). After binding with calcium, calmodulin undergoes a conformational change and alters the activity of specific effectors. Calcium/calmodulin, for example, activates one form of cyclic nucleotide phosphodiesterase. This provides reciprocal interaction of the calcium and cyclic nucleotide second-messenger systems. The activity of some isozymes of adenylyl cyclase is regulated by calcium/calmodulin. The interaction of various second-messenger systems is common and is called **cross talk**.

The activity of several protein kinases is enhanced by calcium/calmodulin. **Glycogen metabolism**, for example, is one process regulated by calcium. **Phosphorylase kinase** is composed of four different types of subunits: α, β, γ, and δ. The δ-subunits, tightly attached to the others, are actually calmodulin. Phosphorylase kinase is activated as calcium binds to the δ-subunit. Calcium triggers both muscle contraction and activation of phosphorylase kinase in parallel. This dual regulation constitutes a part of the molecular logic of the cell. Phosphorylase kinase also is activated by the phosphorylation catalyzed by protein kinase A (Table 15-6). Regulation of glycogen metabolism by calcium and by cAMP demonstrates the convergence of two second-messenger systems on a common function.

Three enzymes have been distinguished by their substrate specificity and are called **calcium/calmodulin-dependent protein kinases I, II,** and **III** (based on the historical order of their discovery). Calcium/calmodulin-dependent protein kinase II is a broad-specificity protein kinase and may play a general regulatory role in several aspects of metabolism. The other two calmodulin-regulated kinases exhibit restricted substrate specificity.

TABLE 15-6. Some Physiologic Substrates of cAMP-Dependent Protein Kinase

Protein Substrate	Effect
Glycogen synthase	Inhibits
Hormone-sensitive lipase	Activates
Phosphatase inhibitor I	Activates
Phospholamban (muscle)	Activates
Phosphorylase kinase	Activates
Pyruvate kinase (liver)	Inhibits
Tyrosine hydroxylase (nerve)	Activates

The Diglyceride Second-Messenger System

The hydrolysis of phosphatidylinositol 4,5-bisphosphate generates two second messengers: **inositol 1,4,5-trisphosphate** and **diglyceride**. Diglyceride, or diacylglycerol, activates its cognate protein kinase, which is called **protein kinase C**. **Protein kinase C** (C refers to calcium) requires phospholipid and calcium for full expression of its activity. Diglyceride increases the affinity of protein kinase C for calcium so that the kinase can function at basal intracellular calcium concentrations (0.1×10^{-6} mol/L). One intriguing finding concerning protein kinase C is that it is specifically activated by tumor promoters, including phorbol esters. Tumor promoters alone do not produce cancers. For tumorigenesis to occur, they must be added after an initiating agent such as benzo[*a*]pyrene. Protein kinase C has been implicated in the regulation of scores of processes, and its postulated actions rival those of protein kinase A.

PROTEIN KINASES

Protein-Serine Kinases

All of the protein kinases mentioned in the previous sections (i.e., protein kinases A, C, and G and calcium/calmodulin-dependent protein kinases) catalyze the phosphorylation of protein-serine residues. In many cases, the substrate proteins and even the residue(s) phosphorylated have been determined. **Protein kinase A** catalyzes the phosphorylation of phosphorylase kinase (and activates it) and glycogen synthase (and inactivates it). A partial list of its other protein substrates is given in Table 15-6.

Less is known about the identity of the physiologic substrates of protein kinase G, but considerable information is available on the substrates of calcium/calmodulin-dependent protein kinase (II) and protein kinase C. The vast majority of phosphate in proteins (approximately 99%) is attached to serine. **Protein-serine kinases** constitute one of the most important means of enzyme and protein regulation by covalent modification in humans. A large number of protein-serine kinases are not regulated by these second messengers. An example already considered in Chapter 6 is **pyruvate dehydrogenase kinase**, which is regulated by acetyl-CoA and NADH in an allosteric fashion. After phosphorylation of pyruvate dehydrogenase, enzyme activity is decreased.

Protein-Tyrosine Kinases

Phosphorylation of protein-tyrosine residues plays an important role in the action of several hormones and growth factors and perhaps in the mechanism of tumorigenesis; for example, the insulin, epidermal growth factor, and platelet-derived growth factor receptors possess protein-tyrosine kinase activity. These are called **receptor protein-tyrosine kinases** (Fig. 15-3). The growth hormone receptor, not a protein-tyrosine kinase, activates a protein-tyrosine kinase. The ***src* gene protein**, a protein that plays a role in the pathogenesis of tumors in some animals, is also a protein-tyrosine kinase. The latter two are called **nonreceptor protein-tyrosine kinases**.

The **insulin receptor** is a tetramer of two α-chains and two β-chains linked by disulfide bridges as α-β-β-α. The α-chains recognize and interact with insulin on the cell exterior. The β-chains interact with the α-chains, which extend into the cell interior where protein-tyrosine kinase activity occurs. The insulin receptor is synthesized as a single chain, and removal of a four amino acid segment from each chain as catalyzed by endoproteases is required to produce the external (α) and internal (β) chains. Two sets of external and internal chains combine to produce the functional form of the

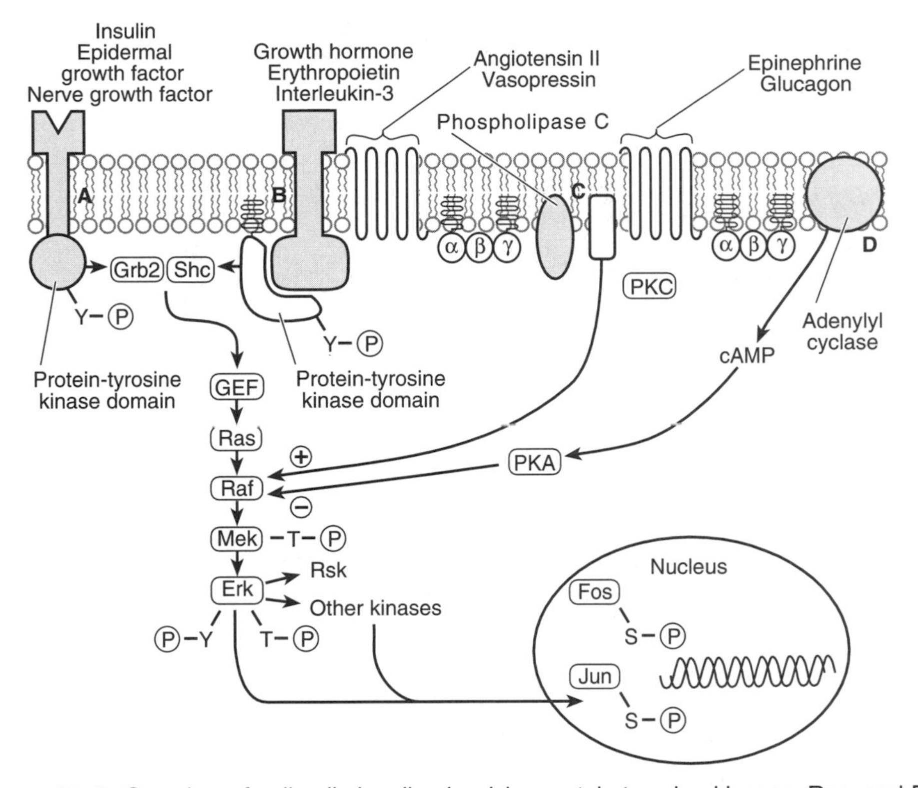

Figure 15-3. Overview of cell-cell signaling involving protein-tyrosine kinases, Ras, and Raf. The letters *A–D* refer to the sequence of steps. cAMP, cyclic adenosine monophosphate; GEF, guanine nucleotide exchange factor; PKA, prekallikrein activator; PKC, protein kinase C; T, threonine; Y, tyrosine.

insulin receptor. After activation of the insulin receptor, each β-chain undergoes an autophosphorylation of an intrinsic tyrosine residue catalyzed by the opposite β-chain.

One mechanism of insulin action involves the binding of **Grb2**, an adaptor protein, to the **protein-tyrosine phosphate** on the insulin receptor. Grb2 contains an SH2 domain (*src* homology domain 2) that binds to protein-tyrosine phosphates. Guanine nucleotide exchange factor (**GEF**) binds to Grb2 via an SH3 domain. This domain recognizes proline-rich regions of proteins. GEF activates Ras, a G-protein. Active **Ras** (with bound GTP) activates **Raf**, a protein kinase, which catalyzes the phosphorylation of MEK, a mitogen-activated/ERK kinase.

MEK is a dual-specificity protein kinase, and MEK catalyzes the phosphorylation of extracellular signal regulated kinases (**ERK**) on specific threonine and tyrosine residues. ERK, in turn, catalyzes the phosphorylation of target proteins and causes their activation. Downstream targets include Fos and Jun, transcription factors. The cascade from the insulin receptor, Ras, the Raf-MEK-ERK cascade, and transcription factor phosphorylation alter gene expression and mediate many insulin effects.

Other growth factor receptors, such as that for epidermal growth factor and platelet-derived growth factor, follow similar protein kinase cascade mechanisms. Growth hormone, prolactin, and several interleukins activate nonreceptor protein-tyrosine kinases. These enzymes can also activate the Raf-MEK-ERK protein kinase cascade. Moreover, numerous other cascades exist.

This brief summary demonstrates that **protein kinases** play a major role in metabolic and genetic regulation. Protein phosphorylation is the most common form of post-translational modification. Moreover, life scientists estimate that the human genome encodes between 1,000 and 3,000 different protein kinases.

Chapter 16

Muscle and Connective Tissue

MUSCLE METABOLISM

Muscle accounts for approximately half the mass of humans. There are three classes of muscle: skeletal, cardiac, and smooth. Skeletal and cardiac muscle are striated in microscopic appearance. This section examines the nature of the proteins that constitute muscle as well as the use of adenosine triphosphate (ATP) as a source of chemical energy for muscle contraction.

Muscle Proteins

The two most abundant proteins in muscle are actin and myosin. **Myosin** is the chief protein of the thick filament, and **actin** is found in the thin filament. Some of their properties are noted in Table 16-1. Actin occurs in most nonmuscle cells, where it is a component of the thin filaments of the cytoskeleton. Myosin contains a globular head [with adenosine triphosphatase (ATPase) activity] and a rodlike tail. The two essential and two regulatory light chains of myosin interact with the two heavy chains of myosin in the globular region.

Role of ATP in Muscle Contraction

The following is a description of the sliding filament theory of muscle contraction. **ATP** (with bound Mg^{2+}) interacts with an actin-myosin complex to displace actin and form a complex with myosin. ATP is hydrolyzed to yield adenosine diphosphate (**ADP**) and **inorganic phosphate (P_i)** and an energized state of myosin (shown diagrammatically in Fig. 16-1). Note that ATP hydrolysis occurs before force generation.

In response to a nerve impulse, **calcium** is released from the sarcoplasmic reticulum, and triggers muscle contraction according to the following scheme. Calcium binds to **troponin C**, and binding produces a conformational change such that tropomyosin rotates out of the path between actin and myosin (Fig. 16-2). Actin then interacts with myosin, the energized myosin moves relative to actin by a ratchetlike mechanism to generate force, and myosin is converted to a deenergized state. The **ATPase cycle** repeats as many as eight times per second to shorten the sarcomere and generate force.

After the cessation of nerve impulses, calcium is sequestered into the sarcoplasmic reticulum by an ATP-dependent process involving a **class P ATPase** (see Chapter 4). The calcium concentration decreases, and troponin reverts to its original conformation. Tropomyosin again blocks the interaction of actin and myosin.

Both skeletal and cardiac muscle, as well as brain, contain **creatine phosphate**. Creatine phosphate serves as a storage form of energy-rich phosphate. Creatine is

TABLE 16-1. Contractile Proteins

Component	Molecular Mass	Structure	Comments
Thick filament			
Myosin	520,000	Two heavy chains, four light chains (two essential and two regulatory)	ATPase heads
Thin filament			
Actin	42,000	Globular monomers form fibrous aggregate	Interacts with myosin to generate force
Tropomyosin	70,000	Two coiled subunits that extend the length of seven actin monomers	Blocks binding of actin to myosin
Troponin	76,000	—	—
Troponin I	21,000	—	Inhibitory
Troponin T	37,000	—	Binds tropomyosin
Troponin C	18,000	—	Binds Ca^{2+}

ATPase, adenosine triphosphatase.

N-methylguanidinoacetate ($HOOCCH_2N(CH_3)C(=NH)N(H)\sim PO_3^{=}$). The P~N, or phosphoramidate, bond is of the high-energy or energy-rich variety. **Creatine phosphokinase** (CPK) catalyzes a reversible reaction between creatine and ATP to form creatine phosphate and ADP. After contraction and the formation of ADP, CPK catalyzes the phosphorylation of ADP to form ATP. **ATP** can serve as a substrate for continued muscle contraction. In exercising muscle, the level of creatine phosphate falls before the decrease in the level of ATP. The level of creatine phosphate returns to its initial value during rest. ATP, formed from substrate-level and oxidative phosphorylation, serves as the source of chemical energy for regenerating creatine phosphate. **CPK** measured in serum samples is commonly elevated after a myocardial infarction and also in individuals with muscular dystrophy and their carrier mothers.

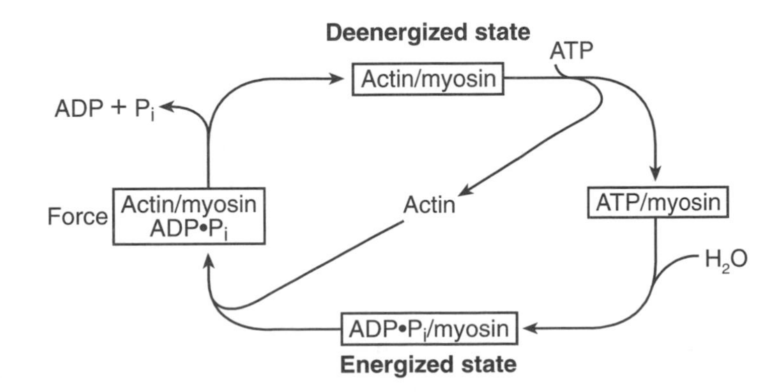

Figure 16-1. The actin/myosin energy transduction cycle. ADP, adenosine diphosphate; ATP, adenosine triphosphate; P_i, inorganic phosphate.

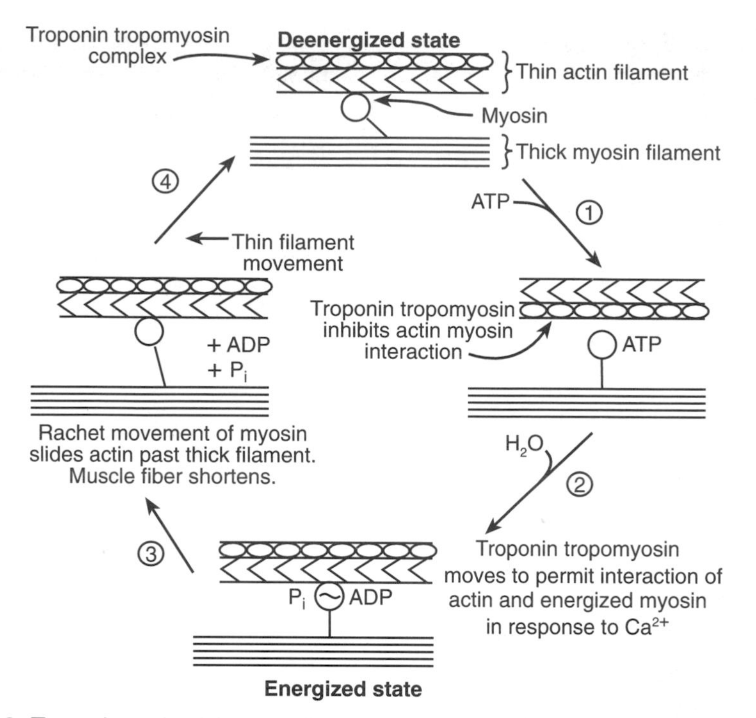

Figure 16-2. Troponin and calcium mediate the interaction of actin and myosin during muscle contraction. The circled numbers represent the sequential steps in muscle contraction described in Fig. 16-1. ADP, adenosine diphosphate; ATP, adenosine triphosphate; P_i, inorganic phosphate.

CONNECTIVE TISSUE

The intercellular space in humans contains an organic matrix rich in **proteoglycans**, which consist of acidic polysaccharides (95% by mass) and proteins (5%). Proteoglycans form the ground substance in which collagen and cells are embedded to form tissues.

Proteoglycans

Proteoglycans are composed of a polysaccharide axis of hyaluronate. Many core proteins emanate laterally from the long, thin hyaluronate axis. A **link protein** stabilizes the **hyaluronate–core protein** complex. Many chondroitin sulfate and keratan sulfate sugar chains are covalently attached to the core proteins and constitute the major mass of the molecule.

There are three types of bonds by which the polysaccharide is covalently linked to the core protein: (1) an *O*-glycosidic bond between **xylose and serine**, (2) an *O*-glycosidic bond between ***N*-acetylgalactosamine and serine or threonine**, and (3) an *N*-glycosidic bond between ***N*-acetylglucosamine and the amide nitrogen of asparagine**.

The components of the **polysaccharide repeating units** are listed in Table 16-2. The following generalizations can be made to simplify the subject. One component of the repeating disaccharide unit consists of either *N*-acetylglucosamine or *N*-acetylgalactosamine, or one of their sulfate derivatives. Except for keratan sulfate (with galactose), the second component consists of a uronic acid salt; it is either D-glucuronate or its 5-epimer L-iduronate. Except for hyaluronate, the polysaccharides of proteoglycans contain *O*-sulfate or *N*-sulfate esters. The donor of sulfate during biosynthesis is **phosphoadenosylphosphosulfate**. The linkages connecting the disaccharides, which

TABLE 16-2. Properties of the Polysaccharide Components of Proteoglycans

Class	Composition of Repeating Unit	Comments
Hyaluronate	Glucuronic acid and *N*-acetylglucosamine	Widely distributed in animal tissues, synovial fluid, and vitreous of the eye
Chondroitin sulfate	Glucuronic acid and *N*-acetylglucosamine as the *O*-4 or *O*-6 sulfate	Polysaccharide chains of 20,000 daltons
Keratan sulfate	Galactose and *N*-acetylglucosamine as the *O*-6 sulfate	Two forms (I and III) differ in bonding to protein
Heparan sulfate	L-Iduronate *O*-2 sulfate and glucosamine *O*-6 and *N*-2 sulfate or acetate	—
Dermatan sulfate	L-Iduronate *O*-2 sulfate and *N*-acetylgalactosamine 4-sulfate; glucuronic acid and *N*-acetylgalactose	Two types of repeating subunit
Heparin	*N*-Acetylglucosamine and L-iduronic acid (90%) or glucuronic acid (10%)	Protein core almost all serine and glycine

are not given here, are specific and determined by the biosynthetic enzyme (specific glycosyltransferases catalyze each addition from a nucleotide sugar). The sulfates and glucuronides account for the acidic nature of this group of complex carbohydrates.

Several inherited **proteoglycan** or **mucopolysaccharide storage diseases** are the result of deficient activities of lysosomal catabolic enzymes. Some of these diseases are listed in Table 16-3.

Collagen

Collagen is the major macromolecule of connective tissue and the most abundant protein in humans and the animal kingdom. It accounts for approximately one-third of

TABLE 16-3. Proteoglycan Storage Diseases [Mucopolysaccharidoses (MPSs)]

Disease	Enzyme Effect	Alternate Designation
Hurler	α-L-Iduronidase	MPS I
Hunter	Iduronate sulfatase	MPS II
Sanfilippo A	*N*-Sulfatase	MPS III A
Sanfilippo B	α-*N*-Acetylglucosamidase	MPS III B
Sanfilippo C	Acetyltransferase	MPS III C
Morquio	*N*-Acetylgalactosamine 6-sulfatase	MPS IV

TABLE 16-4. Major Types of Collagen

Type	Molecular Formula	Location
I	$[\alpha 1(I)]_2\alpha 2(I)$	Skin, tendon, bone
II	$[\alpha(II)]_3$	Cartilage
III	$[\alpha 1(III)]_3$	Skin, blood vessels, uterus
IV	$[\alpha 1(IV)]_3$ and $[\alpha 2(IV)]_3$	Basement membranes
V	$\alpha A(\alpha B)_2$ or $(\alpha A)_3$ and $(\alpha B)_3$	Widespread in small amounts

human protein by mass. Collagen is secreted by fibroblasts; it forms insoluble fibers of high tensile strength. The most distinguishing property of collagen is that it forms a **triple, left-handed helix** made up of three polypeptide chains. There are three amino acid residues per turn. Three left-handed helices combine to form a right-handed super helix, which is long (300 nm) and narrow (1.5 nm).

Another distinguishing feature of collagen is that every third residue is glycine. **Glycine** is the only amino acid small enough to exist at the central core of a triple helix. The sequence of the main body of collagen is thus Gly-X-Y, where X and Y are residues other than glycine. Of the 1,050 residues per chain, approximately 100 of the X residues are proline and 100 Y residues are 4-hydroxyproline. Collagen also contains 5-hydroxylysine residues, which serve as attachment sites for carbohydrate. The collagen helix is stabilized by multiple interchain cross links. The collagens of humans consist of more than one dozen molecules composed of more than 18 genetically distinct α-chains, some of which are given for reference purposes in Table 16-4. **Osteogenesis imperfecta**, or brittle bone disease, involves mutations involving type I collagen, and these mutations generally involve a replacement of glycine by another amino acid.

The synthesis of collagen requires a number of posttranslational reactions similar to those given previously. Collagen synthesis begins intracellularly, and synthesis is completed extracellularly by proteolytic processing. Collagen is synthesized as **preprocollagen** in the rough endoplasmic reticulum. The signal peptide is promptly cleaved, yielding **procollagen**. Prolyl and lysyl hydroxylation are followed by the subsequent glycosylation of several hydroxylysyl residues. Prolyl and lysyl hydroxylases use ascorbate as reductants, and the deficiency of ascorbate explains the pathophysiology of connective tissue defects in **scurvy**.

The formation of the **triple helical structure** occurs in the Golgi. The procollagen triple helix is secreted from the fibroblast. Procollagen amino-terminal protease and procollagen carboxyterminal protease are extracellular enzymes that catalyze the hydrolytic removal of amino-terminal and carboxyterminal fragments, yielding **tropocollagen**. The amino-terminal and carboxyterminal fragments, the proprotein sequences, contain disulfide bonds. Tropocollagen and collagen, however, lack cysteine residues. The tropocollagen molecules form regular, parallel arrays and are stabilized by cross-linking reactions. Lysyl oxidase, an extracellular enzyme, catalyzes the formation of protein-lysyl aldehyde groups, and cross links form between these ε-aldehyde groups and ε-amino groups of other collagen-lysines.

Chapter 17

Hemoglobin, Heme, and Blood Coagulation

HEMOGLOBIN

Blood transports oxygen and carbon dioxide, and consists of **cells** and **plasma**. The liquid remaining when blood plasma clots is the **serum**. The blood clot consists of fibrin, which is derived from fibrinogen. The **hematocrit** is the percent, by volume, of the cellular component to the total volume of blood. Hematocrit reflects the red blood cell mass and is one of the most important laboratory values used in clinical medicine (e.g., an individual with a low hematocrit is anemic).

Erythrocytes are small (7 μm in diameter), biconcave, disc-shaped cells that have lost their nucleus, mitochondria, peroxisomes, and endoplasmic reticulum. They lack mitochondria and derive their adenosine triphosphate (ATP) from anaerobic glycolysis by converting glucose to lactate. Erythrocytes also possess an active pentose phosphate pathway (see Chapter 6). Reduced nicotinamide adenine dinucleotide phosphate (NADPH) generated by this pathway is responsible for the maintenance of reduced glutathione. Glutathione participates in the destruction of hydrogen peroxide in a reaction catalyzed by glutathione peroxidase.

Hemoglobin Structure

The key molecule of oxygen transport is **hemoglobin**. It consists of four polypeptide chains and four heme groups. Hemoglobin is a dimer of dimers, with two chains from the α-family and two chains from the β-family. The four chains are held together by noncovalent attractions. Each chain contains one heme group that binds one molecule of oxygen. Hemoglobin A, the principal hemoglobin in adults, consists of two **α-chains** and two **β-chains**, designated $\alpha_2\beta_2$. The α-chains each contain 141 amino acid residues, and the β-chains each contain 146 amino acid residues. **Hemoglobin A_{1c}** levels result from nonenzymatic modification of hemoglobin A by glucose, and are used to monitor the control of diabetes mellitus. The α-chains of hemoglobin consist of seven helical regions, and the β-chains consists of eight (labeled A through H). There is an invariant histidine in all hemoglobins, called *F8* or the **proximal histidine**, which binds heme. F8 refers to the eighth residue of the F helix.

Genes for the **α-gene** family occur on **chromosome 16**; those for the **β-gene** family occur on **chromosome 11**. Two identical α-genes (α1 and α2) per haploid genome encode for the α-chain of hemoglobin. Both genes are expressed, with the normal diploid genome consisting of four α-genes designated αα/αα. This unusual pairing helps to explain the mechanism for the production of **α-thalassemia**, a genetic disease that leads to a deficiency of α-chains (see Thalassemias). Hemoglobin chain genes are developmentally expressed. α-Chain production begins *in utero*, but β-chain production occurs near to term and postnatally.

Cooperative Oxygen Binding

The concentration dependence of oxygen binding to myoglobin is expressed by a **rectangular hyperbola** (see Fig. 5-9). Myoglobin is a monomer; thus, the rectangular hyperbola signifies oxygen binding to identical sites. The concentration dependence of oxygen binding to hemoglobin is expressed by a **sigmoidal curve**. The sigmoidal curve is diagnostic of positive cooperativity. Physiologically this is important, because hemoglobin can bind and release more of its oxygen cargo over the physiologic range of oxygen tension.

The structure of hemoglobin differs in both the oxygenated and deoxygenated states. The quaternary structure of the oxygenated state is called the **R state** (for relaxed), and the conformation of the deoxygenated state is called the **T state** (for tense). Deoxyhemoglobin is more taut and constrained than oxyhemoglobin because the T form contains eight more salt bridges between subunits than the R form.

As acidity increases, hemoglobin's ability to bind oxygen decreases; protons cause hemoglobin to dump oxygen. The pH dependence of oxygen binding makes physiologic sense and follows the molecular logic of the cell. When erythrocytes pass through active tissues that are producing lactic acid, lactic acid serves as a signal that more oxygen is needed. The increased acidity causes the hemoglobin to release oxygen, a process known as the **Bohr effect**.

Deoxyhemoglobin is a weaker acid (has greater proton affinity) than oxyhemoglobin, where

$$HHb^+ + O_2 \rightleftharpoons HbO_2 + H^+$$

Weaker acid   Stronger acid

The difference in acidity is due to a difference in the pK_a values of the imidazole side chain of one histidine in the α-chains and one histidine in the β-chains. The average pK_a of these **histidines** in the microenvironment of deoxyhemoglobin is slightly basic (7.9), and that for oxyhemoglobin is slightly acidic (6.7). At physiologic pH, each mole of tetrameric of deoxyhemoglobin takes up approximately 0.6 mol of protons.

2,3-Bisphosphoglycerate occurs in the red cell. It binds preferentially to deoxyhemoglobin, therefore, promoting the release of oxygen from hemoglobin (2,3-bisphosphoglycerate forces hemoglobin to dump oxygen). The concentrations of this allosteric regulator of hemoglobin increase when humans at high altitudes adapt to lower oxygen pressure. 2,3-Bisphosphoglycerate is formed from 1,3-bisphosphoglycerate in a reaction catalyzed by bisphosphoglycerate mutase. 2,3-Bisphosphoglycerate is converted to 3-phosphoglycerate in a reaction catalyzed by 2,3-bisphosphoglycerate phosphatase.

DISORDERS RELATED TO ABNORMAL HEMOGLOBIN

Sickle Cell Anemia

Sickle cell anemia causes considerable mortality and morbidity in Africa and in every population in which individuals of African descent have migrated from other areas. Individuals with sickle cell anemia are homozygous for hemoglobin S expression.

Hemoglobin S differs from hemoglobin A by the conversion of β6 glu $\rightarrow$ val. This alteration is caused by a point mutation that results in an A $\rightarrow$ T transversion in the triplet codon for the sixth residue of the β-globin chain. In concentrated hemoglobin solutions that are partially or fully deoxygenated, this mutation leads to polymerization of hemoglobin that causes cell deformity. The formula for sickle cell hemoglobin is $\alpha_2{}^A\beta_2{}^S$. Heterozygotes with sickle cell trait are asymptomatic, accounting for the designation of the disease as autosomal recessive. A single blood cell of a heterozygote contains 60% $\alpha_2{}^A\beta_2{}^A$ and 40% $\alpha_2{}^A\beta_2{}^S$.

Deoxyhemoglobin S aggregates and sickle-shaped cells form because of the point mutation. This morphology leads to the reduced deformability of the red cell and its defective passage through the microcirculation, which, in turn, is responsible for the vascular occlusion. Such structural alterations lead to hemolysis and chronic anemia. Normally, red blood cells spend approximately 1 second in the capillary circulation. Sickling can occur when cells containing hemoglobin S spend a longer period at reduced oxygen concentrations.

Sickle cell anemia is characterized by a lifelong hemolytic anemia, the occurrence of acute or sudden exacerbations called **crises**, and various complications resulting from an increased propensity to infection and the damaging effects of repeated vascular blockage. Sickle cell anemia can be diagnosed *in utero*; however, individuals with sickle cell anemia usually present during the first or second year of life. They exhibit a failure to thrive and have repeated infections caused by *Streptococcus pneumoniae* or *Haemophilus influenzae*.

The **crises** of sickle cell anemia can follow many forms, with chest or abdominal pain, bone infarction, and possibly an underlying infection. Renal disease and retinopathy are associated with long-term disease. Therapy is supportive and includes treatment of infections, analgesics for pain, and transfusions, as required.

Individuals with **sickle cell trait** exhibit greater resistance to malaria than those without the trait, a characteristic that accounts for the continued prevalence of sickle cell anemia in the population. Protection afforded individuals with sickle cell trait is called **heterozygote advantage**.

Thalassemias

Thalassemias are a heterogeneous group of hemoglobinopathies in which the mutation reduces the level of synthesis of the α- or β-chains. There are two main groups of diseases and several clinical pictures that bear specific names: the **α-thalassemias**, in which α-chain synthesis is impaired; and the **β-thalassemias**, in which β-chain synthesis is impaired. Thalassemias are classified as α^+ and α^0 for some or no α-chain production, and β^+ and β^0 for some or no β-chain production.

A normal individual has two identical genes for α-globin on each chromosome 16 for a total of 4 α-globin genes. Possible genotypes indicate that humans can contain anywhere from zero to four normal α-globin genes. Table 17-1 lists the possible genotypes and the approximate α-globin production. If an individual possesses two normal genes, there is a mild anemia, called the **α-thalassemia trait**. Possession of one or no functional genes produces severe disease.

β-Thalassemias and **α-thalassemias** share many of the same characteristics. There is decreased β-globin production and the imbalance of globin chain synthesis, which lead to precipitation of the excess α-chains and hemolytic anemia. The β-chain is important in the postnatal period; onset of β-thalassemia is not apparent until a few months after birth.

Infants with **homozygous β-thalassemia** present with severe anemia before the age of 2 years. They exhibit jaundice, hepatosplenomegaly, and skeletal changes reflecting an increase in bone marrow volume. These infants have **thalassemia major**, which may be of the more severe β^0 or β^+ variety. In the former case, there is no detectable hemo-

TABLE 17-1. Genotypes of α-Thalassemias

Clinical Condition (Severity of Disease)	Number of Functional α-Genes	Genotype	α-Chain Production (%)
Normal	4	αα/αα	100
Silent carrier (nil)	3	αα/α-	75
α-Thalassemia (mild)	2	α-/α- or αα/—	50
Hemoglobin H disease (moderate)	1	α-/—	25
Hydrops fetalis (severe)	0	—/—	0

globin A; in the latter case, some hemoglobin A is detectable. Carriers of one β-thalassemia allele are clinically well and are said to have **thalassemia minor**.

β-Thalassemias are usually due to single nucleotide substitutions rather than deletions. More than 80 different mutations cause β-thalassemia, including almost every conceivable type of abnormality that can reduce the synthesis of a messenger RNA or protein. Mutations include promoter mutations that decrease transcription, cap site and initiator codon mutations, splicing mutations, frameshift mutations, nonsense mutations, and point mutations that produce unstable hemoglobin. In contrast to the α-thalassemias, deletions are a minor cause of the β-thalassemias.

HEME

Biosynthesis

Heme is a **tetrapyrrole** that occurs in hemoglobin, myoglobin, and mitochondrial cytochromes, cytochrome P450, catalase, peroxidases, tryptophan pyrrolase, prostaglandin endoperoxide synthase, and the soluble form of guanylate cyclase. **Hemoglobin** is quantitatively the most abundant heme protein in the body. Heme consists of iron bound to **protoporphyrin IX**.

Protoporphyrin IX is derived from eight molecules of **succinyl-coenzyme A** (**succinyl-CoA**) and eight molecules of **glycine**. Heme synthesis is initiated within the mitochondrion (one step), continues in the cytosol (three steps), and is completed within the mitochondrion (three steps). The first and rate-limiting step in the biosynthesis of heme is catalyzed by **δ-aminolevulinic acid synthase**, a pyridoxal phosphate enzyme. The amino group of glycine forms a Schiff base with pyridoxal phosphate.

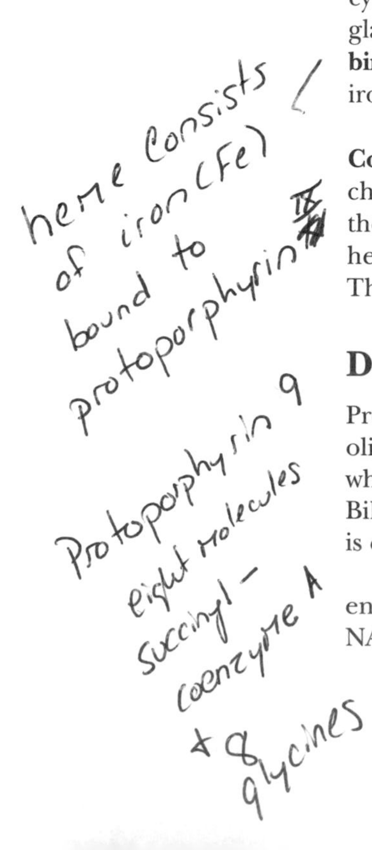

Degradation

Proteins undergo turnover, and the heme liberated during protein catabolism is metabolized. Heme is converted into various bile pigments (not to be confused with bile salts, which are cholesterol derivatives). Key intermediates include **biliverdin** and **bilirubin**. Bilirubin can undergo a conjugation reaction to form bilirubin diglucuronide. It also is converted to urobilin (isolated from urine) and stercobilin (isolated from feces).

The first step in the degradation of heme is catalyzed by **heme oxygenase**, an enzyme of the endoplasmic reticulum. The reactants include 2 mol oxygen, 1 mol NADPH, and heme. This process generates carbon monoxide as well as playing a role in heme degradation. **Carbon monoxide**, which may serve as a second messenger, is analogous to nitric oxide metabolism in the brain (see Chapter 15). Heme is cleaved between the A and B rings to produce **biliverdin**, ferric iron, carbon monoxide, and water. Two atoms of oxygen occur in biliverdin, a third occurs in carbon monoxide, and a fourth occurs in water. The oxidation of organic compounds by molecular oxygen is exergonic and irreversible. Biliverdin reacts with NADPH in a reaction catalyzed by **biliverdin reductase** to form **bilirubin**. Bilirubin, a lipophilic substance, is transported to the liver as a complex with albumin. Bilirubin reacts with two molecules of energy-rich uridine diphosphate~glucuronate to form **bilirubin diglucuronide**, a hydrophilic energy-poor substance. Bilirubin diglucuronide is excreted in the bile.

In the large bowel, bilirubin diglucuronide can be hydrolyzed by bacterial enzymes (β-glucuronidases) to form free bilirubin. Bacteria reduce bilirubin to colorless **urobilinogen**. Most of the fecal urobilinogen is converted to **stercobilin**, a pigment that gives feces its characteristic brown color. A small portion of urobilinogen can be taken up from the gut, transported in the circulation, and excreted in the urine as **urobilin**. Urobilins give urine its characteristic amber color.

When blood contains excessive bilirubin, the sclerae of the eyes and the skin become yellow, a condition known as **jaundice**. Although jaundice is not a disease per se, it does indicate an abnormal condition. Total bilirubin (bilirubin and bilirubin diglucuronide) and bilirubin diglucuronide are commonly measured by clinical chemistry laboratories. Rapidly reacting or soluble bilirubin diglucuronide is called **direct reacting bilirubin**; the insoluble bilirubin is called **indirect reacting bilirubin**, determined by the van den Bergh reaction. The relative proportions of the direct and indirect van den Bergh bilirubin reactions are helpful in pinpointing the mechanisms for producing jaundice.

PORPHYRIAS

Abnormal Porphyrin Metabolism

The porphyrias are a group of **inherited** and **acquired** disorders in which the activities of the enzymes of heme biosynthesis are partially deficient. These rare disorders are classified as hepatic or erythroid, depending on the location of the principal site of the disease. Deficiencies of seven of the eight biosynthetic enzymes have been noted. The only enzyme not associated with porphyria is δ-aminolevulinate synthase, the first enzyme of the pathway.

Acute intermittent porphyria is an autosomal dominant disorder and is the most common form of hepatic porphyria. This disease is due to a 50% reduction of **uroporphyrinogen synthase**, the third enzyme of the pathway. Approximately 90% of individuals with this defect remain asymptomatic throughout life. The symptoms are quite variable, and may involve the peripheral, autonomic, or central nervous system, but usually include abdominal pain. Individuals are not photosensitive. The mechanism for the production of the symptoms is unclear. In the United States, the incidence of this disorder is from 5 to 10 cases per 100,000.

Porphyria cutanea tarda is the most common form of all of the porphyrias. The disorder can be congenital or acquired. Symptoms for the acquired form occur in middle or late life. This disease is due to reduced hepatic **uroporphyrinogen decarboxylase** activity, a cytosolic enzyme that catalyzes the fourth reaction of the pathway. These individuals are photosensitive. The most common clinical finding is cutaneous lesions on the light exposed areas of the hands, arms, and face. There is an increase in uroporphyrin I, formed nonenzymatically from a porphobilinogen metabolite, in the urine.

Lead poisoning results from an assortment of exposures to such things as inhaled fumes, ingested contaminated food, lodged bullets, and lead solubilized from glazes of utensils. The primary source of lead poisoning in children is ingestion of lead-based paint chips. The anemia that accompanies lead poisoning (plumbism) is partially the result of various inhibitory effects of lead on heme biosynthesis. **Porphobilinogen synthase** and ferrochelatase are most sensitive to lead. Typical clinical manifestations are autonomic neuropathy, which causes abdominal pain and ileus (lead colic), and motor neuropathy (lead palsy)—both symptoms produced by advanced lead toxicity. However, insidious nonspecific musculoskeletal and neuropsychiatric complaints are more prevalent.

BLOOD CLOTTING

The balance between initiating and retarding blood clotting is exquisite. Normal hemostasis requires interactions among blood vessels, platelets, monocytes, and blood coagulation proteins. Blood coagulation is initiated by substances in injured tissues and is propagated by a network of serine proteases with trypsin-like specificity. These proteolytic factors contain a **catalytic triad** that consists of serine, histidine, and aspartate residues. This triad is characteristic of serine proteases, including trypsin and chymotrypsin. **Calcium** and **phospholipid** are essential for blood coagulation. The coagulation reactions occur quickly, but remain localized. Natural anticoagulation mechanisms depend on serine proteases. After several days, fibrin clots are lysed by serine proteases and replaced by connective tissue matrix molecules.

There are two pathways that initiate blood clotting: intrinsic (complicated) and extrinsic (simple). The **intrinsic pathway** may be initiated by an abnormal surface provided by damaged endothelium *in vivo* or by glass *in vitro*, and is so named because all components are present in blood; no exogenous component is required to initiate or propagate the reaction. In contrast, the **extrinsic pathway** requires an extravascular component, **thromboplastin** or **factor III**, which results when blood contacts any tissue because of injury. Many tissues express factor III. Both pathways merge at the common pathway (Fig. 17-1). The **common pathway** involves the conversion of prothrombin to active thrombin, a serine protease. Thrombin catalyzes the conversion of fibrinogen to fibrin. These are the two most important steps of the process.

Blood clotting proteins are designated by Roman numerals and common names that are used interchangeably (Table 17-2). The numerals reflect the order of discovery, not their order in the overall process. Note, however, that factor VI is omitted from the blood clotting components. Most hematologists use the terms **fibrinogen**, **prothrombin**, **thromboplastin**, and **calcium** rather than their Roman numbers.

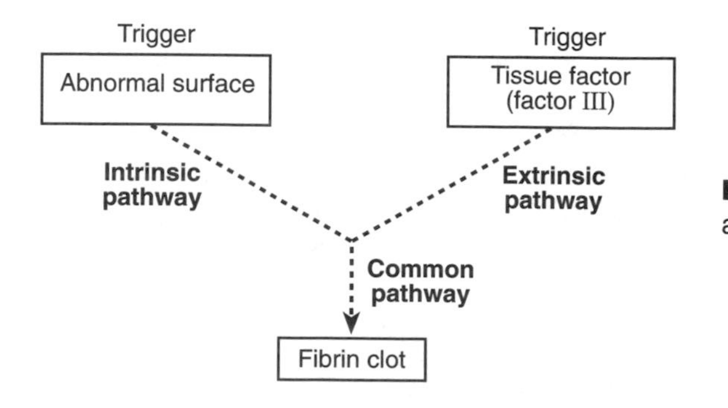

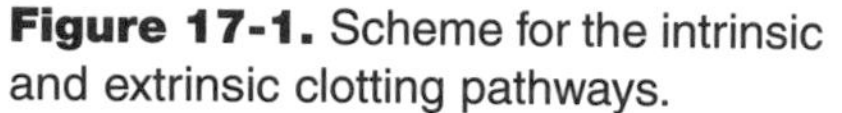
Figure 17-1. Scheme for the intrinsic and extrinsic clotting pathways.

TABLE 17-2. Blood Clotting Factors

Name (Disorder)	Numerical Factor Designation	Site of Synthesis	Serine Protease	γ-Carboxy-glutamate	Ca^{2+} Required	General Properties
Fibrinogen (afibrinogenemia)	I	Liver	–	–	–	Converted to fibrin clot by thrombin-catalyzed proteolysis; stabilized by transglutaminyl transferase (factor XIIIa)
Prothrombin (hypoprothrombinemia)	II	Liver	+	+	+	Acts on fibrinogen and factors V, VII, VIII, XIII by proteolysis
Thromboplastin	III	Most tissues	–	–	–	Auxiliary component in factor VII activation
Calcium	IV	Liver	–	–	—	Required for vitamin K–dependent factors (II, VII, IX, X) and activation of XIII
Proaccelerin	V	Liver	–	–	–	Auxiliary to action of Xa
Proconvertin	VII	Liver	+	+	+	Activates IX and X
Antihemophilic factor (hemophilia)	VIII	Liver	–	–	–	Auxiliary to action of IXa
Christmas factor (Christmas disease)	IX	Liver	+	+	+	Activates X
Stuart factor	X	Liver	+	+	+	Activates prothrombin
Plasma thromboplastin antecedent	XI	Liver	+	–	–	Activates IX
Hageman factor	XII	?	+	–	–	Initiates intrinsic pathway
Fibrin-stabilizing factor	XIII	?	–	–	+	Transglutaminase cross links fibrin
Protein C	XIV	Liver	+	–	–	Inactivates V and VII by proteolysis; anticoagulant
Protein S	—	?	–	–	–	Auxiliary to action of protein C
Prekallikrein	—	Liver	+	–	–	Kallikrein activates XII by proteolysis
High–molecular-weight kininogen	—	Liver	–	–	–	Accessory to kallikrein
von Willebrand's factor (von Willebrand's disease)	VIII antigen	—	–	–	–	Carrier of VIII

The positive (+) and negative (–) signs indicate whether a specific factor possesses serine activity, contains γ-carboxyglutamate (vitamin K–dependent) modifications, or requires calcium for activity.

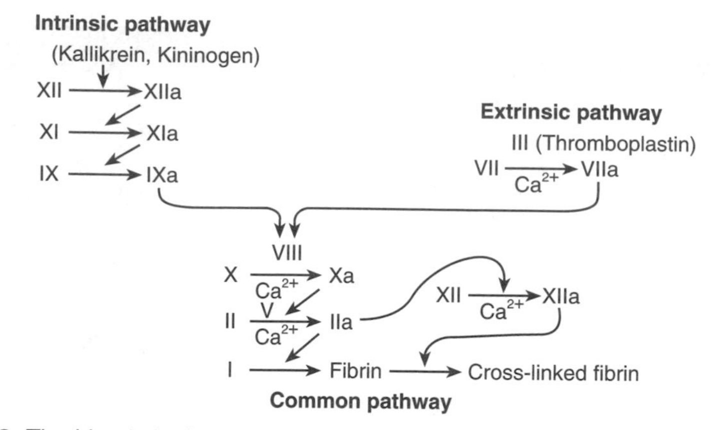

Figure 17-2. The blood clotting cascade of reactions.

Blood clotting involves a series of reactions in which the product of one process initiates a subsequent process; the product of that process then initiates still another process. This scheme is called a **cascade** (Fig. 17-2). The concepts of a blood clotting cascade and protein kinase cascades were formulated simultaneously in the 1960s. Blood clotting involves at least six distinct proteases (i.e., Factors II, VII, IX, X, XI, and XII) with a serine residue at the active site. These serine proteases exhibit trypsin-like specificity (hydrolyzing a peptide bond on the carboxyl side of a basic amino acid residue). Not all such peptide bonds are attacked by the factors, and the enzyme factors operate on selected targets in their specific substrates. During proteolysis, an inactive proenzyme (e.g., factor XII) is converted to an active enzyme, designated by an "a" following the factor number (e.g., factor XIIa). At the end of the blood clotting cascade, prothrombin (e.g., factor II) is converted to thrombin (e.g., factor IIa). The protein substrate for thrombin that results in clot formation is fibrinogen (e.g., factor I). Fibrinogen is converted into a fibrin clot by proteolysis catalyzed by thrombin or factor IIa.

Four factors function as auxiliary components in mediating specific conversions: factors III, IV, V, and VIII. **Factor IV** is calcium, and **factors III, V, and VIII** are proteins. Ca^{2+} is necessary for at least five steps in the clotting cascade: formation of IIa, VIIa, IXa, Xa, and XIIIa. The first four of these protein factors contain several γ-carboxyglutamyl residues that bind calcium. γ-Carboxyglutamate is a product of posttranslational modification and involves a vitamin K–dependent reaction.

Factor VIII, antihemophilic factor, is associated with hemophilia A, an X-linked bleeding disorder. Factor IX is called **Christmas factor**. The name is based on that of a person who had a disease resembling classical hemophilia, but whose biochemical lesion was shown to be different. **Christmas disease** is hemophilia B.

von Willebrand's factor is found in plasma, platelet granules, and subendothelial tissue. von Willebrand's factor performs two functions. It binds to receptors on the platelet surface, and it forms a bridge between the platelet and areas of vascular damage. von Willebrand's factor also binds to and stabilizes factor VIII. A deficiency of any of these three factors produces the same clinical manifestations, demonstrating that the three factors interact during blood clotting.

Blood Clotting Cascade

Hemostasis involves many plasma proteins, platelets, and their interaction with the vascular endothelium. The main points of blood clotting can be summarized as follows. **Thromboplastin**, factor III, initiates the short extrinsic pathway. Factor VII has weak proteolytic activity that is augmented by thromboplastin. Research suggests that factor VIIa, which formally belongs to the extrinsic pathway, can lead to the activation

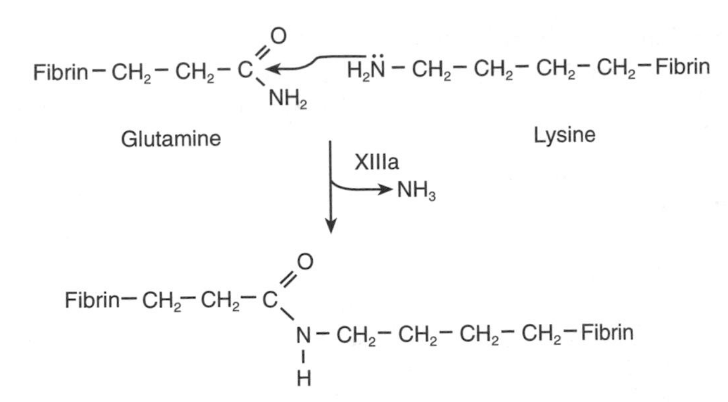

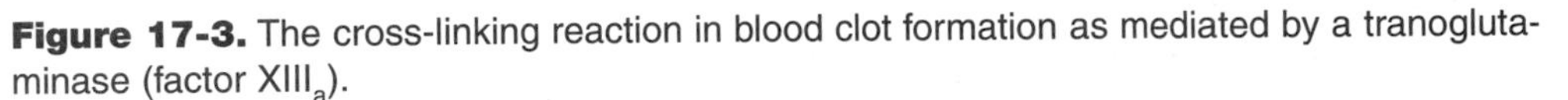
Figure 17-3. The cross-linking reaction in blood clot formation as mediated by a tranoglutaminase (factor XIII_a).

of factors XI and IX of the intrinsic pathway. Rather than having two separate pathways, both pathways collaborate *in vivo* to produce a physiologic response. Factors IXa and VIII (antihemophilic factor) activate factor X. Factors Xa and V activate prothrombin. Thrombin catalyzes the proteolytic conversion of fibrinogen to fibrin.

In the common pathway, thrombin catalyzes the proteolytic conversion of fibrinogen to fibrin. **Fibrinogen** consists of dimers of three different polypeptides $[(A\alpha)_2(B\beta)_2\gamma_2]$ and molecular mass of 340 kd. $A\alpha$ and $B\beta$ each represent a single polypeptide chain. **Thrombin** catalyzes the hydrolysis of specific Arg-Gly bonds in each of the two $A\alpha$-and two $B\beta$-chains to form four small **fibrinopeptides** (two A and two B peptides) and fibrin. Fibrin spontaneously associates to form a loose fibrin clot. Thrombin also activates factor XIII by proteolysis to yield XIIIa. Factor XIIIa, a transglutaminase, catalyzes cross-link formation between glutaminyl and lysyl residues (Fig. 17-3). Although the amide bond of glutamine is not high energy in nature, the amide linkage provides a modicum of activation to energize the cross-linking reaction. Table 17-2 lists the properties of the blood clotting factors.

ANTICOAGULANTS AND FIBRINOLYSIS

Antithrombin III and protein C are anticoagulants in blood. **Antithrombin III** is activated by heparin and binds and inhibits many serine proteases, including thrombin and factor Xa, which accounts for the anticoagulant effects of heparin. Antithrombin III is a suicide inhibitor. When thrombin or factor Xa attack a specific Arg-Ser peptide bond of antithrombin III, a stable one-to-one complex containing the protease and antithrombin III forms. Blockade of the active site of thrombin or factor Xa by antithrombin III produces irreversible inhibition. Heparin increases the rate of formation of the antithrombin-protease complex 1,000-fold.

After tissue repair eliminates the need for a clot, the clot is removed by fibrinolysis. **Plasmin**, a serine protease formed from plasminogen, is responsible for fibrinolysis. Plasmin catalyzes the hydrolysis of fibrin to produce **fibrin degradation products**. Like trypsin, plasmin attacks many peptide bonds on the carboxyl-terminal side of lysine or arginine. Soluble plasmin is inhibited by α_2-plasmin inhibitor. When bound to fibrin, however, plasmin is resistant to α_2-plasmin inhibitor. Binding of plasmin to fibrin or α_2-plasmin inhibitor is mutually exclusive. Besides cross-linking fibrin peptides, factor XIIIa catalyzes the cross-linking of fibrin with α_2-plasmin inhibitor, which causes fibrin to be more resistant to digestion by plasmin. The properties of **α_2-plasmin inhibitor** account for the specificity and efficiency of this protein as the principal inhibitor of fibrinolysis.

Tissue plasminogen activator is a serine protease that catalyzes the conversion of plasminogen to plasmin by hydrolyzing a single peptide bond. Tissue plasminogen activator is synthesized and secreted by endothelium. Like the blood clotting pathway, the plasminogen pathway consists of a cascade. **Streptokinase** is a protein isolated from *Streptococci* and, like tissue plasminogen activator, it initiates fibrinolysis. Streptokinase per se lacks protease activity. It forms a complex with plasminogen and activates its proteolytic activity. Activated plasminogen catalyzes the proteolytic activation of other plasminogen molecules to form plasmin, thereby initiating fibrinolysis. **Urokinase**, a family of serine proteases produced by the kidney and excreted in the urine, functions as a plasminogen activator. It is also found in the plasma. Recombinant tissue plasminogen activator, streptokinase, and urokinase are all under investigation as to their relative efficacy for the treatment of thrombotic diseases.

Heparin, injected intravenously, is a short-acting anticoagulant with a rapid onset of action. **Coumarin** is a vitamin K analogue that inhibits the vitamin K–dependent carboxylation of several blood clotting factors; the coumarins require 12 to 24 hours to elicit an anticoagulant effect.

Blood Clotting Laboratory Tests

The following procedures are used to monitor various components of the blood clotting pathways. **Primary hemostasis**, or the instantaneous plugging of a hole in a blood vessel wall, is achieved by a combination of vasoconstriction and platelet adhesion and aggregation. The formation of a fibrin clot is not required.

Measurement of the **bleeding time** is a sensitive laboratory evaluation of primary hemostatic function. Bleeding time is increased in individuals with von Willebrand's disease and thrombocytopenia. Thrombocytopenia, decreased numbers of platelets, is a common cause of abnormal bleeding. Individuals with hemophilia have a normal bleeding time.

Activated partial thromboplastin time (aPTT) reflects the functional state of the intrinsic pathway. Therapeutic heparin, which is used in the treatment of venous thrombosis, increases aPTT. Patients with hemophilias A and B have increased aPTTs. **Prothrombin time** reflects the functional state of both the extrinsic and the intrinsic pathways. A normal result requires adequate fibrinogen, prothrombin, and factors V, VII, and X. Because the biosynthesis of prothrombin and factors VII, IX, and X requires vitamin K, the prothrombin time is used to monitor individuals administered coumarin.

Genetics of Coagulopathies

Hemophilia A and **hemophilia B** (Christmas disease) are **X-linked bleeding disorders**. Because of defective fibrin formation, affected individuals experience joint and muscle hemorrhage, easy bruising, and prolonged bleeding from wounds. The incidence of both types is approximately 1 in 10,000 individuals, which is similar in different parts of the world and in different races. Due to the decreased reproductive ability of individuals with hemophilia, approximately one-third of the mutant alleles are lost per generation. Because the number of affected individuals has remained constant, approximately one-third of newly diagnosed cases are the result of new mutations. The genes for human factors VIII and IX are near each other on the tip of the long arm of the X chromosome at q26-q27.

The **gene for factor VIII** is approximately 186 kilobases in size and consists of 26 exons and 25 introns. Factor VIII has sequences that are homologous to ceruloplasmin (a copper-containing oxidase in human plasma); however, the significance of this finding is unknown. The human factor VIII gene contains two notable features: an unusually large exon of 3,106 base pairs (average size for exons is 75 to 200 base pairs), and an extremely large intron of 32,000 base pairs between exons 22 and 23. Transcribing such a large gene into RNA may take hours. Liver cells are the main source of circulating factor.

Hemophilia A is due to several point mutations and deletions in the factor VIII gene. Diverse mutations give rise to the same clinical disorder in both hemophilia A and hemophilia B. **Hemophilia B,** or **Christmas disease**, is due to mutations of the factor IX gene. Missense, nonsense, and splicing mutations of factor IX are known. Single nucleotide mutations in the prepropeptide of factor IX also have been characterized. The inability to correctly process the prepropeptide leads to the synthesis of defective molecules. The clinical variation in hemophilia B may be explained by the expression of multiple mutations ranging from no activity to partial activity.

von Willebrand's disease is due to a mutation of the factor gene that is located on or near the tip of the short arm of chromosome 12. More than 20 variants of the disease that are produced by point mutations, insertions, and deletions. In its most common form, von Willebrand's disease is transmitted as an autosomal dominant trait. The clinical course is variable. A mild form seems to reflect a quantitative deficiency of the factor, and a severe form seems to reflect the formation of an abnormal factor.

Individuals with **hemophilia A** and **hemophilia B** have a **prolonged aPPT**. The bleeding time is increased in von Willebrand's disease. These diseases are treated with proteins prepared from human pooled plasma samples obtained from more than 2,000 individual blood donors. Factor IX is stable, but factor VIII and von Willebrand's factor are unstable. The possible and actual transmission of acquired immunodeficiency syndrome by such therapies was recognized in the 1980s. The preparation of these factors by recombinant DNA methods is under development. Factor VIII, produced by recombinant DNA procedures, has been approved for human use with the trade name Recombinate (for recombinate factor VIII).

Chapter 18

Nutrition

ESSENTIAL REQUIREMENTS

The essential requirements for human growth, development, maintenance, and activity are oxygen, energy-yielding metabolic fuels (mainly carbohydrate and fat), protein, vitamins, minerals, and water.

Water

Approximately 70% of the lean body mass is water. Distribution of water in the various compartments is given in Table 18-1. Interstitial fluid is the intercellular fluid exclusive of that in the arteries, veins, and heart (i.e., it is extra-cardiovascular). Plasma volume, which amounts to approximately 5% of the lean body mass, is the extracellular fluid within the cardiovascular system. The latter value is important in calculating intravenous fluid requirements.

Water balance describes the condition that exists when fluid intake is equivalent to output. Representative values for water balance are given in Table 18-2. Wide variations in fluid intake in the normal range are possible. Compensating changes in the urinary output maintain balance under physiologic conditions. **Metabolic water** is produced by oxidative phosphorylation. Protons, electrons, and oxygen react in the terminal step of respiratory chain phosphorylation in a reaction catalyzed by cytochrome oxidase to produce water; hence, the statement that glucose is broken down to carbon dioxide and water. Humans are able to live without oxygen for only a matter of minutes. Water is the next most essential requirement for life. Death from dehydration ensues within several days in the absence of fluid intake.

Caloric and Energy Expenditure

Body weight in adults is determined by the balance between energy expended and energy consumed. If more energy is consumed than expended, then body weight increases. Energy expenditure varies considerably among individuals and is generally greater in children than adults, men than women, and young adults than elderly. Energy output is also increased by activity. A general formula for calculating approximate energy requirements in adults is shown in Table 18-3. The **basal metabolic rate** is the energy necessary for maintaining basal physiologic activities (cardiac output, brain activity, renal output, body temperature, and respiratory function) but does not take into account any additional activities. The energy requirement for basal metabolism is approximately 1 kcal/kg per hour or 24 kcal/kg per day. As noted in Table 18-3, additional energy is expended in the course of daily activities and varies depending on lifestyle. The additional energy required for activity can be much greater than indicated in Table 18-3 for athletes and those individuals in very active occupations. The bioenergetic

TABLE 18-1. Fluid Compartments in Humans

	Percent of Lean Body Mass	Volume (L)/70-kg Body Mass
Total body water	70	49.0
Intracellular compartment	50	35.0
Extracellular compartment	20	14.0
Interstitial fluid	15	10.5
Plasma	5	3.5

TABLE 18-2. Fluid Intake and Output in Humans (mL)

Output	
Expired air	800
Feces	200
Perspiration	400
Urine	1,200
Total	**2,600**
Intake	
Food and beverage	2,300
Metabolic	300
Total	**2,600**

TABLE 18-3. Daily Energy Requirements for Adults

Daily energy = BMR + activity

Expenditure

BMR = weight (kg) × 24 kcal

Activity: Modest = 0.3 × BMR

Moderate = 0.4 × BMR

Heavy = 0.5 × BMR

1 kcal = 1 Calorie of the nutritionist.

BMR, basil metabolic rate

TABLE 18-4. Metabolic Calories Derivable from Foodstuffs

Class	kcal/g
Carbohydrate	4
Protein	4
Fat	9
Ethanol	7

kcal (kilocalorie) is equivalent to the nutritional Calorie. The approximate energy requirement for a 70-kg individual with moderate activity is estimated as follows:

$$\begin{aligned} 70 \times 24 &= 1{,}680 \\ 0.4 \times 1{,}680 &= \underline{672} \\ \text{Total requirement} &= 2{,}352 \text{ kcal or Calories} \end{aligned}$$

Because this is an approximation, the figure can be rounded off to 2,400 kcal.

Energy Sources

The main sources of energy include **carbohydrates**, **lipids**, and **proteins**. The metabolic energy derived from 1 g of each is given in Table 18-4. Note that the caloric value of ethanol is rather substantial. Ethanol is more reduced than carbohydrate, and this accounts for the greater energy yield of ethanol when compared with carbohydrate. Ethanol is oxidized to acetic acid in two NAD^+-dependent steps. Acetic acid is converted into acetylcoenzyme A (acetyl-CoA) and metabolized by the Krebs cycle. Ethanol metabolism increases the reduced nicotinamide adenine dinucleotide/ NAD^+ ratio, and this can increase the conversion of pyruvate to lactate. As a consequence, less pyruvate is available for gluconeogenesis.

When energy expenditure exceeds intake, weight loss ensues. The caloric equivalent of a pound of adipose tissue is approximately 3,500 kcal. This is based on a value of 85% fat, 15% water, and negligible carbohydrate and protein per pound of adipose tissue.

$$1 \text{ lb} \times 454 \text{ g lb}^{-1} \times 0.85 \times 9 \text{ Cal g}^{-1} = 3{,}500 \text{ Cal}$$

Triglyceride is the major storage form of metabolic fuel in humans. Triglycerides are valued as an energy storage form because they have a high energy content and pack into a dense (hydrophobic) compartment.

In the United States, the caloric energy derived from the average diet is as follows: carbohydrates, 46%; fats, 42%; proteins, 12%. Various health authorities suggest that decreasing fats to provide 30% of the energy requirement and increasing carbohydrates to provide 58% of the energy requirement would promote health. Others suggest that decreasing fat consumption even further to provide only 20% of the total energy would be more beneficial.

Essential Human Nutrients

Essential nutrients are those that cannot be synthesized in adequate amounts (if at all) and are required in the diet. These are listed in Table 18-5. The **essential amino acids**

TABLE 18-5. Essential Human Nutrients

Amino Acids	Fatty Acids	Vitamins	Minerals	Other
		Water-soluble	**Bulk**	
Phenylalanine	Linoleic	Thiamine (B_1)	Sodium	Water
Valine	Linolenic	Riboflavin (B_2)	Potassium	Energy
Threonine		Niacin	Phosphate	
Tryptophan		Pyridoxine (B_6)	Magnesium	
Isoleucine		Pantothenate	Calcium	
Methionine		Folate	Chloride	
Histidine[a]		Cobalamin (B_{12})	**Trace**	
Arginine[b]		Ascorbate (C)	Chromium	
Leucine		Biotin[b]	Cobalt[c]	
Lysine		Myonositol[b]	Copper	
		Fat-soluble	Iodine	
		A	Iron	
		D	Molybdenum	
		E	Selenium	
		K	Zinc	
			Fluoride[d]	

[a] *Essential in infants.*

[b] *Human requirement not rigorously established.*

[c] *As vitamin B_{12}.*

[d] *Promotes stronger teeth and bones; whether essential not established.*

can be represented by the acronym PVT TIM *HA*LL (**ph**enylalanine, **v**aline, **t**hreonine, **t**ryptophan, **i**soleucine, **m**ethionine, **h**istidine, **a**rginine, **l**eucine, and **l**ysine; read private Tim Hall). The italicized histidine and arginine indicate that histidine is essential in infants, and the essential requirement of arginine has not been rigorously established. Foods vary in protein quality. This reflects the proportion of essential amino acids in them and their digestibility. In general, animal proteins are of higher quality than plant proteins. The recommended dietary allowance for proteins in adults is 56 g per day. The estimated average consumption of proteins in the United States is 100 g per day. The recommended daily allowance for children per kg is approximately twice that for adults.

The most important worldwide nutritional problem is protein-energy malnutrition in children. **Kwashiorkor** develops in children with adequate energy but insufficient protein intake. **Marasmus** develops in children with both inadequate energy and protein intake.

Linoleic and **linolenic acids** are essential in humans. They cannot be synthesized and therefore are required in the diet. Linoleic acid serves as a precursor of prostaglandins, leukotrienes, thromboxanes, and prostacyclins. The function of linolenic acid is a puzzle. Linoleic acid belongs to the ω6 family of fatty acids. The double bond is found at carbon atom 6 when counting from the ω or methyl end of the fatty acid. Linolenic acid belongs to the ω3-family of fatty acids.

Carbohydrates can be synthesized from glycogenic amino acids. They can also be formed from glycerol derived from triglyceride. Carbohydrates are a common and abundant source of energy. Individuals on a carbohydrate-free or carbohydrate-deficient diet develop ketosis associated with production of the ketone bodies by the pathway noted previously. Therefore, although carbohydrates are not considered essential, their deficiency can produce serious consequences.

The general properties and functions of the vitamins are listed in Table 18-6. Vitamins are generally divided into two classes: water soluble and fat soluble. The **fat-soluble vitamins** include vitamins A, D, E, and K. Except for **vitamin C deficiency** and **scurvy**, deficiencies of a single vitamin are unusual in nature. Deficiencies of two or more vitamins are more characteristic. Pantothenate deficiency, however, is unusual in humans. A deficiency state occurs only during starvation when symptoms of other avitaminoses are manifest.

There are several specific imbalances in micronutrients that are noteworthy (in addition to the common states described in Table 18-6). **Wernicke-Korsakoff syndrome** is a version of thiamine deficiency that is associated with chronic alcoholics. Based on the role of pyridoxal phosphate in heme biosynthesis, vitamin B_6 deficiency can lead to **sideroblastic anemia** (microcytic anemia associated with high serum iron levels). **Goiter** is a well-characterized deficiency in iodine (manifested as an enlarged thyroid) that has largely been eliminated in the United States through the use of iodized salt. **Iron-deficiency anemia**, which is a hypochromic microcytic anemia, is probably the most widespread dietary deficiency disorder in the United States and has resulted in campaigns to supplement dietary intakes of iron in children and pregnant women. **Wilson's disease** (hepatolenticular degeneration) is an abnormal accumulation of copper, while **Menkes' syndrome** is a rare genetic defect in copper transport.

The **fat-soluble vitamins** are associated both with deficiency disorders as well as toxicities caused by their slow excretion and accumulation in the body. This can be problematic in a society that is particularly obsessed with nutritional approaches to health and disease. **Rickets** (children) and **osteomalacia** (adults) are two different manifestations of the same vitamin D deficiency. In children, vitamin D deficiency results in improper mineralization of bone resulting in soft, pliable bones. In adults, it produces demineralization of existing bones, weakening them.

Pernicious anemia is associated with inadequate **vitamin B_{12}**. Pernicious anemia is more commonly related to a failure to absorb vitamin B_{12} than due to inadequate dietary intake. Pernicious anemia develops in individuals who fail to produce a gastric glycoprotein (called **intrinsic factor**) required for vitamin B_{12} (extrinsic factor) absorption. The vitamin B_{12} content of plants is zero (unless the plant is contaminated with bacteria). A deficiency state may therefore occur in individuals on a strict vegetarian diet. Vegetarians and individuals with pernicious anemia often excrete methylmalonate in the urine. This is due to the diminished ability to convert methylmalonyl-CoA to succinyl-CoA.

The interrelationship of **vitamin B_{12}** and **folate** metabolism is complex and incompletely understood. Individuals with pernicious anemia develop both **anemia** and **lesions of the central nervous system**. Folate will correct the anemia but not the neuropathology. The anemia is reversible, but not the nervous system pathology. Therefore, folate is absent from proprietary vitamins to avoid masking the symptomatology of pernicious anemia, which prompts the behavior of seeking medical attention. The advisability of limiting folate in proprietary vitamins is debatable because many people have marginally adequate folate intake. In humans, vitamin B_{12} is necessary for the transfer of the methyl group from methyltetrahydrofolate to homocysteine. It is unclear whether this is the only interrelationship between folate and vitamin B_{12}.

Provitamin D (7-dehydrocholesterol and ergosterol) requires two metabolic transformations (hydroxylations) for conversion to the active form. The first step occurs in the liver and produces **25-hydroxyvitamin D_3**. The second occurs in the kidney and

TABLE 18-6. Major Characteristics of Vitamins

Vitamin	Cofactor	Deficiency State	Biochemical Functions	Comments
Water-soluble group				
Thiamine (B_1)	Thiamine pyrophosphate	Beriberi	Oxidative decarboxylation, pyruvate dehydrogenase (pyruvate → acetyl-CoA) α-Ketoglutarate dehydrogenase (α-Ketoglutarate → succinyl-CoA) Ketoacid dehydrogenase Leucine, isoleucine, valine catabolism Transketolase	—
Riboflavin (B_2)	FAD FMN	—	Oxidative decarboxylation reactions listed under thiamine and/or electron transport	—
Niacin Nicotinic acid Nicotinamide	NAD+	Pellagra	Many hydrogen-transfer redox reactions	Some derived from tryptophan metabolism
Pantothenate	CoA, 4'-phosphopantetheine	—	Acyl transfer Cofactor of acyl carrier protein in fatty acid biosynthesis	Widely distributed (pantothenate) and isolated deficiency unknown
Biotin	Biotinyllysyl	—	Acetyl-CoA carboxylase Propionyl-CoA carboxylase Pyruvate carboxylase	—
Cobalamin (B_{12})	Methylcobalamin 5'-Deoxyadenosylcobalamin	Pernicious anemia	Methylmalonyl-CoA mutase 5-Methyl H_4-folate homocysteine transmethylase	Vitamin B_{12} = extrinsic factor Not found in plants
Folate	Derivatives of H_4-folate	—	One-carbon transfer Thymidylate synthase Purine biosynthesis	—
Ascorbate (C)	Ascorbate	Scurvy	Prolyl and lysyl hydroxylases (collagen) Dopamine β-hydroxylase	Effectiveness in viral disease controversial
Fat-soluble group				
A, retinol	—	Night blindness	Forms 11-*cis* retinal with rhodopsin	Retinal and retinoic acid may play a role in differentiation; possibly beneficial in prevention of some types of cancer

(*continued*)

TABLE 18-6. *(continued)*

Vitamin	Cofactor	Deficiency State	Biochemical Functions	Comments
D,7-Dehydrocholesterol (skin), ergosterol (plants, yeast)	1,25-Dihydroxyvitamin D_3	Rickets in children; osteomalacia in adults	Calcium and phosphate metabolism	25-Hydroxylation in liver, 1-hydroxylation in kidney
E, tocopherol	—	Unknown	Unknown	—
K	—	Bleeding diathesis	Activated blood-clotting factors II, VII, IX, and X	Mediates formulation of γ-carboxyglutamyl protein residues

CoA, coenzyme A; FAD, flavin adenine dinucleotide; FMN, flavin mononucleotide; NAD^+, oxidized form of nicotinamide adenine dinucleotide.

produces **1,25-dihydroxyvitamin D_3** (1,25-dihydroxycholecalciferol). This compound, which plays a pivotal role in calcium metabolism, is transported in the circulation to act on a variety of tissues and organs such as the kidney, bone, and intestine. Vitamin D constitutes both a vitamin and a hormone.

The biochemical role of **vitamin K** in blood clotting has been elucidated. It is necessary for the carboxylation of glutamyl residues in five blood clotting factors (II, VII, IX, X, and XIII). The production of γ-carboxyglutamyl groups results in residues that bind calcium.

Calcium is the predominant mineral found in the body. Besides serving as a structural component of bones and teeth, calcium serves as an important intracellular signalling ion. Recommended treatment of individuals with **osteoporosis** include foodstuffs that are calcium rich, such as dairy products, and calcium salts. Magnesium sulfate injections are used to treat eclampsia.

Iodine is a constituent of thyroid hormone (triiodothyronine and tetraiodothyronine). A deficiency of iodine can lead to an enlarged thyroid gland (goiter). **Graves' disease** and **hyperthyroidism** can be treated with radioiodine. The isotope is concentrated in thyroid gland cells, and radioiodine leads to the (partial) destruction of these cells. Phosphate is a component of many metabolites, DNA, and RNA. Incorporation of radiophosphate into DNA is deleterious to the cell. **Polycythemia vera**, which is characterized by excessive production of red and white blood cells, is treated with radiophosphate.

BIOCHEMISTRY QUESTIONS

DIRECTIONS: Each of the numbered items or incomplete statements in this section is followed by answers or by completions of the statement. Select the ONE lettered answer or completion that is BEST in each case.

1. Niemann-Pick disease is related to the inability of cells to degrade sphingomyelin in visceral organs and white blood cells. The enzyme responsible for this process, sphingomyelinase, is found in which of the following cellular locations?

(**A**) Cytosol
(**B**) Endoplasmic reticulum
(**C**) Golgi complex
(**D**) Lysosome
(**E**) Mitochondrion
(**F**) Nucleus
(**G**) Peroxisome

2. The preferred treatment of an infant with metabolic acidosis and an increased urinary excretion of methylmalonate and homocysteine is with

(**A**) folate
(**B**) niacin
(**C**) pyridoxine
(**D**) thiamine
(**E**) vitamin B_{12}

3. Menkes' syndrome is due to defective intestinal absorption of copper. Patients with this disorder have a deficit in which of the following copper-requiring components of the mitochondrial electron transport chain?

(**A**) Complex I [reduced nicotinamide adenine dinucleotide (NADH)-Q reductase]
(**B**) Complex II (succinate-Q reductase)
(**C**) Complex III (cytochrome reductase)
(**D**) Complex IV (cytochrome oxidase)
(**E**) Complex V [adenosine triphosphate (ATP) synthase]

4. The digitalis receptor is found in which of the following cellular locations?

(**A**) Cytosol
(**B**) Mitochondrion
(**C**) Nucleus
(**D**) Peroxisome
(**E**) Plasma membrane

5. The nonenzymatic glycosylation of proteins is postulated to account for the retinopathy, nephropathy, and neuropathy of diabetes mellitus. This glycosylation reaction involves the reaction of amino groups of proteins and which of the following organic groups found in glucose?

(**A**) Alcohol
(**B**) Aldehyde
(**C**) Alkane
(**D**) Hemiacetal
(**E**) Ketone

6. A serving of chicken tostados contains the following components: protein, 30 g; carbohydrate, 40 g; cholesterol, 50 mg; sodium, 750 mg; fiber, 3 g; fat, 10 g. The approximate percentage of total nutritional calories derived from fat in this serving is

(A) 10%
(B) 15%
(C) 20%
(D) 25%
(E) 30%

7. A 3-month-old infant is found to have cataracts, galactosemia, and galactosuria, without mental deficiency or aminoaciduria. He is not in any distress, and his developmental milestones have been normal. The most likely cause of a defect in galactose metabolism involves which of the following enzymes?

(A) Galactokinase
(B) Galactose-1-phosphate uridyltransferase
(C) β-Galactosidase
(D) Lactase
(E) Uridine diphosphate (UDP)-galactose epimerase

8. The maturation of which of the following receptors requires endoprotease activity that removes an internal segment to produce the mature and functional form?

(A) α-Adrenergic receptor
(B) β-Adrenergic receptor
(C) Glucagon receptor
(D) Insulin receptor
(E) Muscarinic acetylcholine receptor

9. A diet consisting of only plant products can lead to which of the following conditions?

(A) Beriberi
(B) Marasmus
(C) Megaloblastic anemia
(D) Pellagra
(E) Scurvy

10. Action of which of the following enzymes liberates diacylglycerol and inositol-1, 4,5-trisphosphate from phosphatidylinositol 4,5-bisphosphate?

(A) Phospholipase A_1
(B) Phospholipase A_2
(C) Phospholipase C
(D) Phospholipase D

11. Mutations in DNA that alter codons can lead to anomalies in β-globin chain initiation in β-thalassemia. Which of the following codons will initiate this process?

(A) AUG
(B) UAA
(C) UAG
(D) UGA
(E) UUU

12. A deficiency of which one of the following enzymes leaves pyruvate dehydrogenase in a permanently inactivated form?

(A) Phosphoprotein phosphatase 1
(B) Phosphoprotein phosphatase 2A
(C) Protein kinase A
(D) Pyruvate dehydrogenase kinase
(E) Pyruvate dehydrogenase phosphatase

13. A deficiency of which of the following leads to steatorrhea (impaired lipid absorption)?

(A) Bile pigments
(B) Bile salts
(C) Bilirubin
(D) Intrinsic factor
(E) Low-density lipoprotein (LDL) receptor

14. The chief mechanism for the production of hyperglycemia in diabetes mellitus is a consequence of

(A) a failure of glucose to enter liver cells
(B) increased hepatic gluconeogenesis
(C) increased hepatic glycogenolysis
(D) increased hepatic glycolysis
(E) increased muscle glycolysis

15. Actinomycin D, which is used in the treatment of Kaposi's sarcoma and testicular carcinoma, inhibits which of the following metabolic activities?

(A) DNA polymerase α
(B) DNA polymerase γ
(C) RNA capping
(D) RNA polymerase
(E) RNA splicing

16. Which of the following enzymes is sensitive to the oxidative stress produced by a variety of drugs, including antimalarials such as chloroquine?

(A) Alcohol dehydrogenase
(B) Carnitine acyltransferase
(C) Citrate synthase
(D) Glucokinase
(E) Glucose-6-phosphate dehydrogenase

17. The hypocalcemia and hyperphosphatemia of pseudohypoparathyroidism (Albright's disease) is due to a deficiency of a G protein that is activated by the seven-transmembrane segment receptor for parathyroid hormone. The mutations reside in the protein that directly stimulates adenylyl cyclase. The stimulatory protein is

(A) G_s-α
(B) G_s-β
(C) G_s-γ
(D) G_s-$\beta\gamma$
(E) G_i-α

18. The form of hemoglobin that is used to monitor average blood glucose concentrations found in patients with diabetes mellitus is

(**A**) hemoglobin A
(**B**) hemoglobin A_1
(**C**) hemoglobin F
(**D**) hemoglobin H
(**E**) hemoglobin S

19. Inadequate cleavage of trypsinogen to form trypsin leads to diarrhea and failure to thrive. This condition is due to a deficiency of which of the following enzymes?

(**A**) Cathepsin
(**B**) Chymotrypsin
(**C**) Elastase
(**D**) Enterokinase (enteropeptidase)
(**E**) Pepsin

20. Cyanide, one of the most rapid acting poisons in humans, blocks ATP generation by inhibiting

(**A**) complex I of the electron transport chain
(**B**) complex II of the electron transport chain
(**C**) complex III of the electron transport chain
(**D**) complex IV of the electron transport chain
(**E**) complex V of the electron transport chain

21. If one strand of DNA has the sequence 5' GCAT 3', what is the sequence of its complementary strand, written in the 5'-to-3' direction?

(**A**) ATGC
(**B**) AUGC
(**C**) GCAT
(**D**) CGTA
(**E**) CGTU

22. Which of the following food substances lacks cholesterol?

(**A**) Bacon
(**B**) Butter
(**C**) Eggs
(**D**) Orange juice
(**E**) Sausage

23. An insufficient intake of which of the following amino acids leads to a negative nitrogen balance?

(**A**) Asparagine
(**B**) Cysteine
(**C**) Serine
(**D**) Tryptophan
(**E**) Tyrosine

24. Xeroderma pigmentosum is characterized by progressive degenerative changes of sun-exposed portions of the skin and eyes. There is a 200-fold increase in skin cancers. Affected individuals exhibit increased sensitivity to ultraviolet light as a result of thymine dimer formation and related lesions of DNA. Xeroderma pigmentosum is a result of a defect in

(**A**) DNA ligase
(**B**) DNA *N*-glycosylase
(**C**) DNA polymerase γ
(**D**) nucleotide-excision repair
(**E**) repair polymerase

25. Patients with chronic glomerulonephritis who are undergoing renal dialysis often exhibit abnormalities of calcium metabolism resulting from defects in the metabolism of which of the following vitamins?

(**A**) Vitamin A
(**B**) Vitamin B_1
(**C**) Vitamin C
(**D**) Vitamin D
(**E**) Vitamin E

26. Warfarin is an anticoagulant used in the prevention and treatment of venous thrombosis and coronary artery disease. Which of the following actions is associated with this drug?

(**A**) Warfarin mediates the cleavage of tissue plasminogen activator.
(**B**) Warfarin binds to calcium.
(**C**) Warfarin inhibits factor Xa.
(**D**) Warfarin inhibits vitamin K–dependent carboxylation reactions.
(**E**) Warfarin binds to the active site of thrombin.

27. Which of the following metabolites serves as a precursor of heme?

(**A**) Acetoacetyl-coenzyme A (CoA)
(**B**) Mg^{2+}
(**C**) N^5,N^{10}-methylenetetrahydrofolate
(**D**) Succinyl-CoA
(**E**) Valine

28. Which of the following substances maintains ATP levels in cardiac and skeletal muscle but not in liver?

(**A**) 1,3-Bisphosphoglycerate
(**B**) Creatine phosphate
(**C**) The oxidized form of nicotinamide adenine dinucleotide (NAD^+)
(**D**) Phosphoenolpyruvate
(**E**) Uridine diphosphate glucose

29. A 4-day-old infant presents with hypoglycemic convulsions, lactic acidosis, and hepatomegaly. He responds to the administration of glucose and sodium bicarbonate. Galactose infusion increases the plasma glucose concentration. When the infant is in a fed state, a glucagon injection produces hyperglycemia. The most likely enzyme deficiency in this infant is

(**A**) fructose-1-phosphate aldolase
(**B**) fructose-1,6-bisphosphatase
(**C**) galactokinase
(**D**) galactose-1-phosphate uridyltransferase
(**E**) glucose-6-phosphatase

30. Classic hemophilia is due to a deficiency of factor

(**A**) III
(**B**) IV
(**C**) VI
(**D**) VIII
(**E**) IX

31. Which of the following protein kinase activities phosphorylates tyrosine residues in acceptor proteins?

(**A**) Calcium/calmodulin protein kinase II
(**B**) Insulin receptor
(**C**) Protein kinase A
(**D**) Protein kinase C
(**E**) Protein kinase G

32. Which of the following enzymes is a hydrolase?

(**A**) Aldolase
(**B**) Aminoacyl–transfer RNA (tRNA) synthetase
(**C**) Glucose-6-phosphatase
(**D**) Glycogen phosphorylase
(**E**) Hexokinase

33. Folate is important in normal cellular development. A deficiency of folate during pregnancy leads to neural tube defects. The known biochemical role of folate in metabolism is to

(**A**) function in one-carbon transfer reactions
(**B**) mediate decarboxylation reactions
(**C**) participate in transamination reactions
(**D**) serve as an antioxidant

34. The oncogene *trk* is expressed in some colon and thyroid carcinomas and encodes a plasma transmembrane protein that possesses protein kinase activity. This enzyme catalyzes the phosphorylation of which of the following amino acid residues in target substrates?

(**A**) Aspartate
(**B**) Glutamate
(**C**) Glutamine
(**D**) Histidine
(**E**) Tyrosine

35. Which of the following human cells lacks a nucleus and relies on anaerobic glycolysis?

(**A**) *Escherichia coli* cells
(**B**) Mature erythrocytes
(**C**) Mature T cells
(**D**) Short-lived epithelial cells of the gut
(**E**) Terminally differentiated neurons

36. Which of the following bases does NOT normally occur in DNA?

(**A**) Adenine
(**B**) Cytosine
(**C**) Guanosine
(**D**) Thymine
(**E**) Uracil

37. Lovastatin is a widely prescribed drug used in the treatment of hypercholesterolemia. The drug is a competitive inhibitor of the rate-limiting enzyme in cholesterol biosynthesis. Lovastatin

(A) decreases the equilibrium constant of the reaction
(B) decreases the maximum velocity (V_{max}) of the enzyme
(C) increases the apparent Michaelis constant (K_m) of the enzyme for its isoprenoid substrate
(D) increases the standard free energy of the reaction

38. Which of the following enzymes functions in the oxidative segment of the pentose phosphate pathway?

(A) Fructose-1,6-bisphosphatase
(B) Glyceraldehyde-3-phosphate dehydrogenase
(C) 6-Phosphogluconate dehydrogenase
(D) Transaldolase
(E) Transketolase

39. Which of the following amino acid segments most typifies membrane spanning domains in receptors and ion channels?

(A) -Ala-Pro-Asn-Tyr-Gln-Val-
(B) -Ile-Leu-Asp-Glu-Cys-Trp-
(C) -Leu-His-Ala-Gln-Ser-Phe-
(D) -Leu-Val-Phe-Ile-Ala-Gly-Trp-
(E) -Thr-Ala-Val-Lys-Cys-Leu-

40. Which of the following components is a cytoplasmic protein that transiently associates with the ribosome?

(A) A site
(B) P site
(C) Elongation factor-2 (EF-2)
(D) Peptidyltransferase
(E) Ribosomal RNA (rRNA)

41. The elevation of blood ketone bodies seen in hyperketotic hypoglycemia results from a deficiency in which of the following enzymes?

(A) Hydroxymethylglutaryl-CoA lyase
(B) Hydroxymethylglutaryl-CoA reductase
(C) Hydroxymethylglutaryl-CoA synthase
(D) β-Ketothiolase
(E) Succinyl-CoA:acetoacetate CoA-transferase

42. Hemoglobin S, found in people with sickle cell anemia, differs from normal adult hemoglobin at position 6 in the β-chain. This difference in amino acid sequence refers to which of the following attributes of these proteins?

(A) Primary structure
(B) Secondary structure
(C) Tertiary structure
(D) Quaternary structure
(E) Pentameric structure

43. Enalapril, which is used in the treatment of hypertension and cardiac failure, is a competitive inhibitor of angiotensin-converting enzyme. Enalapril

(**A**) binds to an allosteric site to inhibit the reaction
(**B**) binds to the catalytic, or active, site of its target enzyme
(**C**) decreases the equilibrium constant for the enzyme-catalyzed reaction
(**D**) forms a covalent complex with its target enzyme
(**E**) increases the free energy of activation of the reaction

44. Commercial antioxidant and vitamin supplements contain manganese chloride. Manganese is a cofactor for which of the following mitochondrial enzymes?

(**A**) Catalase
(**B**) Cytochrome *c*
(**C**) Cytochrome P-450
(**D**) Dopamine β-hydroxylase
(**E**) Superoxide dismutase

45. Deficiency of which of the following enzymes can lead to a hemolytic anemia resulting from decreased energy metabolism in the mature red blood cell?

(**A**) ATP synthase
(**B**) α-Ketoglutarate dehydrogenase
(**C**) Pyruvate carboxylase
(**D**) Pyruvate dehydrogenase
(**E**) Pyruvate kinase

46. Which one of the following human genes lacks introns?

(**A**) Dystrophin
(**B**) Histone
(**C**) Immunoglobulin G
(**D**) LDL receptor
(**E**) Thyroglobulin

47. The deficiency of which of the following lipid-associated proteins leads to hyperlipemia due to lack of an activator of lipoprotein lipase?

(**A**) A-I
(**B**) B-100
(**C**) C-II
(**D**) D
(**E**) E

48. In humans, the metabolism of which of the following amino acids leads to the production of small amounts of the vitamin nicotinic acid?

(**A**) Cysteine
(**B**) Methionine
(**C**) Serine
(**D**) Tryptophan
(**E**) Valine

49. A 1-week-old infant with normal plasma glucose presents with convulsions, hyperventilation, and lactic acidosis. The most likely cause of this condition is related to a deficiency of which of the following enzymes?

(A) Hexokinase
(B) Phosphofructokinase
(C) Phosphoglucoisomerase
(D) Pyruvate dehydrogenase
(E) Triose phosphate isomerase

50. A 3-month-old infant presents with failure to thrive, vomiting, hepatomegaly, hyperbilirubinemia, acidosis, elevated serum lactate dehydrogenase, and elevated serum glutamate-oxaloacetate transaminase. Her plasma glucose is normal. Infusion of fructose results in hyperglycemia, but galactose infusion does not produce an elevation of blood glucose. The most likely enzyme defect in this patient is

(A) fructose-1,6-diphosphatase
(B) galactokinase
(C) galactose-1-phosphate uridyltransferase
(D) hexokinase
(E) lactase

51. Biological lipid bilayer membranes contain which of the following lipids?

(A) Cholesterol
(B) Cholesteryl ester
(C) Linoleic acid
(D) Palmitic acid
(E) Stearic acid

52. The enzyme that transfers the methyl group from N^5-methyltetrahydrofolate to homocysteine contains which of the following?

(A) Lipoic acid
(B) Nicotinic acid
(C) Pyridoxal phosphate
(D) Thiamine pyrophosphate
(E) Vitamin B_{12}

53. The molar percentage of G in human DNA is 30%. The molar percentage of A is

(A) 10%
(B) 20%
(C) 30%
(D) 40%
(E) 50%

54. Physiologic functions of glucokinase include which one of the following?

(A) Phosphorylation of fructose
(B) Phosphorylation of galactose
(C) Phosphorylation of mannose
(D) Regulation of glucose release into the circulation from the liver
(E) Regulation of insulin release from β-cells of the pancreas

55. Restriction fragment length polymorphisms, which are used in the diagnosis of genetic diseases, are observed on

(**A**) C_ot curves
(**B**) Eastern blots
(**C**) Northern blots
(**D**) Southern blots
(**E**) Western blots

56. Which of the following proteins is a serine protease?

(**A**) Antithrombin III
(**B**) Factor VIII
(**C**) Fibrinogen
(**D**) Heparin
(**E**) Tissue plasminogen activator

57. An inborn error of metabolism that is associated with lactic acidosis, secondary carnitine deficiency, and excretion of β-hydroxy dicarboxylic acids is due to a deficiency of which of the following enzymes?

(**A**) Carnitine acyltransferase II
(**B**) β-Hydroxyacyl-CoA dehydrogenase
(**C**) β-Ketothiolase
(**D**) Long-chain fatty acyl-CoA dehydrogenase
(**E**) Succinyl-CoA:acetoacetate CoA-transferase

58. Which one of the following genetic disorders is due to a defect in lysosomal metabolism?

(**A**) Ehlers-Danlos syndrome
(**B**) Familial hypercholesterolemia
(**C**) Hurler's disease (type I mucopolysaccharidosis)
(**D**) McArdle's disease
(**E**) Muscular dystrophy

59. Which of the following amino acids is both glycogenic and ketogenic?

(**A**) Glutamine
(**B**) Histidine
(**C**) Methionine
(**D**) Phenylalanine
(**E**) Serine

60. The enzyme glucose-6-phosphate dehydrogenase has an altered electrophoretic mobility at pH 8.6 caused by the mutation of a valine residue to which one of the following amino acids?

(**A**) Alanine
(**B**) Asparagine
(**C**) Glutamate
(**D**) Proline
(**E**) Tyrosine

61. Hereditary phosphofructokinase deficiency is characterized by a mild hemolytic anemia and myopathy. Noninvasive nuclear magnetic resonance studies show an elevation of which of the following metabolites?

(**A**) 1,3-Bisphosphoglycerate
(**B**) Dihydroxyacetone phosphate
(**C**) Fructose 6-phosphate
(**D**) Glyceraldehyde 3-phosphate
(**E**) Phosphoenolpyruvate

62. The negative logarithm of acid ionization constant of the physiologic phosphate buffer system is 6.8. If the concentration of $H_2PO_4^{1-}$ is 5.0 mmol/L at pH 6.8 in a urine sample, the concentration of HPO_4^{2-} is

(**A**) 0.5 mmol/L
(**B**) 1.0 mmol/L
(**C**) 5.0 mmol/L
(**D**) 6.8 mmol/L
(**E**) 50.0 mmol/L

63. The standard free energy for the conversion of 2-phosphoglycerate to 3-phosphoglycerate and for the conversion of fructose 1,6-bisphosphate to glyceraldehyde 3-phosphate and dihydroxyacetone phosphate is positive. When fluoride is present in bacterial cells, which of the following conditions prevails?

(**A**) Aldolase is inhibited.
(**B**) Glyceraldehyde-3-phosphate dehydrogenase is inhibited.
(**C**) Glyceraldehyde 3-phosphate and dihydroxyacetone phosphate accumulate.
(**D**) Phosphoglycerate kinase is inhibited.
(**E**) 3-Phosphoglycerate accumulates.

64. Following the oxygenation of hemoglobin, the α- and β-subunits move relative to each other. This describes a change in which of the following structural aspects of the protein?

(**A**) Covalent
(**B**) Primary
(**C**) Secondary
(**D**) Tertiary
(**E**) Quaternary

65. Tay-Sachs disease is due to a deficiency of hexosaminidase A. This enzyme is an example of which of the following classes of enzymes?

(**A**) Hydrolase
(**B**) Ligase
(**C**) Lyase
(**D**) Oxidoreductase
(**E**) Transferase

66. Hereditary triose phosphate isomerase deficiency is the most devastating of the glycolytic enzyme deficiencies involving anemia, neurologic dysfunction, and cardiomyopathy. People with this disorder exhibit elevated levels of which of the following metabolites?

(**A**) 1,3-Bisphosphoglycerate
(**B**) Dihydroxyacetone phosphate
(**D**) 2-Phosphoglycerate
(**C**) 3-Phosphoglycerate
(**E**) Phosphoenolpyruvate

67. The hyperammonemia found in patients with pyruvate carboxylase deficiency is due to which of the following mechanisms?

(**A**) Decreased aspartate levels, which lead to impaired urea production in liver
(**B**) Decreased glutamine levels, which lead to impaired asparagine synthesis
(**C**) Decreased utilization of ATP
(**D**) Impaired conversion of pyruvate to fructose 1,6-bisphosphate
(**E**) Impaired conversion of pyruvate to alanine

68. Which of the following components of the electron transport chain reacts physiologically with oxygen?

(**A**) Cytochrome aa_3
(**B**) Cytochrome *b*
(**C**) Cytochrome *c*
(**D**) Coenzyme Q
(**E**) NADH dehydrogenase

69. A 2-week-old infant presents with lethargy and hepatomegaly. His plasma carnitine is elevated, the majority of which is combined with C_{14} myristate. The most likely metabolic defect in this infant is which of the following?

(**A**) 2,4-Dienoyl-CoA reductase
(**B**) Long-chain fatty acyl-CoA dehydrogenase
(**C**) Long-chain hydroxyacyl-CoA dehydrogenase
(**D**) Medium-chain fatty acyl-CoA dehydrogenase
(**E**) Short-chain fatty acyl-CoA dehydrogenase

70. The increase in triglyceride synthesis in response to insulin injection in people with diabetes mellitus involves which of the following reactions?

(**A**) Activation of hormone-sensitive lipase
(**B**) Reaction of *S*-adenosylmethionine with phosphatidylcholine
(**C**) Reaction of 1,2-diglyceride with acyl-CoA
(**D**) Reaction of 1,2-diglyceride with cytidine diphosphate-choline
(**E**) Reaction of palmitoyl-CoA and serine

71. Patients with lactic acidosis and which of the following enzyme defects are treated with thiamine?

(**A**) Cytochrome oxidase
(**B**) Glucose-6-phosphatase
(**C**) Medium-chain fatty acyl-CoA dehydrogenase
(**D**) Pyruvate carboxylase
(**E**) Pyruvate dehydrogenase

72. A patient with a partial gastrectomy can develop pernicious anemia due to a deficiency in the production of which of the following substances?

(**A**) Elongation factor
(**B**) Extrinsic factor
(**C**) Factor VIII
(**D**) Intrinsic factor
(**E**) Release factor

73. Physiologic exercise leads to the breakdown of muscle glycogen and an elevation of serum lactate. Neither of these actions occurs in people with McArdle's disease. Which of the following muscle enzymes is defective in McArdle's disease?

(**A**) Branching enzyme
(**B**) Debranching enzyme
(**C**) Glycogen phosphorylase
(**D**) Glycogen synthase
(**E**) Uridinediphosphoglucose pyrophosphorylase

74. A 42-year-old man presents with ruddy cyanosis, erythrocytosis, normal red blood cell morphology, and a hemoglobin concentration of 19 g/dL (normal, <16 g/dL). This condition is duc to a dcficicncy of which of the following enzymes?

(**A**) Bisphosphoglycerate mutase
(**B**) 2,3-Bisphosphoglycerate phosphatase
(**C**) 1,6-Fructose bisphosphatase
(**D**) Hexokinase
(**E**) Phosphofructokinase

75. The anabolism of pyruvate involves which of the following pathways?

(**A**) Conversion to acetyl-CoA
(**B**) Conversion to carbon dioxide and water
(**C**) Conversion to glucose via oxaloacetate
(**D**) Conversion to lactate

76. A 9-month-old infant presents with hepatomegaly and hypoketotic hypoglycemia. Her plasma total and free carnitine levels are normal. Mass spectral analysis shows no abnormal fatty acid metabolites in the urine. The most likely defect in this infant is

(**A**) carnitine/acylcarnitine translocase deficiency
(**B**) carnitine-palmitoyltransferase I deficiency
(**C**) carnitine-palmitoyltransferase II deficiency
(**D**) electron transfer flavoprotein deficiency
(**E**) long-chain 3-hydroxyacyl-CoA dehydrogenase deficiency

77. The conversion of isocitrate to α-ketoglutarate requires

(**A**) biotin
(**B**) CoA
(**C**) lipoate
(**D**) NAD^+
(**E**) thiamine

78. Leber hereditary optic neuropathy is a maternally inherited form of adult-onset blindness due to missense mutations of mitochondrially encoded proteins in complex I or III of the electron transport chain. The electron transport chain of oxidative phosphorylation is located in the

(**A**) endoplasmic reticulum
(**B**) inner membranous space
(**C**) inner mitochondrial membrane
(**D**) mitochondrial matrix
(**E**) outer mitochondrial membrane

79. While sprinting, a runner generates an increase in serum lactate, and ATP is broken down to adenosine diphosphate (ADP) and then to adenosine monophosphate (AMP). 5'-AMP is an allosteric activator of which of the following enzymes?

(A) Glucose-6-phosphatase
(B) Glycogen phosphorylase
(C) Glycogen synthase
(D) Phosphoenolpyruvate carboxykinase
(E) Pyruvate carboxylase

80. The principal function of the pentose phosphate pathway in humans is

(A) ATP production
(B) NAD^+ use
(C) reduced nicotinamide adenine dinucleotide phosphate production
(D) 1,5-ribulose bisphosphate production
(E) thioredoxin reduction

81. Patients with lactic acidosis and which of the following enzyme defects are treated with a low carbohydrate diet?

(A) Fructose-1,6-bisphosphatase
(B) Glucose-6-phosphatase
(C) Phosphoenolpyruvate (PEP) carboxykinase
(D) Pyruvate carboxylase
(E) Pyruvate dehydrogenase

82. A deficiency in the activity of long-chain acyl-CoA dehydrogenase used during the β-oxidation of fatty acids might be corrected by the ingestion of which of the following?

(A) Carnitine
(B) Fumarate
(C) Niacin
(D) Riboflavin
(E) Thiamine pyrophosphate

83. During starvation, ketone bodies are synthesized in the

(A) adipose tissues
(B) brain
(C) liver
(D) muscles
(E) thymus

84. Administration of insulin to a diabetic patient promotes glycogen storage via which of the following mechanisms?

(A) Activation of branching enzyme
(B) Activation of phosphofructokinase
(C) Activation of protein kinase A
(D) Inhibition of glycogen phosphorylase
(E) Inhibition of UDP-glucose pyrophosphorylase

85. The lock-and-key model of enzyme catalysis states that

(A) substrates bind to and activate an enzyme so that it can catalyze conversion of other substrate molecules
(B) the enzyme binds the substrate that alters it in such a way as to increase its inherent free energy
(C) the enzyme has a conformation and forms an active site that is complementary to the structure of the substrates
(D) the enzyme locks the product in a conformation that prevents it from being converted back to substrate

86. Which of the following enzymes catalyzes the hydrolysis of proteins on the carboxyl terminal side of lysine and arginine?

(A) Carboxypeptidase B
(B) Chymotrypsin
(C) Elastase
(D) Pepsin
(E) Trypsin

87. A defect in which of the following enzymes can lead to neonatal lactic acidosis?

(A) Arginase
(B) 1,3-Bisphosphoglycerate mutase
(C) Phosphofructokinase
(D) Pyruvate dehydrogenase kinase
(E) Pyruvate dehydrogenase phosphatase

88. A 61-year-old woman has a peripheral neuropathy, macrocytic anemia, and decreased vitamin B_{12} absorption. Gas chromatographic analysis of her urine demonstrates methylcitrate. This metabolite is a result of which of the following reactions?

(A) Acetyl-CoA with methyloxaloacetate
(B) Citrate with methyltetrahydrofolate
(C) Methylmalonyl-CoA with malonate
(D) Methylmalonyl-CoA with water
(E) Propionyl-CoA with oxaloacetate

89. Administration of insulin to a diabetic patient leads to an increase in lipogenesis by which of the following mechanisms?

(A) Activation of hormone-sensitive lipase
(B) Activation of protein kinase A
(C) Induction of acetyl-CoA carboxylase
(D) Induction of PEP carboxylase
(E) Repression of glucose-6-phosphate dehydrogenase

90. The most common form of sphingolipidosis is Gaucher's disease. The accumulation of excess glucosylcerebroside is due to which of the following conditions?

(A) Deficiency of ceramide-UDP glucose transferase
(B) Deficiency of glucosylceramide β-glucosidase
(C) Deficiency of 3-ketosphinganine reductase
(D) Deficiency of 3-ketosphinganine synthetase
(E) Deficiency of sphinganine acyltransferase

91. Which of the following is a risk factor for atherosclerosis and cardiovascular disease?

(**A**) Elevated high-density lipoprotein (HDL) levels
(**B**) Elevated LDL levels
(**C**) Subnormal LDL levels
(**D**) Total plasma cholesterol <200 mg/dL

92. A 2-week-old infant presents with vomiting, lethargy, and convulsions without evidence of an infectious process. Laboratory examination reveals hyperammonemia without acidosis. The infant is diagnosed with a hereditary deficiency of glutamate *N*-acetyltransferase. *N*-acetylglutamate is an allosteric activator of which of the following enzymes?

(**A**) Arginase
(**B**) Argininosuccinate lyase
(**C**) Argininosuccinate synthetase
(**D**) Carbamoyl-phosphate synthetase-I
(**E**) Ornithine transcarbamoylase

93. Gastric proton/potassium adenosinetriphosphatase (ATPase) is the antigen that leads to autoimmune type A gastritis and then to pernicious anemia. This enzyme is homologous to the sodium/potassium ATPase, or sodium pump. The pumping mechanism involves the formation of phosphate covalently attached to which of the following amino acid residues?

(**A**) Aspartate
(**B**) Cysteine
(**C**) Glutamate
(**D**) Histidine
(**E**) Serine

94. Which one of the following fatty acids is classified as an essential human nutrient?

(**A**) Linoleate
(**B**) Oleate
(**C**) Palmitate
(**D**) Palmitoleate
(**E**) Stearate

95. Metachromatic leukodystrophy is associated with the accumulation of sulfated galactosylceramide. The biosynthesis of this compound involves which of the following reactions?

(**A**) *N*-acetylmannosamine with phosphoenolpyruvate
(**B**) Cytidine monophosphate (CMP)-*N*-acetylneuraminate with a galactose residue of cerebroside
(**C**) Inorganic sulfate with a galactose residue of cerebroside
(**D**) Phosphoadenosylphosphosulfate with a galactose residue of cerebroside
(**E**) UDP-glucose with a galactose residue of cerebroside

96. A 2-week-old infant presents with vomiting, lethargy, and convulsions without evidence of an infectious process. Laboratory examination reveals hyperammonemia without acidosis and an elevated urinary orotic aciduria. This clinical picture is best explained by a deficiency of which of the following enzymes?

(**A**) Arginine succinate lyase
(**B**) Arginine succinate synthase
(**C**) Carbamoyl-phosphate synthetase-I
(**D**) Ornithine transcarbamoylase
(**E**) Uridine monophosphate (UMP) synthase

97. Infants fed milk formula deficient in vitamin B_6 develop seizures. Which of the following hypotheses best explains the development of seizures? Vitamin B_6 is a cofactor of

(A) cystathionine synthase
(B) glutamate decarboxylase
(C) glutamate-oxaloacetate transaminase
(D) glycogen phosphorylase
(E) serine dehydratase

98. The increase in glucagon action that occurs in diabetes mellitus results in the elevation in the concentration of hepatic

(A) acetyl-CoA
(B) ATP
(C) citrate
(D) cyclic AMP
(E) NADH

99. During a 100-m sprint (under conditions of anaerobic glycolysis), there is increased conversion of glucose into lactic acid. The increase in the rate of glycolysis is related to an increase in intracellular

(A) acetyl-CoA
(B) AMP
(C) ATP
(D) citrate
(E) NAD^+

100. The most likely enzyme defect in a patient with hepatocellular dysfunction and maleylacetoacetate in the urine is

(A) fumarylacetoacetate hydrolase
(B) homogentisate dioxygenase
(C) *p*-hydroxyphenylpyruvate dioxygenase
(D) phenylalanine hydroxylase

101. The most likely diagnosis in a patient with metabolic acidosis, normal plasma homocysteine, and elevated methylmalonic acid in the urine is

(A) biotinidase deficiency
(B) lipoamide dehydrogenase deficiency
(C) 3-methylglutaconyl-CoA hydratase deficiency
(D) methylmalonyl-CoA mutase
(E) short-chain acyl-CoA dehydrogenase deficiency

102. The nonenzymatic glycosylation of proteins is postulated to account for the retinopathy, nephropathy, and neuropathy of diabetes mellitus. This glycosylation reaction involves the reaction of amino groups of proteins and glucose. The amino acid residue found in proteins whose R-group or side chain contributes to the amino function is which of the following amino acids?

(A) Arginine
(B) Lysine
(C) Serine
(D) Threonine
(E) Tyrosine

103. A 4-day-old infant presents with hypoglycemic convulsions, lactic acidosis, and hepatomegaly. He responds to the administration of glucose and sodium bicarbonate. When the infant is in a fed state, a glucagon injection fails to produce hyperglycemia. The most likely enzyme deficiency in this infant is

(**A**) fructose-1,6-bisphosphatase
(**B**) fructose-1-phosphate aldolase
(**C**) glucose-6-phosphatase
(**D**) phosphoenolpyruvate carboxykinase
(**E**) pyruvate carboxylase

104. Glutamate occurs at position 6 (from the amino terminus) in the β-chain of normal adult hemoglobin. A variety of amino acid replacements have been found in mutant hemoglobins, including hemoglobin S. Many of the original mutant forms were detected by electrophoresis. Substitution of which of the following residues for glutamate alters the charge of hemoglobin ($\alpha_2\beta_2$) by four units at pH 7 to yield hemoglobin C?

(**A**) Alanine
(**B**) Arginine
(**C**) Cysteine
(**D**) Leucine
(**E**) Tyrosine

105. Fructose-2,6-bisphosphate levels are elevated in diabetes mellitus, resulting from the unopposed action of glucagon and the cyclic AMP second-messenger system. Fructose-2,6-bisphosphate activates which of the following liver enzymes by an allosteric mechanism?

(**A**) Fructose-1,6-bisphosphatase
(**B**) Hexokinase
(**C**) Lactate dehydrogenase
(**D**) Phosphofructokinase
(**E**) Pyruvate kinase

106. Which of the following components participates in fatty acid biosynthesis but not in fatty acid β-oxidation?

(**A**) Acetyl-CoA
(**B**) Acyl carrier protein
(**C**) Mitochondria
(**D**) NAD^+
(**E**) Water

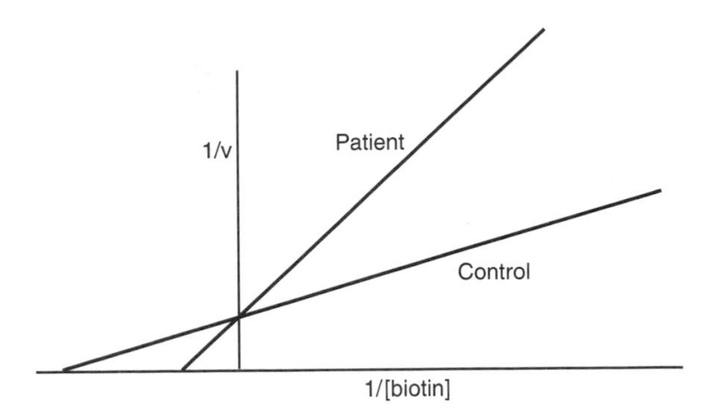

107. The above figure shows biotin holocarboxylase synthetase activity as a function of biotin concentration in samples taken from a control individual and a patient with metabolic acidosis. These data indicate that

(A) the K_m of the patient's enzyme is larger than that of the control
(B) the K_m of the patient's enzyme is smaller than that of the control
(C) the V_{max} for the patient's enzyme is equal to that of the control
(D) the V_{max} for the patient's enzyme is larger than that of the control

108. Acute ethanol intake results in an increase of $NADH/NAD^+$. This condition can inhibit gluconeogenesis by influencing which of the following enzyme-catalyzed reactions?

(A) Fructose-1,6-bisphosphatase
(B) Glucose-6-phosphatase
(C) Lactate dehydrogenase
(D) PEP carboxykinase
(E) Pyruvate carboxylase

109. A 32-year-old woman presents with pallor and easy fatigability. Laboratory examination reveals hypochromic microcytic anemia. A deficiency of which one of the following substances is the most likely cause anemia in this patient?

(A) Folate
(B) Intrinsic factor
(C) Iron
(D) Niacin
(E) Vitamin B_{12}

110. The substitution of which of the following amino acids for lysine is a conservative amino acid substitution?

(A) Arginine
(B) Glycine
(C) Proline
(D) Serine
(E) Tryptophan

111. Aspirin is an irreversible inhibitor of which of the following enzymes?

(A) Adenylyl cyclase
(B) Cyclooxygenase
(C) Phosphatidylinositol phosphate 3-kinase
(D) Protein kinase A
(E) Protein kinase C

112. *Streptococcus mutans*, the causative agent of dental caries, generates which of the following acids that dissolves dental enamel?

(**A**) Acetic acid
(**B**) Fumaric acid
(**C**) Lactic acid
(**D**) Propionic acid
(**E**) Succinic acid

113. Amyl nitrate converts the ferrous iron of hemoglobin to the ferric iron of methemoglobin. Which of the following occurs when amyl nitrate is administered clinically to treat cyanide poisoning?

(**A**) Adverse effects of cyanide are ameliorated by dilation of the coronary arteries.
(**B**) Carbon monoxide forms a complex with cyanide in a process accelerated by ferric iron.
(**C**) Cyanide ion binds to methemoglobin iron.
(**D**) Metabolism of cyanide as catalyzed by rhodanese is increased.
(**E**) Oxygen-carrying capacity of the blood is increased.

114. The predominant nitrogen metabolite excreted by humans is

(**A**) ammonium ion
(**B**) creatinine
(**C**) fecal protein
(**D**) urea
(**E**) uric acid

115. Phosphoribosyl pyrophosphate aminotransferase is inhibited allosterically by which of the following metabolites?

(**A**) Allopurinol ribonucleoside monophosphate
(**B**) ATP
(**C**) CMP
(**D**) Inosine triphosphate
(**E**) UMP

116. Which of the following enzymes alters the supercoiling of DNA and is a target of numerous cancer chemotherapeutic agents?

(**A**) DNA gyrase
(**B**) DNA ligase
(**C**) DNA topoisomerase II
(**D**) Primase
(**E**) SSB (single-strand binding) protein

117. Which of the following transcription factors is a potential target for cancer chemotherapeutic agents?

(**A**) *Eco*RI
(**B**) Myc
(**C**) TFID
(**D**) TFIID
(**E**) TFIIIA

118. A nutritional deficiency of which of the following amino acids inhibits protein synthesis by limiting a required substrate?

(**A**) Alanine
(**B**) Glycine
(**C**) Isoleucine
(**D**) Serine
(**E**) Tyrosine

119. Which of the following enzymes is rate-limiting and a target for drugs in the treatment of hypercholesterolemia?

(**A**) Hepatic hydroxymethylglutaryl (HMG)-CoA reductase
(**B**) HMG-CoA synthase
(**C**) Isopentenyl pyrophosphate isomerase
(**D**) Mevalonate kinase
(**E**) Phosphomevalonate kinase

120. Phorbol ester, a tumor promoter, activates which of the following enzymes?

(**A**) Mitogen activated protein (MAP) kinase
(**B**) Protein kinase A
(**C**) Protein kinase C
(**D**) Protein kinase G
(**E**) Src gene protein kinase

121. The activity of which of the following enzymes is decreased in scurvy?

(**A**) Aromatic acid decarboxylase
(**B**) Malate dehydrogenase
(**C**) Prolyl hydroxylase
(**D**) Pyruvate dehydrogenase
(**E**) Vitamin K-epoxide reductase

Questions 122 and 123

A 40-year-old man with a history of alcoholism develops a low-grade fever, cough, and night sweats. A chest X-ray reveals hilar adenopathy. A diagnosis of tuberculosis is made. The patient begins a course of antibiotics, including rifamycin.

122. The action of rifamycin on bacterial metabolism is to

(**A**) inhibit the initiation of protein synthesis
(**B**) inhibit the initiation of RNA synthesis
(**C**) produce misreading of the genetic code
(**D**) inhibit ribosomal transpeptidase activity
(**E**) inhibit the ribosomal translocation reactions

123. The patient is found to have acquired immunodeficiency syndrome and is treated with a combination of drugs that includes azidothymidine (AZT). The action of this nucleotide on the human immunodeficiency virus (HIV) is to

(**A**) inhibit HIV messenger RNA synthesis
(**B**) inhibit HIV protease
(**C**) inhibit HIV reverse transcriptase
(**D**) inhibit translation of HIV messenger RNA (mRNA)
(**E**) produce misreading of the genetic code

124. Nitric oxide is a free radical that functions in a variety of physiologic processes in nerve, muscle, and other cells. The action of nitric oxide is to directly activate

(A) calcium/calmodulin stimulated protein kinase
(B) guanylate cyclase
(C) protein kinase C
(D) protein phosphatase-1
(E) the nicotinic cholinergic receptor

125. A 52-year-old woman develops skin flushing and diarrhea. Urinalysis reveals an elevated excretion of 5-hydroxyindoleacetic acid (5-HIAA). A diagnosis of carcinoid syndrome is made. 5-HIAA is derived from which of the following substances?

(A) Histidine
(B) Homogentisate
(C) Phenylalanine
(D) Tryptophan
(E) Tyrosine

126. The metabolic aberration in diabetes mellitus is related to the increased action of glucagon, which is not restrained by insulin. Glucagon affects the human liver in which of the following ways?

(A) Glucagon increases hexokinase activity.
(B) Glucagon increases pyruvate kinase activity.
(C) Glucagon increases phosphofructokinase activity.
(D) Glucagon increases the levels of cyclic AMP.
(E) Glucagon increases the levels of fructose-2,6-bisphosphate.

127. Which of the following neurotransmitters is inactivated at the synapse by hydrolysis?

(A) Acetylcholine
(B) Dopamine
(C) Glutamate
(D) Norepinephrine
(E) Serotonin

128. Which of the following cancer chemotherapeutic agents inhibits dihydrofolate reductase?

(A) Actinomycin D
(B) Asparaginase
(C) Cytosine arabinoside
(D) Hydroxyurea
(E) Methotrexate

129. A defect in a plasma membrane transport protein for which of the following substances leads to an inborn error of metabolism?

(A) Carbon dioxide
(B) Carnitine
(C) Nitric oxide
(D) Oxygen
(E) Urea

130. The mutation that is responsible for sickle cell anemia is

(A) elimination of a stop codon
(B) an insertion in the promoter
(C) a nonsense mutation
(D) a point mutation
(E) a result of aberrant splicing

131. A 4-day-old infant presents in the emergency department with lethargy, convulsions, and acidosis. Urinalysis reveals the presence of isovalerylglycine. The most likely diagnosis in this patient is

(A) biotin deficiency
(B) β-ketothiolase deficiency (2-methylacetoacetyl CoA thiolase deficiency)
(C) isovaleryl CoA dehydrogenase deficiency
(D) maple syrup urine disease
(E) propionyl-CoA carboxylase deficiency

132. Which of the following amino acids has only a single codon?

(A) Alanine
(B) Glycine
(C) Histidine
(D) Tryptophan
(E) Tyrosine

133. The lysosomal storage diseases such as Gaucher's disease, Tay-Sachs disease, and Hurler's disease are due to a

(A) defect in catabolism
(B) defect in cytochrome P-450
(C) deficiency of ω-3 fatty acids
(D) defect in electron transport
(E) defect in metabolite biosynthesis

134. Cystic fibrosis is due to a defect in the transport of

(A) bicarbonate
(B) chloride
(C) phosphate
(D) potassium
(E) sodium

135. A pregnant patient with edema, hypertension, and proteinuria is treated for eclampsia with which of the following substances?

(A) Aldosterone
(B) Calcium chloride
(C) Enalapril
(D) Magnesium sulfate
(E) Potassium chloride

136. A 33-year-old woman has a 3-month history of progressive weakness. A diagnosis of myasthenia gravis is established. The pathogenesis of this disorder is due to a decrease in which of the following muscle components?

(A) Muscle actin
(B) Muscle ATP
(C) Muscle calcium
(D) Muscle myosin
(E) Nicotinic acetylcholine receptors

137. Fasting hypoglycemia in medium-chain acyl-CoA dehydrogenase deficiency is due to decreased acetyl-CoA concentrations and decreased activity of which of the following enzymes?

(**A**) Fumarase
(**B**) Lactate dehydrogenase
(**C**) Malate dehydrogenase
(**D**) Phosphoenolpyruvate carboxykinase
(**E**) Pyruvate carboxylase

138. The substrates for ω-oxidation include long- and medium-chain fatty acids. Defects in the metabolism of long- and medium-chain fatty acids result in the formation of dicarboxylic acids, which are excreted in the urine. ω-Oxidation occurs in which of the following cellular locations?

(**A**) Cytosol
(**B**) Endoplasmic reticulum
(**C**) Inner mitochondrial membrane
(**D**) Outer mitochondrial membrane
(**E**) Peroxisome

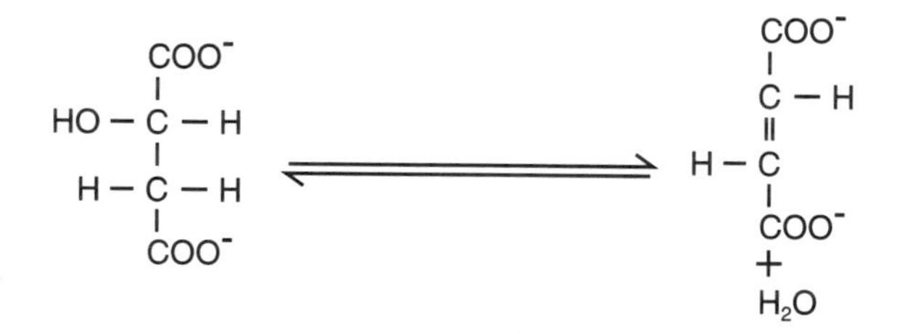

139. The reaction illustrated in the above diagram is catalyzed by which of the following enzymes?

(**A**) Hydrolase
(**B**) Ligase
(**C**) Lyase
(**D**) Oxidoreductase
(**E**) Transferase

140. A 70-year-old man presents with back pain and osteoblastic lesions, which are observed by X-ray, and an elevated serum acid phosphatase. The most likely diagnosis of this patient is

(**A**) cancer of the prostate
(**B**) osteomalacia
(**C**) osteoporosis
(**D**) Paget's disease of the bone
(**E**) rickets

141. Which of the following substances is generally elevated in serum within 12 hours of a myocardial infarction?

(**A**) Alkaline phosphatase
(**B**) Creatine phosphokinase
(**C**) Myoinositol
(**D**) Myokinase
(**E**) Sodium/potassium ATPase

142. Using an ATP yield of 2.5 per mole of NADH and 1.5 per mole of flavin adenine dinucleotide ($FADH_2$), determine the yield of ATP per mole of acetyl-CoA oxidized to carbon dioxide and water by the Krebs cycle and oxidative phosphorylation.

(**A**) 7.5
(**B**) 8.5
(**C**) 10.0
(**D**) 11.0
(**E**) 12.0

143. The class of histones that occurs in the linker region of nucleosomes is

(**A**) H1
(**B**) H2A
(**C**) H2B
(**D**) H3
(**E**) H4

144. A person with elevated plasma concentrations of lactate, pyruvate, propionate, and 3-hydroxyisovalerate is a candidate for treatment with

(**A**) biotin
(**B**) niacin
(**C**) pantothenate
(**D**) pyridoxine
(**E**) thiamine

145. The time-dependent cleavage of one amino-terminal lysine from one of the chains of cardiac creatine phosphokinase differentiates the MB1 isozyme (without lysine) from the MB2 isozyme (with lysine). As a result of this modification, creatine phosphokinase recently released from a myocardial infarct can be differentiated from background enzyme by which one of the following criteria?

(**A**) The MB2 isozyme is more negatively charged than the MB1 isozyme.
(**B**) The MB2 isozyme migrates anodally (toward the negatively charged pole) when compared with the MB1 isozyme.
(**C**) The MB1 and MB2 isozymes bear equivalent charges at pH 7.
(**D**) Recently released enzyme is more negatively charged than the background enzyme.
(**E**) Recently released enzyme migrates cathodally (toward the positively charged pole) when compared with the background enzyme.

146. The most likely diagnosis in a newborn with acute encephalopathy, hypoglycemia, and ketonuria and elevated plasma leucine, isoleucine, and valine but normal plasma ammonium ion is

(**A**) arginase deficiency
(**B**) long-chain acyl-CoA dehydrogenase deficiency
(**C**) maple syrup urine disease
(**D**) methylmalonic aciduria
(**E**) mitochondrial complex II deficiency

147. The most likely diagnosis in a newborn with acute encephalopathy, metabolic acidosis, normal urinary ketones, normal plasma ammonium ion, normal plasma carnitine, and the absence of organic aciduria is

(**A**) arginase deficiency
(**B**) argininosuccinate synthetase deficiency
(**C**) long-chain acyl-CoA deficiency
(**D**) maple syrup urine disease
(**E**) mitochondrial complex III deficiency

148. A field worker presents to the emergency department in respiratory distress with emesis, diarrhea, and muscle fasciculation following exposure to Sarin, an insecticide with an action similar to diisopropylfluorophosphate. The action of diisopropylfluorophosphate and Sarin is due to inhibition of which of the following proteins?

(**A**) Acetylcholinesterase
(**B**) Choline acetyltransferase
(**C**) Nicotinic receptor
(**D**) Norepinephrine transporter
(**E**) Tyrosine hydroxylase

149. Malonyl-CoA decreases fatty acid oxidation by inhibiting which of the following reactions?

(**A**) Acyl-CoA dehydrogenase
(**B**) Carnitine acyltransferase I
(**C**) Enoyl-CoA hydratase
(**D**) Fatty acyl-CoA ligase
(**E**) Hydroxyacyl-CoA dehydrogenase

150. The most likely enzyme defect in a patient with metabolic acidosis and increased urinary lactic, α-ketoglutaric, α-ketoisocaproic, α-keto-β-methylvaleric, and α-ketoisovaleric aciduria is

(**A**) branched-chain ketoacid dehydrogenase
(**B**) holocarboxylase synthetase
(**C**) lipoamide dehydrogenase
(**D**) mevalonate kinase
(**E**) propionyl-CoA carboxylase

151. An elevation of which of the following lipoproteins leads to hypercholesterolemia with little or no elevation of triglyceride?

(**A**) Chylomicrons
(**B**) Intermediate-density lipoprotein
(**C**) LDL
(**D**) Very-low-density lipoprotein (VLDL)

152. A 74-year-old woman develops tremor and a shuffling gait over a period of several months. This disorder is related to a deficiency in the metabolism of which of the following amino acids?

(**A**) Arginine
(**B**) Glutamate
(**C**) Histidine
(**D**) Tryptophan
(**E**) Tyrosine

153. Complex biochemical machinery is required for replication because of which of the following properties of DNA and DNA polymerases?

(**A**) DNA polymerases catalyze 3' to 5' polymerization reactions, and duplex DNA is parallel.
(**B**) DNA polymerases catalyze 5' to 3' polymerization reactions, and duplex DNA is antiparallel.
(**C**) DNA polymerases catalyze 5' to 3' polymerization reactions, and DNA is double stranded.
(**D**) DNA polymerases catalyze 5' to 3' polymerization reactions, and duplex DNA is parallel.
(**E**) DNA polymerases catalyze both 5' to 3' and 3' to 5' polymerization reactions, and duplex DNA is antiparallel.

154. Which one of the following enzymes catalyzes the combination of Okazaki fragments?

(**A**) DNA gyrase
(**B**) DNA ligase
(**C**) DNA polymerase α
(**D**) DNA topoisomerase I
(**E**) DNA topoisomerase II

155. Unlike DNA polymerases, RNA polymerases

(**A**) can elongate a primer oligonucleotide
(**B**) can initiate synthesis of their polynucleotide product
(**C**) generate inorganic pyrophosphate
(**D**) occur within the mitochondrion
(**E**) use a DNA template

156. A deficiency of long-chain acyl-CoA dehydrogenase leads to acidosis without ketosis. The lack of ketogenesis is due to which of the following mechanisms?

(**A**) Failure to convert hydroxymethylglutaryl-CoA to ketone bodies
(**B**) Failure of fatty acid to enter the mitochondrion
(**C**) Failure to obtain acetyl-CoA by β-oxidation
(**D**) Failure to obtain reducing equivalents for the electron transport chain
(**E**) Failure to provide reducing equivalents for acetoacetate reduction

157. Which one of the following enzymes is required for the synthesis of heterogeneous nuclear RNA in humans?

(**A**) Reverse transcriptase
(**B**) RNA ligase
(**C**) RNA polymerase I
(**D**) RNA polymerase II
(**E**) RNA polymerase III

158. Pyrazinamide exhibits liver toxicity. Which of the following clinical chemistry tests can be used to assess liver damage?

(**A**) Amylase
(**B**) Creatine phosphokinase
(**C**) γ-Glutamyl transpeptidase and aspartate aminotransferase activity
(**D**) Serum cholesterol
(**E**) Serum glucose

159. The component of protein synthesis machinery that directly uses the chemical energy of ATP is

(**A**) aminoacyl-tRNA ligases
(**B**) eukaryotic elongation factor 1 (eEF1)
(**C**) eukaryotic elongation factor 2 (eEF2)
(**D**) eukaryotic initiation factor 2 (eIF2)
(**E**) eukaryotic release factor (eRF)

160. Mutations of the signal peptide sequence can alter the function of which of the following proteins?

(**A**) Blood clotting factor IX
(**B**) Estrogen
(**C**) EF1
(**D**) Hexokinase
(**E**) Pyruvate dehydrogenase

161. Which of the following receptors contains seven transmembrane segments?

(**A**) Epidermal growth factor
(**B**) Estrogen
(**C**) Glucagon
(**D**) Insulin
(**E**) Nicotinic acetylcholine

162. The receptor for which of the following hormones binds to DNA and alters the transcription of target genes?

(**A**) Calcitonin
(**B**) Parathyroid hormone
(**C**) Thyrotropin-releasing hormone (TRH)
(**D**) Thyrotropin (thyroid-stimulating hormone)
(**E**) Thyroxine

163. Fluorouracil, a prodrug used in cancer chemotherapy, leads to the inhibition of which of the following enzymes?

(**A**) Dihydrofolate reductase
(**B**) DNA polymerase β
(**C**) Epidermal growth factor receptor protein-tyrosine kinase
(**D**) Phosphoribosyl pyrophosphate amido transferase
(**E**) Thymidylate synthase

164. Omeprazole, an inhibitor of the proton/potassium-ATPase, is used for the treatment of peptic ulcers. The proton/potassium-ATPase is a class

(**A**) A ATPase that secretes acid across apical membranes
(**B**) B ATPase that secretes acid across basil membranes
(**C**) F ATPase that secretes acid across membranes
(**D**) P ATPase that secretes acid across plasma membranes
(**E**) ATPase that secretes acid across vesicular membranes

165. People with polycythemia vera can be treated with radioactive

(**A**) calcium
(**B**) cobalt
(**C**) iodine
(**D**) phosphate
(**E**) radon

166. The increased serum triglyceride levels that occur in diabetes mellitus result from which of the following mechanisms?

(**A**) Decreased insulin activity promotes fatty acid transport into adipocytes, which promotes serum triglyceride synthesis.
(**B**) Decreased insulin activity promotes glucose uptake into hepatocytes, which provides additional carbohydrate substrate for lipogenesis.
(**C**) Increased availability of fatty acids, which results from enhanced activity of hormone-sensitive lipase, drives hepatic triglyceride and VLDL formation.
(**D**) Increased lipoprotein lipase activity, which decreases the liberation of triglyceride from VLDL.

167. The conversion of 1 mole of ethanol to carbon dioxide and water results in the liberation of how many moles of NADH and $FADH_2$?

(**A**) 2
(**B**) 4
(**C**) 5
(**D**) 6
(**E**) 8

168. Insulin increases the synthesis of which of the following proteins?

(**A**) Fructose-1,6-bisphosphatase
(**B**) Glucose-6-phosphatase
(**C**) Glycogen synthase
(**D**) Hormone-sensitive lipase
(**E**) Phosphoenolpyruvate carboxykinase

169. The most abundant protein in humans by mass is

(**A**) albumin
(**B**) collagen
(**C**) hemoglobin
(**D**) myosin

170. Acetyl-CoA carboxylase is located in the which of the following cellular locations?

(**A**) Cytosol
(**B**) Lysosome
(**C**) Mitochondrion
(**D**) Nucleus
(**E**) Plasma membrane

171. The functional form of the LDL receptor, which is deficient in familial hypercholesterolemia, is located in the

(**A**) endoplasmic reticulum
(**B**) Golgi complex
(**C**) mitochondrion
(**D**) nucleus
(**E**) plasma membrane

172. Wernicke's encephalopathy, which occurs in people who chronically abuse ethanol, is due to a deficiency of

(**A**) ascorbate
(**B**) folate
(**C**) pantothenate
(**D**) riboflavin
(**E**) thiamine

173. The intake of which of the following amino acids decreases the requirement for methionine in the diet by a process called sparing?

(**A**) Alanine
(**B**) Asparagine
(**C**) Cysteine
(**D**) Lysine
(**E**) Serine

174. What is the net production of ATP during the conversion of 1 molecule of glucose to 2 molecules of pyruvate?

(**A**) 1
(**B**) 2
(**C**) 3
(**D**) 6
(**E**) 32

175. The concentration of hydrogen ions in a patient with acidosis (pH 7.3) compared with the concentration of hydrogen ions at physiologic pH is

(**A**) decreased by 0.1 mol/L
(**B**) decreased by 1×10^{-7} mol/L
(**C**) decreased by 1×10^{-8} mol/L
(**D**) increased by 1×10^{-7} mol/L
(**E**) increased by 1×10^{-8} mol/L

176. The α-helix found in proteins such as hemoglobin and myoglobin is an example of which of the following properties of a protein?

(**A**) Primary structure
(**B**) Secondary structure
(**C**) Tertiary structure
(**D**) Quaternary structure

177. Ketogenic diets are thought to be effective in the treatment of certain forms of epilepsy. Which of the following compounds is the most ketogenic?

(**A**) Alanine
(**B**) Lactate
(**C**) Leucine
(**D**) Serine
(**E**) Threonine

178. Which of the following molecules lacks discrete ends?

(**A**) mRNA
(**B**) Mitochondrial DNA
(**C**) tRNA
(**D**) The X chromosome
(**E**) The Y chromosome

179. Approximately half of all human cancers involve mutations in which of the following tumor suppressor genes?

(A) *abl*
(B) *p53*
(C) *ras*
(D) *src*
(E) *RB1*

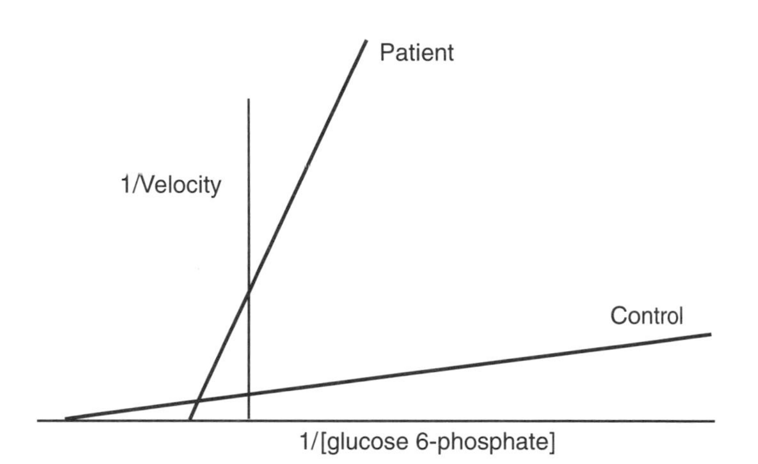

180. A woman was given the drug chloroquine before a trip to a region where malaria is endemic. She develops a hemolytic anemia. Analysis of her red blood cell glucose-6-phosphate dehydrogenase activity and that of a control yield the above data. From the double reciprocal plot shown above, you can conclude that

(A) the K_m of the patient's enzyme is less than that of the control subject, and the V_{max} of the patient's enzyme is greater than that of the control subject
(B) the K_m of the patient's enzyme is greater than that of the control subject, and the V_{max} of the patient's enzyme is less than that of the control subject
(C) the K_m of the patient's enzyme is greater than that of the control subject, and the V_{max} of the patient's enzyme is greater than that of the control subject
(D) the K_m of the patient's enzyme is less than that of the control subject, and the V_{max} of the patient's enzyme is less than that of the control subject

181. The chief excitatory neurotransmitter in the human brain is

(A) dopamine
(B) γ-aminobutyrate
(C) glutamate
(D) norepinephrine
(E) serotonin

182. The catalytic triad of serine proteases consists of serine, glutamate, and

(A) arginine
(B) aspartate
(C) glutamine
(D) histidine
(E) lysine

183. Creatine phosphokinase consists of a dimer made up of M (muscle) and B (brain) subunits. What is the total number of possible isozymes?

(**A**) 1
(**B**) 2
(**C**) 3
(**D**) 4
(**E**) 5

184. Which of the following clinical chemistry determinations is the most sensitive indicator of early alcoholic liver disease?

(**A**) Alkaline phosphatase
(**B**) Aspartate aminotransferase
(**C**) Direct bilirubin
(**D**) γ-Glutamyltranspeptidase
(**E**) Lactate dehydrogenase

185. Humans and some bacteria both contain catalase. This enzyme catalyzes the breakdown of which of the following substances?

(**A**) $FADH_2$
(**B**) Hydrogen peroxide
(**C**) Hypochlorite
(**D**) Methemoglobin
(**E**) Superoxide

186. Which of the following enzyme-catalyzed reactions of glycolysis is reversible and also participates in gluconeogenesis?

(**A**) Hexokinase
(**B**) Phosphofructokinase
(**C**) Phosphoglycerate kinase
(**D**) Pyruvate kinase

187. The conversion of 1 mole of glucosyl residues from glycogen to 2 moles of pyruvate in reactions involving glycogen phosphorylase and glycolysis results in the net formation of how many moles of ATP?

(**A**) 1
(**B**) 2
(**C**) 3
(**D**) 4
(**E**) 5

188. A deficiency of pyruvate kinase in erythrocytes causes hemolytic anemia. This enzyme catalyzes the following reaction: ATP + pyruvate $\rightleftharpoons$ ADP + phosphoenolpyruvate. Which of the following statements describes this reaction?

(**A**) The products contain more high-energy bonds than the reactants.
(**B**) The reactants contain more high-energy bonds than the products.
(**C**) The reaction is bidirectional *in vivo.*
(**D**) The reaction proceeds from right to left under standard conditions.

189. After an injection of insulin in a diabetic patient with hyperglycemia, there is a prompt normalization of plasma glucose. Plasma glucose returns to normal as a result of the recruitment of which of the following transport proteins to the plasma membrane?

(A) GLUT1
(B) GLUT2
(C) GLUT3
(D) GLUT4
(E) GLUT5

190. Osteogenesis imperfecta, or brittle bone disease, involves mutations of type I collagen. These mutations often involve which one of the following amino acids?

(A) Alanine
(B) Glycine
(C) Hydroxylysine
(D) Hydroxyproline
(E) Methionine

191. A 41-year-old woman who is 5-ft 1-in. tall and weighs 155 lb presents with right upper quadrant pain, followed by the onset of jaundice. Laboratory studies reveal an increase in which of the following substances that produced jaundice in this patient?

(A) Bile acids
(B) Bilirubin
(C) Stercobilin
(D) Urobilin
(E) Urobilinogen

192. Which of the following enzymes is an integral membrane protein?

(A) Citrate synthase
(B) Isocitrate dehydrogenase
(C) α-Ketoglutarate dehydrogenase
(D) Succinate dehydrogenase
(E) Succinate thiokinase

193. Quantitatively, the predominant reaction that generates ammonium ion is catalyzed by which of the following enzymes?

(A) Adenosine deaminase
(B) L-Amino acid oxidase
(C) Asparaginase
(D) Aspartate aminotransferase
(E) Glutamate dehydrogenase

194. Which of the following metabolites is the stoichiometric substrate of the Krebs tricarboxylic acid cycle?

(A) Acetyl-CoA
(B) Carbon dioxide
(C) Citrate
(D) Oxaloacetate
(E) Pyruvate

195. A deficiency of which of the following enzymes leads to an increase in 2,3-bisphosphoglycerate in red cells?

(**A**) Aldolase
(**B**) Hexokinase
(**C**) Phosphoglucoisomerase
(**D**) Pyruvate kinase
(**E**) Triose phosphate isomerase

196. Hypoglycemia is characteristic of many forms of glycogen storage disease. Glycogen storage disease due to the deficiency of which of the following enzymes is NOT characterized by hypoglycemia?

(**A**) Amylo-1,6-glucosidase (debranching enzyme)
(**B**) Amylo-1,6-transglucosidase (branching enzyme)
(**C**) Glucose-6-phosphatase
(**D**) Liver phosphorylase
(**E**) Liver phosphorylase kinase

197. The carbon and nitrogen atoms of which of the following amino acids are incorporated into the purine ring system?

(**A**) Aspartate
(**B**) Glycine
(**C**) Glutamate
(**D**) Histidine
(**E**) Lysine

198. The normal sequence between exon 12 and intron 12 of hexosaminidase A has the following sequence: ...TG/GT..., where the slash (/) represents the exon/intron junction. A person is found to have the following sequence at the corresponding site: ...TG/CT... Which of the following statements best describes the genetic defect in this individual?

(**A**) A four-base insertion leads to a frameshift mutation.
(**B**) A one-base deletion leads to a frameshift mutation.
(**C**) The mutation interferes with RNA splicing and intron 12 is not removed.
(**D**) The mutation results in the insertion of a stop codon at this position.
(**E**) The mutation occurs in the intron and is silent.

199. Which of the following mechanisms explains the occurrence of methylmalonic aciduria in a person with a mutation in the leader sequence of methylmalonyl-CoA mutase?

(**A**) Leader sequence mutations eliminate the amino acids from the protein that occurs in the substrate binding site.
(**B**) The enzyme is inactive because of a lack of requisite protein phosphorylation.
(**C**) The enzyme is not synthesized because of abnormal RNA splicing.
(**D**) The enzyme is not synthesized because of a lack of mRNA translation.
(**E**) The normal reaction fails to occur because of a lack of enzyme translocation to its normal cellular location.

200. Which of the following enzymes catalyzes the physiologic formation of hydrogen peroxide?

(**A**) Catalase
(**B**) Cytochrome oxidase
(**C**) Glutathione peroxidase
(**D**) Myeloperoxidase
(**E**) Superoxide dismutase

201. A mutation of an A to G in the promoter of the factor IX gene can lead to hemophilia. Promoters determine the

(**A**) origins of replication
(**B**) polyadenylation site of RNA
(**C**) splice sites for RNA processing
(**D**) start site for RNA synthesis
(**E**) stop site for RNA synthesis

202. ADP-ribosylation reactions are involved in the action of diphtheria toxin and cholera toxin. The metabolite that donates the ADP-ribosyl group is

(**A**) ADP
(**B**) ATP
(**C**) CoA
(**D**) flavin adenine diphosphate (FAD)
(**E**) NAD^+

203. The signal recognition particle receptor is found in which of the following subcellular locations?

(**A**) Endoplasmic reticulum
(**B**) Mitochondrion
(**C**) Nucleus
(**D**) Plasma membrane
(**E**) Ribosome

204. One of the following carbohydrates is important in targeting proteins to the lysosome. Mutations that interfere with the attachment of this moiety to appropriate proteins lead to I-cell disease, which is associated with

(**A**) CMP-*N*-acetylneuraminate
(**B**) dolichol diphosphate glucose
(**C**) dolichol monophosphate glucose
(**D**) mannose 6-phosphate
(**E**) UDP-glucose

205. The nitrogen of nitric oxide is derived directly from which of the following metabolites?

(**A**) Arginine
(**B**) Aspartate
(**C**) Carbamoyl phosphate
(**D**) Glutamate
(**E**) Glutamine

206. The SH2 domain of Grb2, a signal transduction protein, enables Grb2 to

(**A**) bind to leucine zipper proteins
(**B**) bind to phosphoserine residues of heterotrimeric G proteins
(**C**) bind to phosphotyrosine residues of the insulin receptor
(**D**) bind to the promoter regions of target genes
(**E**) bind to zinc finger proteins

207. Extracellular signal regulated kinases or MAP kinases have which of the following properties?

(**A**) They are activated following phosphorylation of histidine.
(**B**) They are activated following phosphorylation of serine.
(**C**) They are activated following phosphorylation of threonine.
(**D**) They are activated following phosphorylation of tyrosine.
(**E**) They are activated following phosphorylation of both tyrosine and threonine.

208. Which of the following oncogene products participates in signal transduction and is active when it binds to guanosine triphosphate?

(A) Bcl-2
(B) ErbB2
(C) Myc
(D) Ras
(E) Src

209. The leading cause of blindness throughout the world is

(A) cataracts
(B) diabetes mellitus
(C) retinitis pigmentosum
(D) vitamin A deficiency

210. Angiotensin II, an important regulator of arterial pressure, has the following structure: Asp-Arg-Val-Tyr-Ile-His-Pro-Phe. Its net charge at pH 7 is approximately

(A) –1
(B) –0.5
(C) 0
(D) +0.5
(E) +1

211. Alcoholism is often associated with thiamine deficiency. As a result, which one of the following enzymes has been studied in people with this disorder?

(A) Glucose-6-phosphate dehydrogenase
(B) Lactonase
(C) 6-Phosphogluconate dehydrogenase
(D) Transaldolase
(E) Transketolase

212. Failure to phosphorylate fructose (essential fructosuria) or galactose (galactokinase deficiency) does not lead to mental retardation and other pathologic sequelae. The benign nature of these disorders is due to the

(A) cleavage and metabolism of these two sugars by aldolase
(B) failure to absorb these two compounds from the gut
(C) isomerization of these two sugars to glucose and subsequent catabolism initiated by glucokinase
(D) lack of accumulation of charged intracellular metabolites
(E) use of hexokinase to initiate the metabolism of these two sugars

213. The reaction of the Krebs cycle that is unidirectional and forces the cycle to also be unidirectional is mediated by which of the following enzymes?

(A) Aconitase
(B) Isocitrate dehydrogenase
(C) α-Ketoglutarate dehydrogenase
(D) Succinate dehydrogenase
(E) Succinate thiokinase

214. Which one of the following enzymes catalyzes a hydrolysis reaction?

(A) Glycogen branching enzyme
(B) Glycogen debranching enzyme
(C) Glycogen phosphorylase
(D) Glycogen phosphorylase kinase
(E) Glycogen synthase

215. Which of the following mutations would most seriously disrupt the conformation of a protein?

(**A**) Alteration of an amino acid at the carboxyl terminus
(**B**) Introduction of a proline residue into an α-helix
(**C**) Substitution of glutamate for aspartate
(**D**) Substitution of serine for threonine

216. Oral administration of ion exchange resins is used in the treatment of hypercholesterolemia. Therapy is directed toward the elimination of which of the following?

(**A**) Bilirubin
(**B**) Cholestanol
(**C**) Coprostanol
(**D**) Deoxycholate
(**E**) Mevalonate

217. A deficiency of α-ketoglutarate dehydrogenase is one cause of lactic acidosis. Which of the following cofactors is NOT associated with the α-ketoglutarate dehydrogenase reaction?

(**A**) Biotin
(**B**) CoA
(**C**) FAD
(**D**) Lipoic acid
(**E**) NAD^+

218. Deficiency of which of the following enzymes does NOT result in hypoglycemia in infants?

(**A**) Fructose-1,6-bisphosphatase
(**B**) Glucose-6-phosphatase
(**C**) Hepatic phosphorylase
(**D**) Muscle phosphorylase
(**E**) Pyruvate carboxylase

219. Which of the following compounds is NOT a normal constituent of DNA and its presence in DNA leads to instability?

(**A**) Adenosine deoxyriboside
(**B**) Cytosine arabinoside
(**C**) Guanine deoxyriboside
(**D**) 5-Methylcytosine deoxyriboside
(**E**) Thymine deoxyriboside

220. Pancreatic enzyme replacement, which is used in the treatment of patients with cystic fibrosis, involves each of the following enzymes EXCEPT

(**A**) amylase
(**B**) lipase
(**C**) pepsin
(**D**) trypsin

DIRECTIONS: Each set of questions in this section consists of a list of three to five options followed by several numbered items. For each numbered item, select the ONE lettered option that is most closely associated with it. Each lettered option may be selected once, more than once, or not at all.

Questions 221–223

(**A**) Chloramphenicol
(**B**) Cycloheximide
(**C**) Puromycin
(**D**) Streptomycin
(**E**) Tetracycline

Match each mechanism of action below with the appropriate drug.

221. Produces mistakes in bacterial translation (misreading of the genetic code)

222. Inhibits bacterial peptidyltransferase activity

223. Causes premature chain termination during protein synthesis in prokaryotes and eukaryotes

Questions 224 and 225

(**A**) Adenosine deaminase
(**B**) Glucose-6-phosphatase
(**C**) Hexosaminidase A
(**D**) Hypoxanthine-guanine phosphoribosyltransferase
(**E**) Phenylalanine hydroxylase

Match each description of an enzyme deficiency with the appropriate enzyme.

224. Deficiency associated with combined immunodeficiency disease

225. Deficiency associated with Lesch-Nyhan syndrome

Questions 226 and 227

(**A**) Albumin
(**B**) Chylomicrons
(**C**) HDL
(**D**) LDL
(**E**) VLDL

Match each description below with the correct transporter.

226. Transports triglyceride from liver to other tissues

227. Transports fatty acids from adipose tissue to the liver

Questions 228–230

(**A**) DNA
(**B**) 5S RNA
(**C**) mRNA
(**D**) rRNA
(**E**) tRNA

Match each property below with the appropriate nucleic acid.

228. Capped by 7-methylguanosine

229. CCA at the 3' end

230. In humans, contains a polyA tail

Questions 231–233

(**A**) DNA polymerase α
(**B**) DNA polymerase β
(**C**) DNA polymerase γ
(**D**) DNA polymerase δ
(**E**) DNA polymerase ε

Match each action below with the corresponding form of DNA polymerase.

231. Catalyzes mitochondrial DNA elongation reactions

232. Activated by proliferating cell nuclear antigen

233. Synthesizes the lagging DNA strand in humans

Questions 234–237

(**A**) eIF1
(**B**) eIF2
(**C**) eIF3
(**D**) eIF4E
(**E**) eRF
(**F**) eEF1α
(**G**) eEF1βγ
(**H**) eEF2

Match each description of eukaryotic activity with its corresponding protein synthesis factor.

234. Target of diphtheria toxin

235. Binds the specific methionine tRNA that is implanted into the P site of the ribosome

236. Mediates the translocation of the peptidyl-tRNA/mRNA complex from the A site to the P site

237. Participates in the hydrolysis of peptidyl-tRNA

Questions 238–240

(**A**) Cytosol
(**B**) Lysosome
(**C**) Mitochondrion
(**D**) Nucleus
(**E**) Peroxisome
(**F**) Plasma membrane
(**G**) Rough endoplasmic reticulum
(**H**) Smooth endoplasmic reticulum

For each description of a biochemical process, select the most appropriate cellular component.

238. Responsible for the oxidation of very long-chain fatty acids (>C18)

239. Site of functional glucocorticoid receptor

240. Site of action of β-glucosidase (defective in Gaucher's disease), an enzyme with an acid pH optimum

Questions 241–244

(**A**) Calcium
(**B**) Chloride
(**C**) Cobalt
(**D**) Copper
(**E**) Fluoride
(**F**) Iodide
(**G**) Iron
(**H**) Magnesium
(**I**) Phosphorus
(**J**) Potassium
(**K**) Radon
(**L**) Sodium

For each characteristic listed below, select the most appropriate element.

241. The transport of this substance from the outside to the inside of heart cells is inhibited by digitalis.

242. Essential nutrient used in the treatment of osteoporosis.

243. An excess of this substance leads to hepatolenticular degeneration.

244. The radioactive form of this element is used in treating Graves' disease.

Questions 245–248

(**A**) Alanine
(**B**) Cysteine
(**C**) Glutamate
(**D**) Glycine
(**E**) Leucine
(**F**) Lysine
(**G**) Methionine
(**H**) Proline
(**I**) Serine
(**J**) Tryptophan

For each biochemical feature listed below, select the most appropriate amino acid.

245. Occurs at every third position in the collagen triple helix; mutations that alter this residue can produce Ehlers-Danlos syndrome

246. This residue in proteins involved in blood clotting can undergo vitamin K–dependent carboxylation.

247. Metabolism of this amino acid results in the formation of serotonin.

248. This amino acid is transported from muscle to liver as part of the Cori cycle.

Questions 249–252

(**A**) Acetyl-CoA carboxylase
(**B**) Alcohol dehydrogenase
(**C**) Carnitine acyltransferase-I
(**D**) Citrate synthase
(**E**) Fructose-1,6-bisphosphatase
(**F**) Glucokinase
(**G**) Glucose-6-phosphatase
(**H**) Glucose-6-phosphate dehydrogenase
(**I**) Glycogen phosphorylase
(**J**) Glycogen synthase
(**K**) Hexokinase
(**L**) Hormone-sensitive lipase
(**M**) Isocitrate dehydrogenase
(**N**) Malate dehydrogenase
(**O**) Phosphofructokinase
(**P**) Phosphoglycerate kinase
(**Q**) Pyruvate carboxylase
(**R**) Pyruvate dehydrogenase kinase
(**S**) Pyruvate kinase
(**T**) Succinate dehydrogenase

For each property listed below, select the corresponding enzyme.

249. Enzyme inhibited by fructose 2,6-bisphosphate

250. Rate-limiting enzyme of the Krebs citric acid cycle

251. Following phosphorylation by protein kinase A, this enzyme is activated.

252. Rate-limiting enzyme for lipogenesis

Questions 253–256

(**A**) Adrenocorticotropic hormone
(**B**) Aldosterone
(**C**) Angiotensin II
(**D**) Epinephrine
(**E**) Glucagon
(**F**) Insulin
(**G**) Progesterone
(**H**) Renin
(**I**) Thyrotropin
(**J**) TRH
(**K**) Triiodothyronine

For each characteristic listed below, select the corresponding hormone.

253. This hormone is a tripeptide produced by the hypothalamus.

254. Increased levels of this hormone produce the increased pigmentation in patients with Addison's disease.

255. This circulating protein is a protease.

256. The receptor for this hormone possesses protein-tyrosine kinase activity.

Questions 257–259

(**A**) Alkaptonuria
(**B**) Argininemia
(**C**) Hyperammonemia
(**D**) Hyperphenylalaninemia
(**E**) Tyrosinemia

For each enzyme deficiency described below, select the corresponding condition.

257. Dihydropteridine reductase deficiency

258. Tyrosine aminotransferase deficiency

259. Homogentisate oxidase deficiency

Questions 260–263

(**A**) Acetyl-CoA
(**B**) Adenosine triphosphate
(**C**) *S*-Adenosylmethionine
(**D**) CMP-*N*-acetylneuraminate
(**E**) Farnesyl pyrophosphate
(**F**) Guanosine diphosphate-mannose
(**G**) N^5-Methyltetrahydrofolate
(**H**) Myristoyl-CoA
(**I**) Palmitoyl-CoA
(**J**) Phosphoadenosylphosphosulfate
(**K**) UDP-glucose

For each biochemical property listed below, select the corresponding metabolite.

260. The allosteric effector of pyruvate carboxylase that is required for gluconeogenesis

261. An intermediate in cholesterol biosynthesis

262. Failure to form this metabolite is one cause of homocystinuria.

263. A precursor for sphinganine biosynthesis

Questions 264–267

(**A**) E_t
(**B**) k_{cat}
(**C**) k_{cat}/K_m
(**D**) K_d
(**E**) K_m
(**F**) V_{max}
(**G**) Competitive inhibition
(**H**) Noncompetitive inhibition
(**I**) Suicide inhibition

For each description associated with enzyme kinetics or inhibition, select the corresponding constant or type of inhibition.

264. Specificity constant

265. The substrate concentration that gives half-maximal velocity

266. A form of inhibition that gives identical K_m values but reduced V_{max} values (compared to reactions in the absence of inhibitor)

267. A form of inhibition overcome at saturating concentrations of substrate

Questions 268–274

- **(A)** Ascorbic acid
- **(B)** Biotin
- **(C)** Cobalamin (vitamin B_{12})
- **(D)** Dehydrocholesterol
- **(E)** Folate
- **(F)** Niacin
- **(G)** Pantothenate
- **(H)** Protein
- **(I)** Retinol
- **(J)** Tetrahydrofolate
- **(K)** Thiamine
- **(L)** Tocopherol
- **(M)** Vitamin K

Match each of the following situations with the corresponding dietary vitamin/nutrient deficiency listed above.

268. Rickets

269. Impaired dopamine β-hydroxylase activity

270. Kwashiorkor

271. Pellagra

272. Beriberi

273. Night blindness

274. Extended bleeding times

Questions 275–277

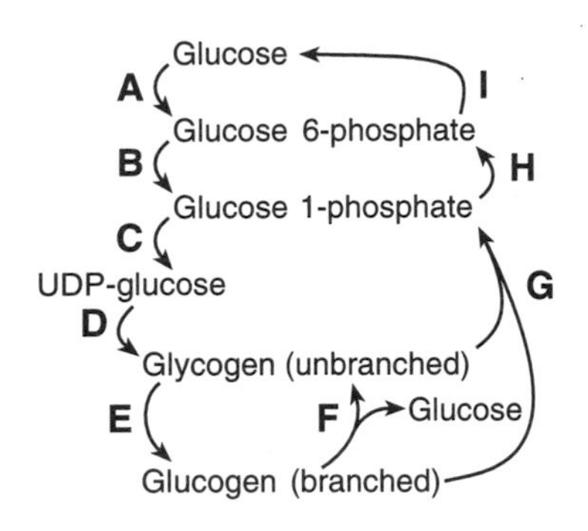

For each description of a deficiency or reaction, select the correct letter on the figure showing the step in glycogen metabolism.

275. A deficiency of this enzyme leads to von Gierke's disease or type I glycogen storage disease.

276. A deficiency of this process in skeletal muscle leads to type X glycogen storage disease, where glycogen has fewer than normal α-1,6-glycosidic linkages.

277. Hydrolysis of inorganic pyrophosphate helps to pull this reaction forward.

Questions 278–282

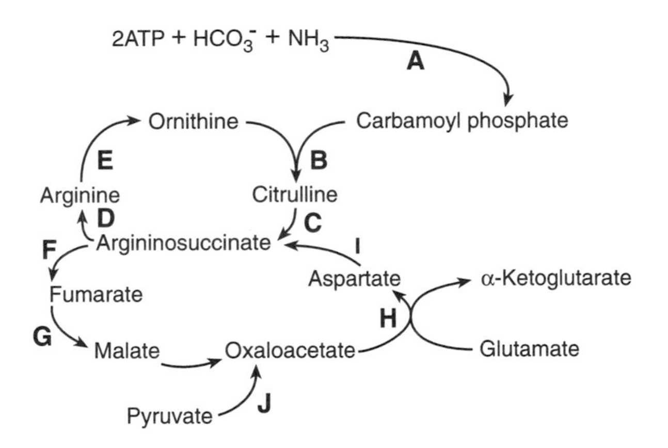

For each defect or deficiency described below, select the correct letter on the figure showing the urea cycle.

278. A defect in this step leads to hyperammonemia and orotic aciduria.

279. This reaction requires pyridoxal phosphate (vitamin B_6) as cofactor.

280. ATP furnishes the chemical energy for this reaction (ATP → AMP + inorganic pyrophosphate).

281. A deficiency of this hydrolase leads to hyperammonemia and hypercitrullinemia.

282. A deficiency of this enzyme leads to hyperammonemia and acidosis.

Questions 283–290

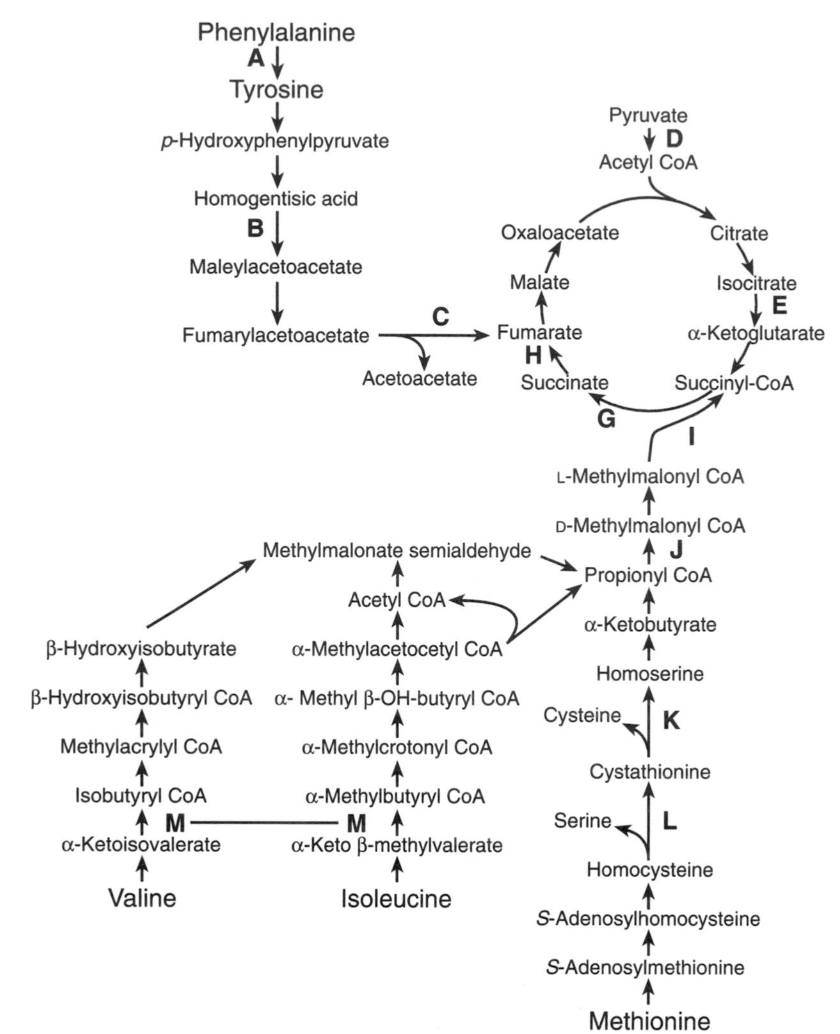

For each deficiency, defect, or reaction listed below, indicate the correct letter on the above figure showing the step involved.

283. Approximately half of the individuals with a deficiency of this enzyme respond to pyridoxine.

284. This reaction is activated by ADP.

285. A deficiency of this enzyme leads to hepatorenal tyrosinemia with the attendant formation of succinylacetone, an abnormal metabolite.

286. A defect in this enzyme results in urine that becomes pigmented black.

287. A failure to synthesize tetrahydrobiopterin leads to a deficiency in this enzyme activity.

288. This reaction requires biotin as an enzyme prosthetic group.

289. This reaction is inhibited by malonate.

290. This reaction is accompanied by substrate-level phosphorylation.

Questions 291–293

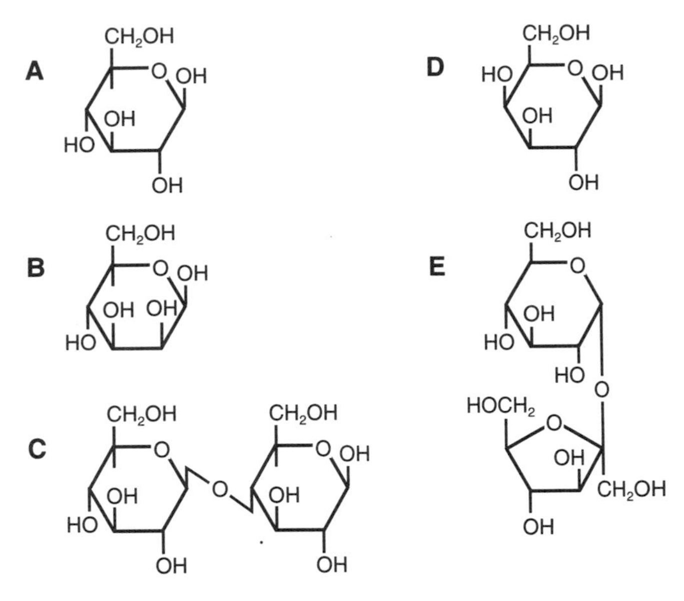

Using the above figure, select the most appropriate compound that possesses the specified features.

291. This sugar is oxidized by alkaline copper and glucose oxidase and accounts for the reducing sugar found in the urine of people with diabetes mellitus.

292. This sugar is oxidized by alkaline copper, not by glucose oxidase, and can occur in the urine of individuals with galactosemia.

293. This sugar is NOT oxidized by alkaline copper nor by glucose oxidase.

Questions 294–296

(**A**) Arginine
(**B**) Cysteine
(**C**) Glutamate
(**D**) Proline
(**E**) Serine
(**F**) Tryptophan

Match each description below with the appropriate amino acid.

294. An α-imino acid

295. Synthesized from α-ketoglutarate in one step

296. Contains an aromatic side chain

Questions 297–299

(**A**) Actin
(**B**) Factor Ia
(**C**) Factor IIa
(**D**) Factor IIIa
(**E**) Lysyl hydroxylase
(**F**) Lysyl oxidase
(**G**) Myosin
(**H**) Prolyl hydroxylase
(**I**) Troponin C

For each of the descriptions below, select the appropriate protein.

297. Functions as the ATPase that sustains muscle contraction

298. A serine protease

299. Functions extracellularly during the cross-linking of tropocollagen molecules to form mature collagen

BIOCHEMISTRY ANSWERS AND DISCUSSION

1—D (Chapter 1) Lysosomes contain a variety of enzymes that catalyze the hydrolysis of sphingolipids. Niemann-Pick disease is a lysosomal disease resulting from a deficiency of the enzyme sphingomyelinase.

2—E (Chapters 7 and 9) Vitamin B_{12} has two functions in humans. It mediates the conversion of methylmalonyl-coenzyme A (CoA) to succinyl-CoA, and it participates in the reaction of methyltetrahydrofolate and homocysteine to form methionine and dihydrofolate.

3—D (Chapter 6) Complex IV (cytochrome oxidase) contains copper and iron. Complexes I, II, and III contain iron only.

4—E (Chapter 15) The sodium/potassium adenosinetriphosphatase (ATPase) is an integral plasma membrane enzyme to which digitalis and its congeners bind. As a result of partial inhibition, intracellular sodium accumulates and is exchanged for extracellular calcium by an antiporter. The elevated calcium concentration augments muscle contraction and plays a therapeutic role in the action of digitalis.

5—B (Chapter 6) The reactive aldehyde of glucose forms a bond with the alcohol of C^4 to form a closed ring. This ring form is in equilibrium with the open-chain form, which possesses the reactive aldehyde that can react with proteins.

6—D (Chapter 18) The total number of calories is 370 (4×30/g protein + 4×40/g carbohydrate + 9×10/g fat). The percentage derived from fat is $90/370 \approx 24.3\%$.

7—A (Chapter 6) In a person with a deficiency of galactose-1-phosphate uridyltransferase, toxic levels of galactose 1-phosphate accumulate, leading to galactosemia, galactosuria, mental retardation, and a variety of other symptoms. With a deficiency of galactokinase, there is no accumulation of charged intracellular metabolites, and the symptoms are less severe. The galactose is excreted in the urine as a reducing sugar.

8—D (Chapter 15) The insulin receptor is synthesized as single chains; a removal of a four-amino-acid segment from each chain, as catalyzed by endoproteases, is required to produce the external (α) chains and internal (β) chains, which are linked by disulfide bonds. Two α-chains are also linked by disulfide bridges to yield the mature receptor, such as β-α-α-β in nature.

9—C (Chapter 18) Plants are unable to synthesize vitamin B_{12}. The diet of strict vegans can be deficient in this substance, which leads to a megaloblastic anemia.

10—C (Chapter 7) Phospholipase C mediates the hydrolysis of phosphatidylinositol 4,5-bisphosphate to yield diglyceride and inositol 1,4,5-trisphosphate.

11—A (Chapter 13) AUG, which corresponds to methionine, is the initiating codon. AUG also codes for internal methionine residues in proteins.

12—E (Chapter 6) Phosphorylation of pyruvate dehydrogenase, as catalyzed by pyruvate dehydrogenase kinase, is inhibitory. A deficiency of pyruvate dehydrogenase phosphatase leaves pyruvate dehydrogenase in its permanently inhibited form.

13—B (Chapter 7) The lipids in bile, including bile salts, solubilize dietary lipids and aid in their absorption. In the absence of emulsifying agents such as bile salts, fats are not absorbed but are excreted in feces.

14—B (Chapter 6) In the absence of insulin, the effects of glucagon predominate. Glucagon results in an induction of the enzymes of gluconeogenesis, which accounts for the observed hyperglycemia. Glucose enters liver cells (but not muscle cells) in a fashion independent of insulin.

15—D (Chapter 12) Actinomycin D binds to DNA and inhibits RNA polymerase activity and transcription. The machinery required for replication can overcome the actions of actinomycin D. RNA capping and splicing are independent of DNA.

16—E (Chapter 6) Many common forms of glucose-6-phosphate dehydrogenase are inhibited by antimalarials such as chloroquine. After inhibition, hemolysis can result.

17—A (Chapter 15) Parathyroid hormone binds to its receptor, and the hormone-receptor complex activates G_s by mediating the exchange of guanosine triphosphate (GTP) for guanosine diphosphate (GDP) bound to G_s-α. G_s-α dissociates from G_s-$\beta\gamma$, and G_s-α stimulates adenylyl cyclase activity.

18—B (Chapter 17) The aldehyde form of glucose reacts nonenzymatically with hemoglobin A (normal adult hemoglobin) to form glycohemoglobin, or hemoglobin A_1.

19—D (Chapter 5) Enterokinase, or enteropeptidase, catalyzes the hydrolysis of specific peptide bonds in inactive trypsinogen to yield active trypsin. Trypsin can then mediate the activation of additional trypsinogen.

20—D (Chapter 6) Cyanide binds to copper ion of cytochrome oxidase, which is found in complex IV of the electron transport chain.

21—A (Chapter 11) The complement of 5' GCAT 3' is 5' ATGC 3'. T base pairs with A, A base pairs with T, C base pairs with G, and G base pairs with C. Note that the two strands are antiparallel, going in opposite directions.

22—D (Chapter 7) Animals, but not plants, are able to synthesize cholesterol. Plant products, such as orange juice, are cholesterol free.

23—D (Chapter 2) An insufficient intake of any essential amino acid leads to negative nitrogen balance. Gelatin, for example, is low in essential amino acids, and it is devoid of tryptophan; a diet with protein derived solely from gelatin would produce negative nitrogen balance. Recall the mnemonic: PVT TIM HALL.

24—D (Chapter 11) Xeroderma pigmentosum is due to a defect in nucleotide-excision repair.

25—D (Chapters 2 and 18) To form the active metabolite, vitamin D undergoes 1-hydroxylation in the liver and 25-hydroxylation in the kidney.

26—D (Chapter 18) Vitamin K functions in the reduced nicotinamide adenine dinucleotide phosphate (NADPH)–dependent carboxylation of specific glutamate residues in calcium-binding proteins, including those of the blood coagulation cascade. This carboxylation is inhibited by warfarin and other vitamin K antagonists.

27—D (Chapter 17) Porphyrin and heme are synthesized from eight molecules of succinyl-CoA and eight molecules of glycine. Ferrous iron is added during the last step of synthesis.

28—B (Chapter 16) Creatine phosphate contains an energy-rich P~N linkage, and creatine phosphokinase catalyzes its reaction with adenosine diphosphate (ADP) to form adenosine triphosphate (ATP) and creatine. Like skeletal and cardiac muscle, the brain also contains creatine phosphokinase.

29—B (Chapter 6) Hypoglycemia and lactate acidosis in newborns can result from impaired gluconeogenesis. This condition could be the result of a deficiency of fructose-1,6-bisphosphatase or glucose-6-phosphatase. For glucagon to exert its hyperglycemic effect requires functional glucose-6-phosphatase, and this rules this deficiency unlikely. A definitive diagnosis of bisphosphatase deficiency requires the measurement of enzyme activity in a liver biopsy.

30—D (Chapter 17) Classic hemophilia (hemophilia A) is due to a deficiency of factor VIII. Christmas disease (hemophilia B) resembles classical hemophilia clinically, but Christmas disease (named after a patient) is due to a deficiency of factor IX.

31—B (Chapter 15) The insulin receptor is made up of two extracellular regions and two intracellular regions. After insulin interacts with the extracellular domain, protein-tyrosine kinase activity of the intracellular domain is activated.

32—C (Chapter 6) Glucose-6-phosphatase, which catalyzes the hydrolysis of glucose 6-phosphate to form glucose and inorganic phosphate, is an example of a hydrolase. A deficiency of this enzyme is responsible for von Gierke's disease, a type I glycogen storage disease that is characterized by hypoglycemia, lactic acidemia, and elevated blood triglycerides.

33—A (Chapter 9) Tetrahydrofolate derivatives function in a variety of one-carbon transfer reactions. *S*-Adenosylmethionine also participates in one-carbon transfer reactions. Because folate is important in normal cellular development, deficiency during pregnancy leads to neural tube defects.

34—E (Chapter 12) Most plasma membrane protein kinases catalyze the phosphorylation of protein-tyrosine residues in acceptor proteins.

35—B (Chapters 1 and 17) The mammalian red blood cell (erythrocyte) lacks a nucleus, mitochondria endoplasmic reticulum, and other membranous organelles. Although the bacteria *Escherichia coli* lacks a nucleus, it is prokaryotic and uses oxidative phosphorylation. Mature T cells, short-lived epithelial cells of the gut, and terminally differentiated neurons have high energy requirements and use oxidative phosphorylation (and also contain nuclei).

36—E (Chapter 10) Uracil is a component of RNA and does not normally occur in DNA. The spontaneous hydrolysis of cytosine in DNA yields uracil. Uracil is recognized by repair enzymes as abnormal, and uracil in DNA is replaced by repair enzymes.

37—C (Chapter 5) Competitive inhibitors do *not* alter the maximum velocity (V_{max}) of enzymes as do noncompetitive and uncompetitive inhibitors. Competitive inhibitors increase the apparent Michaelis constant (K_m) of an enzyme for their substrate analogue. Higher concentrations of substrate are thereby required to achieve half-maximal velocity.

38—C (Chapter 6) Glucose-6-phosphate dehydrogenase, lactonase, and 6-phosphogluconate dehydrogenase are the three enzymes that make up the oxidative portion of the pentose phosphate pathway. Recall that dehydrogenases remove hydrogens or oxidize a substance.

39—D (Chapter 2) The peptide segment -Leu-Val-Phe-Ile-Ala-Gly-Trp- is composed exclusively of hydrophobic amino acids, and it will most effectively interact with the hydrophobic environment of the intramembrane space. -Ala-Pro-Asn-Tyr-Gln-Val-, -Ile-Leu-Asp-Glu-Cys-Trp-, -Leu-His-Ala-Gln-Ser-Phe-, and -Thr-Ala-Val-Lys-Cys-Leu- contain charged and/or uncharged hydrophilic (polar) substituents.

40—C (Chapter 13) Elongation factor-1, elongation factor-2, initiation factors, and release factors are cytosolic proteins that associate transiently with the ribosome.

41—E (Chapter 7) Succinyl-CoA:acetoacetate CoA-transferase deficiency results in ketosis due to the normal synthesis of ketone bodies, but, because of the enzyme defect, acetoacetate cannot be metabolized and builds up in the body fluids.

42—A (Chapter 2) The primary structure of a protein refers to its amino acid sequence.

43—B (Chapter 5) Competitive inhibitors, such as enalapril, bind to the same portion of the active site as their homologous substrate.

44—E (Chapter 2) Manganese is the cofactor of mitochondrial superoxide dismutase, and copper and zinc are cofactors for cytosolic superoxide dismutase.

45—E (Chapter 6) The mature red blood cell uses glycolysis exclusively for ATP production. Of the enzymes listed (α-ketoglutarate dehydrogenase, pyruvate carboxylase, pyruvate dehydrogenase, and pyruvate kinase), only pyruvate kinase occurs in the Embden-Myerhof glycolytic pathway.

46—B (Chapter 12) Histone is unusual because it is not polyadenylated, and its gene lacks introns.

47—C (Chapter 7) C-II is the apolipoprotein that activates lipoprotein lipase. A-I activates lecithin-cholesterol acyltransferase. B-100 is required for recognition of the low-density lipoprotein (LDL) receptor. D is a structural protein, and E is recognized by both LDL and chylomicron remnant receptors.

48—D (Chapter 8) Tryptophan is the only amino acid that is metabolized to form a vitamin in humans. The amount of nicotinic acid formed, however, is minor. Niacin remains an essential human dietary constituent.

49—D (Chapter 6) A deficiency of the enzymes of the glycolytic pathway decreases the rate of pyruvate and lactate production. Once pyruvate is formed, its chief route of catabolism in aerobic cells is via the pyruvate dehydrogenase reaction, which feeds acetyl-CoA into the Krebs tricarboxylic acid cycle. In the absence of pyruvate dehydrogenase, pyruvate is diverted to lactic acid.

50—C (Chapter 6) The inability of galactose infusion to result in hyperglycemia could be due to a defect in galactose metabolism of glucose-6-phosphatase. That fructose infusion leads to hyperglycemia indicates that both glucose-6-phosphatase and aldolase B are functional. A deficiency of galactose-1-phosphate uridyltransferase explains the findings and indicates that the patient has classical galactosemia.

51—A (Chapter 7) Cholesterol, but not cholesteryl esters, occurs in the lipid bilayers of biological membranes. Free fatty acids do not occur in significant amounts in lipid bilayers.

52—E (Chapter 9) The transfer of the methyl group from methyltetrahydrofolate to homocysteine requires vitamin B_{12}. The other reaction in humans that requires vitamin B_{12} is the conversion of D-methylmalonyl-CoA to succinyl-CoA, as catalyzed by a mutase. The reduction of methylenetetrahydrofolate to methyltetrahydrofolate is irreversible. In pernicious anemia, methyltetrahydrofolate accumulates and cannot be metabolized due to a deficiency of vitamin B_{12}. This is the "methyl trap" hypothesis.

53—B (Chapter 11) A + T + G + C = 100%. A = T; G = C. G = 30%, then C = 30%. A + T = 100–60 = 40%. A = T = 20%.

54—E (Chapter 6) Glucokinase, which possesses a high K_m value for glucose (≈10 mmol/L), is found chiefly in the liver and in the β-cells of the pancreas. The liver enzyme catalyzes the phosphorylation of glucose to initiate its metabolism. The pancreatic enzyme also catalyzes this reaction. Because of its high K_m (in contrast to hexokinase with a very low K_m value of approximately 30 μmol/L), glucokinase can function as a sensor for glucose levels and participate in the regulation of insulin secretion.

55—D (Chapter 14) Restriction fragment length polymorphisms are obtained from DNA that differ in the location of restriction enzyme sites due to differences in the sequence of the bases. After digestion with restriction enzymes, the resulting DNA is electrophoresed, and the resulting fragments are resolved by length. In Southern blots (named after the originator), DNA is electrophoresed. In Northern blots, RNA is resolved on the electrophoretic gels.

56—E (Chapter 17) Tissue plasminogen activator is a serine protease that catalyzes the hydrolysis of plasminogen to plasmin. Plasmin, in turn, is a protease that catalyzes the hydrolysis of fibrin.

57—B (Chapter 7) A defect in β-hydroxyacyl-CoA dehydrogenase results in the excretion of substrate byproducts, β-hydroxyacids.

58—C (Chapter 16) Hurler's disease (type I mucopolysaccharidosis) is a lysosomal disease due to a deficiency of L-iduronidase. This disorder is characterized by hepatomegaly, splenomegaly, skeletal deformities, enlarged tongue, and joint stiffness.

59—D (Chapter 8) Phenylalanine is metabolized to acetoacetate and fumarate. Acetoacetate, a ketone body, is ketogenic, and fumarate is glycogenic.

60—C (Chapter 2) The R group of valine is uncharged at pH 8.6. Glutamate and aspartate have negatively charged R groups at pH 7. Except for glutamate, the other amino acids listed (i.e., alanine, asparagine, proline, and tyrosine) lack a charge.

61—C (Chapter 6) The substrate that phosphofructokinase operates on, fructose 6-phosphate, is elevated in hereditary phosphofructokinase deficiency.

62—C (Chapter 3) If a urine sample has a concentration of $H_2PO_4^{1-}$ of 5.0 mmol/L at pH 6.8, then the concentration of HPO_4^{2-} will be 5.0 mmol/L. The pK is the pH at which the acidic ($H_2PO_4^{1-}$) and basic (HPO_4^{2-}) components of a buffer are present at equal concentration.

63—E (Chapter 6) Inhibition of enolase by fluoride was important in elucidating the glycolytic pathway. As a consequence, 2-phosphoglycerate and 3-phosphoglycerate accumulate, but the equilibrium favors the accumulation of 2-phosphoglycerate.

64—E (Chapter 2) The interaction of protein subunits represents the quaternary structure.

65—A (Chapter 5) Hexosaminidase A catalyzes the lysis or cleavage of a covalent bond of its substrate and is a hydrolase. Most of the degradative reactions catalyzed by lysosomal enzymes are hydrolytic reactions.

66—B (Chapter 6) Triose phosphate isomerase catalyzes the conversion of dihydroxyacetone phosphate to glyceraldehyde 3-phosphate.

67—A (Chapter 8) As a result of decreased activity of pyruvate carboxylase, there is decreased conversion of pyruvate to oxaloacetate. Low levels of this four-carbon metabolite lead to decreased levels of aspartate, one donor of nitrogen atoms used in urea biosynthesis.

68—A (Chapter 6) Cytochrome aa_3, or cytochrome oxidase, reacts with oxygen to form metabolic water. The other components of the electron transport chain may form reactive and adventitious intermediates (superoxide, hydrogen peroxide) with oxygen.

69—B (Chapter 7) Long-chain fatty acids are metabolized in peroxisomes to myristoyl-CoA. Metabolites are translocated to the mitochondrion, where further metabolism is impaired. Fatty acyl-CoA is converted enzymatically to C_{14} carnitine derivatives, which leak out of the cell into the circulation. This process leads to a secondary carnitine deficiency within cells. Lower molecular weight metabolites accumulate in medium-chain and short-chain fatty acyl-CoA dehydrogenase deficiency. Long-chain (C_{16} and C_{18}) hydroxyacyl-carnitines accumulate in the plasma in long-chain hydroxyacyl-CoA dehydrogenase deficiency, and unsaturated fatty acyl-CoA accumulates in 2,4-dienoyl-CoA reductase deficiency. Medium-chain fatty acyl-CoA dehydrogenase is the most common inborn error in this group of diseases. C_6 and C_8 compounds accumulate in medium-chain deficiency, and C_4 compounds accumulate in short-chain fatty acid deficiency. The 2-*trans*-4-*cis*-C10:2 accumulates in 2,4-dienoyl-CoA reductase deficiency.

70—C (Chapter 7) Triglyceride synthesis involves the reaction of glycerol phosphate with two moles of acyl-CoA. Phosphate is hydrolyzed from the resulting phosphatidate to yield 1,2-diglyceride, which reacts with acyl-CoA to form triglyceride. Insulin inhibits hormone-sensitive lipase. The reaction of S-adenosylmethionine with phosphatidylethanolamine yields lecithin; the reaction of 1,2-diglyceride with cytidine diphosphate-choline also yields lecithin. The reaction of palmitoyl-CoA with serine is the first step in sphingolipid synthesis.

71—E (Chapter 6) Of the enzymes listed (i.e., cytochrome oxidase, glucose-6-phosphatase, medium-chain fatty acyl-CoA dehydrogenase, pyruvate carboxylase, and pyruvate dehydrogenase), only pyruvate dehydrogenase uses thiamine. It also contains lipoamide and flavin adenine dinucleotide (FAD) as prosthetic groups, but administration of these compounds (FAD as riboflavin) has no significant effect on patients with pyruvate dehydrogenase deficiency.

72—D (Chapter 18) Intrinsic factor, the stomach glycoprotein, is required for the absorption of vitamin B_{12}, or extrinsic factor. The combined intrinsic-extrinsic factor is absorbed in the ileum.

73—C (Chapter 6) Glycogen phosphorylase catalyzes the cleavage of α-1,4-glycosidic bonds by phosphate (phosphorolysis).

74—A (Chapter 17) Bisphosphoglycerate mutase catalyses the conversion of 1,3-bisphosphoglycerate to 2,3-bisphosphoglycerate, an allosteric effector for hemoglobin. In the absence of the allosteric effector, higher than normal concentrations of oxygen are required to saturate hemoglobin. Low oxyhemoglobin levels lead to both increased hemoglobin production and increased hemoglobin levels, as is seen in this patient.

75—C (Chapter 6) Conversion of pyruvate to oxaloacetate and then to glucose is a biosynthetic or anabolic pathway.

76—B (Chapter 7) Unlike patients with defects in β-oxidation, patients with carnitine-palmitoyltransferase I deficiency have normal plasma and tissue carnitine levels and do not have dicarboxylic aciduria or other abnormalities in their urinary organic acid profiles. Avoidance of prolonged fasting is successful in managing these patients. The enzyme defect corresponds to a form that is expressed in liver but not in muscle.

77—D (Chapter 6) Isocitrate dehydrogenase, which catalyzes the conversion of isocitrate to α-ketoglutarate, requires the oxidized form of nicotinamide adenine dinucleotide (NAD^+) as an oxidant. Unlike pyruvate dehydrogenase and α-ketoglutarate dehydrogenase, this enzyme does not require thiamine, lipoate, or CoA. Biotin is a cofactor of ATP-dependent carboxylation reactions, and not decarboxylation reactions.

78—C (Chapter 6) The electron transport chain and the ATP synthase of oxidative phosphorylation are located in the inner mitochondrial membrane.

79—B (Chapter 6) 5'-Adenosine monophosphate (AMP) serves as a signal that ATP levels are submaximal. 5'-AMP stimulates glycogenolysis by activating glycogen phosphorylase. 5'-AMP inhibits gluconeogenesis by inhibiting fructose-1,6-bisphosphatase.

80—C (Chapter 6) The primary function of the pentose phosphate pathway is to generate NADPH for reductive biosynthesis, especially lipid and steroid biosynthesis. A secondary function is to provide ribose 5-phosphate for nucleotide biosynthesis.

81—E (Chapter 6) High carbohydrate intake provides ample substrate for pyruvate dehydrogenase. Pyruvate cannot be metabolized normally with attendant lactic acidosis, therefore, patients are placed on low carbohydrate diets. In contrast, gluconeogenesis is impaired in defects involving fructose-1,6-bisphosphatase, glucose-6-phosphatase, phosphoenolpyruvate carboxykinase, and pyruvate dehydrogenase, and high carbohydrate diets are used.

82—D (Chapter 7) The first step in the β-oxidation of fatty acids within the mitochondrion is catalyzed by acyl-CoA dehydrogenase; this enzyme contains FAD (from riboflavin). Acyl-CoA dehydrogenase is found in the inner mitochondrial membrane, and this dehydrogenase passes its electrons to coenzyme Q. As a result of this process, 1.5 moles of ATP will result from oxidative phosphorylation.

83—C (Chapter 7) Only liver mitochondria contain the enzymes necessary for ketone body biosynthesis, including 3-hydroxy-3-methylglutaryl-CoA (HMG-CoA) lyase.

84—D (Chapter 6) Insulin action leads to dephosphorylation of glycogen phosphorylase and a concomitant decrease in phosphorylase activity via a complex protein kinase cascade. Branching enzyme and uridine diphosphate (UDP)-glucose pyrophosphorylase are not regulatory enzymes. Phosphofructokinase and protein kinase A activities are decreased as a result of insulin action.

85—C (Chapter 5) The lock-and-key hypothesis states that specific amino acids within the active site of an enzyme create a unique environment for binding substrate molecules. Consequently, the enzyme increases the local concentrations of substrates, orients them in the proper manner, and/or puts sufficient strain on specific components to allow the specific reaction to take place at a higher rate. Enzymes do not alter free energies or the ultimate equilibrium state of a reaction.

86—E (Chapter 5) Trypsin catalyzes the hydrolysis of substrate peptides on the carboxyterminal side of lysine and arginine (basic amino acids).

87—E (Chapter 6) Lactic acidosis occurs when pyruvate dehydrogenase is less active and pyruvate catabolism is diminished. After phosphorylation and inhibition of pyruvate dehydrogenase by its kinase, a failure to dephosphorylate pyruvate dehydrogenase results in a chronically inactive enzyme. This situation occurs when there is deficient pyruvate dehydrogenase phosphatase activity.

88—E (Chapter 9) Methylmalonyl-CoA is converted to succinyl-CoA by a vitamin B_{12}–dependent mutase. In vitamin B_{12} deficiency, there is an accumulation of methylmalonyl-CoA. Its precursor, propionyl-CoA, also builds up, and a portion reacts with oxaloacetate to produce methylcitrate, an abnormal metabolite.

89—C (Chapter 9) The rate-limiting step in fatty acid biosynthesis is catalyzed by acetyl-CoA carboxylase, and this enzyme is induced by insulin as part of the lipogenic response. Glucose-6-phosphate dehydrogenase is also induced by insulin.

90—B (Chapter 7) Gaucher's disease is due to the inability to degrade glucosylceramide by β-glucosidase, a lysosomal enzyme. The other activities mentioned are enzymes in the pathway of glucosylceramide biosynthesis.

91—B (Chapter 7) Elevated levels of LDL, or "bad cholesterol," are a risk factor for cardiovascular disease. LDL transports cholesterol from liver to extrahepatic tissues. High-density lipoprotein (HDL) transports cholesterol from extrahepatic cells to the liver. Elevated levels of HDL are associated with decreased incidence of cardiovascular disease. Hypercholesterolemia (>200 mg/dL) is also a risk factor for atherosclerosis and cardiovascular disease.

92—D (Chapter 8) Carbamoyl-phosphate synthetase-I is the rate-limiting enzyme for urea biosynthesis and requires *N*-acetylglutamate as an allosteric activator. The other enzymes listed (i.e., arginase, argininosuccinate lyase, argininosuccinate synthetase, and ornithine transcarbamoylase) participate in urea synthesis.

93—A (Chapter 15) ATP donates a phosphoryl group to an aspartate to form an energized protein. Protons are pumped from parietal cells, and a less energy-rich protein is formed. The pump translocates potassium to the cell interior, and the aspartylphosphate is cleaved by hydrolysis.

94—A (Chapter 7) Of the fatty acids listed (i.e., linoleate, oleate, palmitate, palmitoleate, and stearate), only linoleate is essential. Linolenic acid is the other essential fatty acid.

95—D (Chapter 16) Phosphoadenosylphosphosulfate is the universal biological donor of sulfate. Inorganic sulfate lacks the chemical free energy to react with an acceptor.

96—D (Chapter 9) Hyperammonemia without acidosis is generally due to a defect that occurs in the urea cycle. With ornithine transcarbamoylase deficiency, mitochondrial carbamoyl phosphate cannot react with ornithine. It escapes from the mitochondrion and reacts with aspartate in the cytosol in the pathway for pyrimidine synthesis. The product that accumulates is orotic acid. Infants with uridine monophosphate deficiency have hereditary orotic aciduria, and they exhibit signs and symptoms of a megaloblastic anemia.

97—B (Chapter 15) Vitamin B_6 is a cofactor for all of the enzymes listed (i.e., cystathionine synthase, glutamate decarboxylase, glutamate-oxaloacetate transaminase, glycogen phosphorylase, serine dehydratase). However, glutamate decarboxylase catalyzes the biosynthesis of γ-aminobutyrate (GABA), the chief inhibitory neurotransmitter in the brain, and a deficiency in the synthesis of GABA is postulated to lead to seizures.

98—D (Chapter 6) Glucagon interacts with its seven-transmembrane segment plasma membrane receptor to activate adenylyl cyclase (via a stimulatory 6 protein), which leads to increased formation of cyclic AMP.

99—B (Chapter 6) AMP activates phosphofructokinase allosterically to increase the rate of glycolysis. A decrease in the concentration of ATP during maximal exertion leads to an increase in ADP and AMP. Increased reduced nicotinamide adenine dinucleotide (NADH) drives the pyruvate to lactate conversion.

100—A (Chapter 8) Fumarylacetoacetate hydrolase is the last enzyme in the catabolic pathway for phenylalanine and tyrosine. A deficiency leads to the excretion of maleylacetoacetic acid, a precursor of fumarylacetoacetate, in the urine.

101—D (Chapter 7) Methylmalonyl-CoA mutase is the enzyme that catalyzes the conversion of substrate to succinyl-CoA in a vitamin B_{12}–dependent reaction.

102—B (Chapter 2) The aldehyde group of glucose can react with amino groups nonenzymatically, and the side chain of lysine contributes groups with which glucose can react.

103—C (Chapter 6) Hypoglycemia and lactic acidosis can result from a defect in gluconeogenesis. The hyperglycemic response to glucagon requires functional glucose-6-phosphatase for the sugar to leave the liver, and the lack of a hyperglycemic response indicates that the deficiency involves this enzyme.

104—B (Chapter 2) Two glutamate residues on two β-chains of hemoglobin bear a total of two negative charges (one per glutamate); to increase the charge by four units requires replacement of each glutamate by a residue bearing a positive charge. Arginine is a residue that bears a positive charge at physiologic pH.

105—D (Chapter 6) Fructose 2,6-bisphosphate, which is the most important physiologic regulator of the Embden-Meyerhof glycolytic pathway, activates liver phosphofructokinase. Fructose 2,6-bisphosphate also *inhibits* hepatic fructose 1,6-bisphosphatase.

106—B (Chapter 7) Acyl carrier protein serves as a swinging arm in transporting acyl derivatives from active site to active site of the fatty acid synthase complex. NAD^+ and the mitochondrion participate in fatty acid oxidation. Acetyl-CoA and water participate in both fatty acid synthesis and oxidation.

107—A (Chapter 5) The plot is a Lineweaver-Burke double reciprocal plot, and larger values of x and y correspond to smaller values of the substrate concentration and the velocity. The V_{max} for the patient's enzyme equals that of the control, but the K_m of the patient's enzyme is larger than that of the control (the x value is larger). It requires a higher concentration of biotin to produce a velocity equal to one-half of the V_{max}.

108—C (Chapter 18) The increased $NADH/NAD^+$ ratio favors the conversion of pyruvate to lactate, thereby decreasing substrate sources for gluconeogenesis.

109—C (Chapter 18) The most common cause of a hypochromic microcytic anemia is iron deficiency. The anemia is treated with oral ferrous sulfate and with vitamin C (ascorbate). Ascorbate is a reductant that maintains the iron in the reduced, or ferrous, form; reduced iron is the absorbable form.

110—A (Chapter 2) The R-group of arginine, like lysine, bears a positive charge at physiologic pH. Substitutions of arginine for lysine are considered to be conservative.

111—B (Chapter 7) Cyclooxygenase, an enzyme in the pathway for prostaglandin and prostacyclin synthesis, is irreversibly inhibited by aspirin in a reaction that involves acetylation of an essential serine residue.

112—C (Chapter 6) *Streptococcus mutans* is a bacterium found in the oral cavity that converts glucose to lactic acid by glycolysis. This bacterium lacks the Krebs cycle and oxidative phosphorylation activity; it generates its energy (ATP) by glycolysis and must convert pyruvate to lactate to regenerate NAD^+.

113—C (Chapter 6) Cyanide binds to methemoglobin produced by the action of amyl nitrate; however, this decreases the oxygen-binding capacity of hemoglobin to spare the essential oxidative phosphorylation.

114—D (Chapter 8) The chief nitrogen metabolite in humans is urea. The ammonium ion is toxic.

115—A (Chapter 9) Phosphoribosyl pyrophosphate aminotransferase, the rate-limiting enzyme in purine biosynthesis, is inhibited by purine monophosphates and the allopurinol metabolite, allopurinol ribonucleoside monophosphate.

116—C (Chapter 11) DNA topoisomerase II cleaves two strands of the DNA duplex to alter its supercoiling. DNA topoisomerase I cleaves one strand of the DNA duplex to alter its supercoiling.

117—B (Chapter 12) Myc is a transcription factor that is overexpressed in some cancers and is a potential target for cancer chemotherapeutic agents.

118—C (Chapter 8) A deficiency of any essential amino acid can lead to inhibition of protein synthesis. Of the amino acids listed (alanine, glycine, isoleucine, serine, tyrosine), isoleucine is the essential amino acid (PVT TIM HALL).

119—A (Chapter 7) HMG-CoA reductase is the target for the statin compounds that are used in the treatment of hypercholesterolemia. After enzyme inhibition, cholesterol levels decline, and HMG-CoA reductase and LDL receptor are thereby induced. The decline in cholesterol is related to increased transport of cholesterol in cells via LDL receptor-mediated endocytosis. The other enzymes listed (i.e., HMG-CoA synthase, isopentenyl pyrophosphate isomerase, mevalonate kinase, and phosphomevalonate kinase) are in the pathway for cholesterol biosynthesis.

120—C (Chapter 15) Protein kinase C, which is activated physiologically by diglyceride, is also activated by phorbol esters.

121—C (Chapter 16) The three enzymes in human metabolism that use ascorbate (vitamin C) as a reactant include prolyl hydroxylase, lysyl hydroxylase (both involved in collagen synthesis), and dopamine β-hydroxylase (involved in catecholamine biosynthesis).

122—D (Chapter 12) Rifamycin inhibits the initiation of bacterial RNA synthesis by interacting with the β-subunit of the $\alpha_2\beta\beta'\sigma$-initiation complex.

123—C (Chapter 12) Azidothymidine is a deoxynucleotide whose therapeutic mode of action is to inhibit human immunodeficiency virus reverse transcriptase.

124—B (Chapter 15) Nitric oxide binds to the heme of soluble guanylate cyclase and stimulates cyclic guanosine monophosphate formation.

125—E (Chapter 15) 5-Hydroxyindolacetate is derived from tryptophan via serotonin.

126—D (Chapter 6) Glucagon interacts with its seven-transmembrane segment receptor, which activates G_s, which in turn activates adenylate cyclase, leading to an increase in the level of cyclic AMP. This leads to the activation of protein kinase A. The enzyme protein kinase A then catalyzes the phosphorylation of fructose-6-phosphate-2-kinase/fructose-2,6-bisphosphatase and leads to decreased levels of fructose 2,6-bisphosphate. Protein kinase A also catalyzes the phosphorylation and inactivation of hepatic pyruvate kinase.

127—A (Chapter 15) Of the low-molecular-weight neurotransmitters, acetylcholine is the only one that is inactivated enzymatically by hydrolysis.

128—E (Chapter 9) Methotrexate is a folate analogue that is a competitive inhibitor of dihydrofolate reductase.

129—B (Chapter 7) A deficiency of a carnitine transporter leads to a progressive cardiomyopathy and skeletal muscle weakness that begins at 2 to 4 years of age.

130—D (Chapter 17) The mutation that is responsible for sickle cell anemia is a point mutation. It converts a codon in the β-chain from a glutamate residue to a valine residue.

131—C (Chapter 8) This is a rare condition and is due to the deficiency of isovaleryl CoA dehydrogenase, which catalyzes the conversion of isovaleric acid to 3-methylcrotonoate in the leucine degradative pathway.

132—D (Chapter 13) Tryptophan and methionine are the only two amino acids with a single codon.

133—A (Chapter 1) The lysosomes contain a variety of hydrolase enzymes that catalyze catabolic reactions, or the breakdown of large molecules to small molecules.

134—B (Chapter 1) Cystic fibrosis is due to a defect in chloride transport; the gene product is cystic fibrosis transmembrane conductance regulator.

135—D (Chapter 18) Magnesium sulfate injections lead to less excitable neuronal cells and help prevent seizures, which can accompany eclampsia.

136—E (Chapter 15) Myasthenia gravis is an autoimmune disease with antibodies formed against the nicotinic cholinergic receptors found in skeletal muscle.

137—E (Chapter 6) Pyruvate carboxylase requires acetyl-CoA as an allosteric effector for the expression of its activity. Pyruvate carboxylase catalyzes the first reaction in the conversion of pyruvate to glucose.

138—B (Chapter 7) ω-Oxidation, which involves NADPH, oxygen, and cytochrome P-450, occurs in the endoplasmic reticulum. Free fatty acids, not thioesters, are the substrates. Under physiologic conditions, the decarboxylate is thioesterified in an ATP-dependent process, and the resulting compound can be metabolized by β-oxidation.

139—C (Chapter 5) Lyases are enzymes that convert one reactant into two products that contain one more double bond than occurs in the reactant.

140—A (Chapter 5) Elevated serum acid phosphatase in adult males is nearly pathognomonic for cancer of the prostate.

141—B (Chapter 5) Creatine phosphokinase, an enzyme found in cardiac and skeletal muscle, is characteristically elevated within 12 hours of a myocardial infarction, or heart attack.

142—C (Chapter 6) The yield of ATP per mole of acetyl-CoA oxidized to carbon dioxide and water by the Krebs cycle and oxidative phosphorylation is 3 moles of NADH (isocitrate dehydrogenase, α-ketoglutarate dehydrogenase, malate dehydrogenase) for a total of 7.5 moles ATP, 1 mole reduced flavin adenine dinucleotide ($FADH_2$) (succinate dehydrogenase) for 1.5 moles ATP, and 1 ATP equivalent (GTP from the succinate thiokinase reaction) yields a total of $7.5 + 1.5 + 1 = 10$ ATP per mole of acetyl-CoA oxidized.

143—A (Chapter 11) Histone H1 occurs in the linker region of nucleosomes; H2A, H2B, H3, and H4 occur in the nucleosome.

144—A (Chapter 8) Pyruvate, propionate, and 3-hydroxyisovalerate carboxylase contain biotin as a prosthetic group.

145—B (Chapter 2) The MB2 isozyme with its extra lysine (when compared with MB1 or the background isozyme) migrates anodally (toward the negatively charged pole) when compared with MB1.

146—C (Chapter 8) The three amino acids listed (i.e., leucine, isoleucine, and valine) are branched-chain amino acids. A deficiency of branched-chain ketoacid dehydrogenase results in maple syrup urine disease.

147—E (Chapters 6 and 8) A deficiency of mitochondrial complex III results in acidosis as a result of pyruvic acid and lactic acid build-up. Arginase deficiency and argininosuccinate synthetase deficiency are not associated with acidosis. Maple syrup urine disease and long-chain acyl-CoA deficiency result in organic aciduria.

148—A (Chapter 15) Agents such as diisopropylfluorophosphate and Sarin react with an active-site serine of acetylcholinesterase to produce irreversible inhibition.

149—B (Chapter 7) Malonyl-CoA, the main substrate for fatty acid biosynthesis, inhibits carnitine acyltransferase I, a cytosolic enzyme that is required for the translocation of fatty acyl groups into the mitochondrion from the cytosol.

150—C (Chapters 6 and 8) Lipoamide dehydrogenase occurs in pyruvate dehydrogenase (accounting for increased lactic acid), α-ketoglutarate dehydrogenase (accounting for increased α-ketoglutaric acid), and branched-chain ketoacid dehydrogenase (accounting for increased α-ketoisocaproic, α-keto-β-methylvaleric, and α-ketoisovaleric aciduria).

151—C (Chapter 7) An elevation of LDL leads to hypercholesterolemia with little or no elevation of triglyceride. The elevation of chylomicrons or very-low-density lipoproteins (VLDL) leads to an elevation of triglyceride. The elevation of LDL or intermediate-density lipoprotein leads to elevations of both cholesterol and triglyceride.

152—E (Chapter 15) Tremor and shuffling gain in an elderly person are commonly due to Parkinson's disease. This disorder is related to the loss of neurons.

153—B (Chapter 11) Duplex DNA is antiparallel, and DNA polymerases catalyze only 5' to 3' polymerization reactions. Complex replication machinery is required to produce antiparallel DNA, and it requires activities for both leading and lagging strand biosynthesis.

154—B (Chapter 11) DNA ligase catalyzes the linking together of Okazaki fragments or fragments of DNA that are synthesized discontinuously as the lagging strand.

155—B (Chapter 12) Unlike DNA polymerases, RNA polymerases can initiate the synthesis of their polynucleotide product without a primer.

156—C (Chapter 7) The β-oxidation of fatty acids provides acetyl-CoA, the building block of the ketone bodies.

157—D (Chapter 12) RNA polymerase II is the enzyme that catalyzes the formation of heterogeneous nuclear RNA (hnRNA)/messenger RNA (mRNA) in humans.

158—C (Chapter 5) Elevations of serum γ-glutamyl transpeptidase and aspartate aminotransferase can be used to monitor abnormal liver function.

159—A (Chapter 13) Aminoacyl-transfer RNA (tRNA) ligases catalyze the linking of their corresponding amino acid and tRNA in an ATP–dependent process [ATP → AMP + inorganic pyrophosphate (PP_i)].

160—A (Chapter 17) Blood clotting factor IX is synthesized in the liver and is secreted into the circulation to become functional. Mutations that prevent proteolytic processing of the signal sequence that is required for protein secretion can result in a nonfunctional protein.

161—C (Chapter 6) The glucagon receptor contains seven transmembrane segments. The activated receptor activates adenylyl cyclase.

162—E (Chapter 15) The receptor for thyroxine is a protein that binds to DNA and alters the transcription of target genes.

163—E (Chapter 9) Fluorouracil is a prodrug that is converted to deoxyfluorouridine triphosphate, which is a suicide inhibitor of thymidylate synthase.

164—D (Chapter 4) The proton/potassium-ATPase is a class P ATPase that secretes acid across plasma membranes. The P refers to the phospho-aspartyl intermediate that occurs during the translocation process.

165—D (Chapter 18) People with polycythemia vera (excess red and white blood cells) can be treated with radioactive phosphate.

166—C (Chapter 7) The increased availability of fatty acids as a result of enhanced activity of hormone-sensitive lipase drives hepatic triglyceride and VLDL formation.

167—D (Chapters 6 and 18) The conversion of 1 mole of ethanol to acetic acid results in the generation of 2 moles of NADH (from alcohol dehydrogenase and acetaldehyde dehydrogenase). The conversion of acetate to acetyl-CoA results in the liberation of 3 additional moles of NADH (isocitrate dehydrogenase, α-ketoglutarate dehydrogenase, and malate dehydrogenase) and 1 mole of $FADH_2$. The net synthesis of NADH and $FADH_2$ is $2 + 3 + 1 = 6$.

168—C (Chapter 6) Glycogen synthase is induced by insulin and promotes glycogenesis.

169—B (Chapter 2) Collagen is the most abundant protein in the human body.

170—A (Chapter 7) Acetyl-CoA carboxylase, the rate-limiting enzyme in fatty acid biosynthesis, is located in the cytosol.

171—E (Chapter 7) The functional form of the LDL receptor, which is deficient in familial hypercholesterolemia, is located in the plasma membrane. After the recognition and binding of the receptor with ligands on the lipoprotein particle, the coated pit is taken up by endocytosis to form a coated vesicle. The contents of the coated vesicles are delivered to lysosomes, and the receptor is recycled. The lysosomes catalyze the hydrolysis of cholesteryl ester to produce free cholesterol, which is released to the cytosol. The lysosomes also catalyze the hydrolysis of the proteins derived from the coated vesicles.

172—E (Chapter 18) Wernicke's encephalopathy is due to a deficiency of thiamine. The syndrome consists of ataxia, tremors, and disturbances in ocular motility.

173—C (Chapter 8) The intake of cysteine, a sulfur-containing amino acid, decreases the requirement for methionine, also a sulfur-containing amino acid.

174—B (Chapter 6) Two molecules of ATP are consumed (hexokinase and phosphofructokinase), and four molecules of ATP are produced (two at the phosphoglycerate kinase reaction and two at the pyruvate kinase reaction). The net synthesis of ATP is 4 – 2 = 2.

175—E (Chapter 3) The hydrogen ion concentration equals 10^{-pH}. The difference between acidotic and normal conditions is $10^{-7.3}$ minus $10^{-7.4}$ or 5×10^{-8} mol/L minus 4×10^{-8} mol/L . Therefore, acid is increased by 1×10^{-8} mol/L .

176—B (Chapter 2) Amino acid sequence and primary protein sequence are synonymous. The secondary structure represents the folding of the primary amino acid sequences into structures (α-helix and β-pleated sheets). The organization of these structures relative to each other is tertiary structure, and the assembly of subunits is quaternary structure.

177—C (Chapter 6) Leucine is catabolized to acetoacetyl-CoA and acetyl-CoA and is entirely ketogenic. Alanine, lactate, serine, and threonine are glycogenic.

178—B (Chapter 11) Mitochondrial DNA is circular and lacks discrete ends. Bacterial DNA, such as that of *E. coli,* is also circular.

179—B (Chapter 11) Mutations of the *p53* gene occur in approximately half of all human cancers. *RB1* is also a tumor suppressor gene, but mutations occur in only a small percentage of human cancers.

180—B (Chapter 5) The K_m of the patient's enzyme is greater than that of the control subject, and the V_{max} of the patient's enzyme is less than that of the control subject. Recall that the y-intercept is $1/V_{max}$ and that the x-intercept is $-1K_m$.

181—C (Chapter 15) Glutamate is the predominant excitatory neurotransmitter in the human brain.

182—D (Chapter 17) The catalytic triad of serine proteases consists of histidine, serine, and glutamate.

183—C (Chapter 5) There are three possible enzyme forms for creatine phosphokinase: MM (found in skeletal muscle), MB (found in cardiac muscle), and BB (found in the brain).

184—D (Chapter 5) γ-Glutamyltranspeptidase is the most sensitive indicator of early alcoholic liver disease.

185—B (Chapter 6) Catalase catalyzes the conversion of hydrogen peroxide to oxygen and water.

186—C (Chapter 6) The reaction catalyzed by phosphoglycerate kinase is reversible and participates in both glycolysis and gluconeogenesis.

187—C (Chapter 6) The conversion of the glucosyl residue from glycogen to glucose 6-phosphate does not involve the expenditure of ATP. The phosphofructokinase reaction uses 1 molecule ATP, and the phosphoglycerate kinase reaction generates two ATP molecules for each of the two triose molecules; the pyruvate kinase reaction also generates two ATP molecules for each of the two triose molecules derived from the six carbon glucosyl residue. Thus, the net formation of ATP is –1 + 2 + 2 = +3.

188—D (Chapter 4) Although the number of high-energy bonds is the same in the reactants and products of this reaction, the large standard free energy of hydrolysis of phosphoenolpyruvate makes the reaction proceed from right to left *in vivo* and *in vitro.*

189—D (Chapter 6) GLUT4 translocase, which occurs in skeletal muscle and adipose tissue, is recruited to the plasma membrane in response to insulin.

190—B (Chapter 16) Every third amino acid in the triple helical region of collagen is glycine, and mutations that involve the mutation of glycine to any other residue in these positions can produce osteogenesis imperfecta.

191—B (Chapter 17) There is an increase in bilirubin due to obstruction of the bile ducts, which is caused by gallstones. Urobilinogen, urobilin, and stercobilin are formed in the gut. Bile acids are synthesized in liver but are not pigmented, but chronic blockade of the bile ducts can lead to pruritus (itching), which is due to elevated levels of circulating bile salts.

192—D (Chapter 6) All of the enzymes of the Krebs tricarboxylic acid cycle are hydrophilic and occur within the mitochondrial matrix, except for succinate dehydrogenase, which is an integral membrane protein.

193—E (Chapter 8) Glutamate dehydrogenase is the predominant enzyme that catalyzes the formation of ammonium ions.

194—A (Chapter 6) Acetyl-CoA is the stoichiometric substrate that is consumed by the action of the Krebs cycle. Oxaloacetate is the regenerating substrate that functions catalytically.

195—D (Chapter 6) In pyruvate kinase deficiency, there is an accumulation of each of the glycolytic intermediates up to phosphoenolpyruvate. There is also an increased conversion of 1,3-bisphosphoglycerate to 2,3-bisphosphoglycerate.

196—B (Chapter 6) Branching enzyme deficiency results in normal amounts of liver glycogen but with very long outer branches. Hypoglycemia is not part of the clinical picture of this disorder because the enzymes that catalyze glycogen breakdown and glucose release are functional.

197—B (Chapter 9) The two carbon atoms and one nitrogen atom of glycine are incorporated into the purine ring system in an ATP-dependent process.

198—C (Chapter 12) The GT sequence at intron boundaries is invariant, and such mutations interfere with normal splicing. The person who has the sequence ...TG/CT... at the corresponding site has Tay-Sachs disease, which results from diminished expression of hexosaminidase A.

199—E (Chapter 13) The leader sequence of methylmalonyl-CoA mutase targets the enzyme to the mitochondrion, and mutations of the leader sequence result in a failure of the enzyme to be translocated into the mitochondrion.

200—E (Chapter 4) Superoxide dismutase catalyzes the conversion of superoxide and protons to form hydrogen peroxide and oxygen.

201—D (Chapter 12) Promoters are DNA sequences that determine the start site for RNA synthesis.

202—E (Chapter 13) Diphtheria toxin and cholera toxin catalyze the reaction of their target proteins with NAD^+ to form an ADP-ribosylated protein and nicotinamide.

203—A (Chapter 13) The signal recognition particle receptor is found in the endoplasmic reticulum. The targeted proteins are inserted into membranes or are secreted from the cell.

204—D (Chapter 13) The enzymes within the lysosome contain *N*-linked oligosaccharides with terminal mannose 6-phosphate. A deficiency of *N*-acetylglucosaminyl-1-phosphotransferase produces I-cell disease.

205—A (Chapter 15) The nitrogen of nitric oxide (NO) is derived directly from arginine. The other reactants include oxygen and NADPH, and the products include nitric oxide, citrulline, water, and the oxidized form of nicotinamide adenine dinucleotide phosphate.

206—C (Chapter 15) The SH2 domains of proteins such as Grb2 enable them to bind to protein-tyrosine phosphates.

207—E (Chapter 15) Extracellular signal regulated kinase (ERK), or mitogen activated protein kinase (MAP kinase), is unusual in that it is activated by phosphorylation of both threonine and tyrosine as catalyzed by MEK (*M*AP kinase or *E*RK kinase *K*inase).

208—D (Chapter 15) Ras is an oncogene that is active when it is bound to GTP. The proto-oncogene Ras can readily hydrolyze GTP to GDP and inorganic phosphate (P_i) in the presence of a GTPase activating protein (GAP). Oncogenic Ras has mutations such that it has diminished GTPase activity even in the presence of GAP.

209—D (Chapter 6) Although diabetes mellitus is the leading cause of blindness in the United States, the leading cause of blindness worldwide is caused by vitamin A deficiency.

210—C (Chapter 2) The amino terminus and the R-group of arginine are positively charged and the carboxy terminus and R-group of aspartate are negatively charged, for a net charge of zero. The R-group of histidine is approximately 6 and, at pH 7, approximately 90% of the histidine residues of a group of molecules are unchanged.

211—E (Chapter 6) The cofactor for transketolase, which mediates the transfer of two-carbon fragments, is thiamine pyrophosphate.

212—D (Chapter 6) The compounds fructose and galactose are not phosphorylated and, thus, they do not accumulate in cells. Both sugars are excreted in the urine. Failure to metabolize them is not of great consequence in affected individuals.

213—C (Chapter 6) The α-ketoglutarate dehydrogenase reaction is unidirectional and forces the cycle to proceed in the clockwise direction as usually written.

214—B (Chapter 6) Glycogen debranching enzyme catalyzes a hydrolysis reaction. A deficiency of this enzyme leads to type III glycogen storage disease (Cori's disease).

215—B (Chapter 2) Introduction of a proline residue into an α-helix causes disruption of the helix; proline cannot contribute to α-helix formation because the nitrogen of proline in peptide linkage is not available for hydrogen bond formation, and rotation about the α carbon is constrained.

216—D (Chapter 7) The chief metabolites of cholesterol are bile salts, including deoxycholate, a negatively charged substance. Bile salts are secreted in bile, with more than 95% being reabsorbed in the terminal ileum (enterohepatic circulation). When bile salts bind to nonabsorbable resins, they are excreted in feces, which leads to reduced body stores of cholesterol.

217—A (Chapter 6) Biotin is *not* a cofactor of α-ketoglutarate dehydrogenase; rather, it is a cofactor for ATP-dependent carboxylation reactions. CoA, FAD, lipoic acid, and NAD^+ covalently attached to enzymes and thiamine pyrophosphate are the five cofactors associated with α-ketoglutarate dehydrogenase, pyruvate dehydrogenase, and α-ketobutyrate dehydrogenase.

218—D (Chapter 6) Muscle phosphorylase deficiency is not a cause of hypoglycemia because muscle does not serve as a direct source of circulating glucose.

219—B (Chapter 11) Cytosine arabinoside can be incorporated into DNA (from the corresponding triphosphate), with its presence in DNA leading to instability. Cytosine arabinoside is an antimetabolite that is used for treating acute lymphocytic and acute myelogenous leukemias.

220—C (Chapter 5) Pancreatic enzyme replacement includes amylase, lipase, and trypsin; pepsin is a stomach enzyme.

221—D (Chapter 13) Streptomycin produces mistakes in bacterial translation (e.g., incorporation of leucine in response to a phenylalanine codon). Streptomycin also inhibits the initiation of bacterial protein synthesis.

222—A (Chapter 13) Chloramphenicol inhibits bacterial peptidyltransferase activity. The toxicity of chloramphenicol in humans is postulated to be due to inhibition of the mitochondrial protein synthesis machinery at the peptidyltransferase step.

223—C (Chapter 13) Puromycin is a phenylalanine-tRNA analogue that results in premature chain termination of the nascent polypeptide chain in both prokaryotes and eukaryotes.

224—A (Chapter 9) Adenosine deaminase deficiency impairs nucleotide metabolism in immune cells, leading to severe immune deficiency.

225—D (Chapter 9) Hypoxanthine-guanine phosphoribosyltransferase is deficient in patients with Lesch-Nyhan disease. This disorder is associated with hyperuricemia. Lesch-Nyhan disease and its severity provided the first indication of the importance of the salvage pathway of purines in humans.

226—E (Chapter 7) VLDL is synthesized in the liver and is responsible for the transport of triglyceride from the liver to other tissues.

227—A (Chapter 7) Hormone-sensitive lipase in adipose tissue catalyzes the hydrolysis of triglyceride to fatty acids and glycerol. The free fatty acids are transported as a complex with albumin.

228—C (Chapter 12) mRNA is derived from hnRNA by several processing reactions. Processing includes capping the 5' end with 7-methylguanosine, the adding of up to 200 adenylate residues at the 3' end, and excising introns by splicing reactions.

229—E (Chapter 13) All tRNAs end with CCA at the 3' end. The amino acid forms a covalent high-energy bond with the A residue at the 3' end. A single enzyme exists for replacing all three nucleotides at the 3' end of tRNA in the event that they are removed.

230—C (Chapter 12) hnRNA is processed to form mRNA. A part of this processing (posttranscriptional modification) involves the addition of the polyA tail. ATP is the nucleotide donor. The polyA tail appears to increase the stability of mRNA. These modifications of mRNA occur only in eukaryotes.

231—C (Chapter 11) DNA polymerase γ catalyzes mitochondrial DNA replication. Mitochondrial DNA mutations are more prevalent than nuclear DNA mutations due to the lack of a DNA repair system.

232—D (Chapter 11) DNA polymerase δ and proliferating cell nuclear antigen are responsible for leading strand biosynthesis.

233—A (Chapter 11) DNA polymerase α is responsible for lagging strand biosynthesis.

234—H (Chapter 13) Eukaryotic elongation factor 2 (eEF2) is the protein that is ADP-ribosylated in a reaction catalyzed by diphtheria toxin.

235—B (Chapter 13) Eukaryotic initiation factor 2 binds initiator methionine-tRNA; this methionine-tRNA then is implanted into the P-site of the ribosome. Other aminoacyl-tRNAs are implanted into the A site of the ribosome.

236—H (Chapter 13) eEF2 binds GTP, and this elongation factor mediates peptidyl-tRNA/mRNA translocation from the A site to the P site in preparation for the next elongation reaction.

237—E (Chapter 13) Eukaryotic release factor (eRF) participates in the hydrolysis of peptidyl-tRNA to yield tRNA and a polypeptide. This activity occurs when a termination codon is recognized in the A site and peptidyl-tRNA is in the P site of the ribosome.

238—E (Chapter 7) Very long-chain fatty acids are oxidized initially in peroxisomes, and then the partially oxidized, but shorter, acids are translocated to the mitochondrion for complete degradation.

239—D (Chapter 15) Receptors for steroid hormones are functional in the nucleus, where they alter transcription rates.

240—B (Chapter 1) Lysosomes contain a variety of enzymes with acid pH optima that catalyze the hydrolysis of a variety of carbohydrate and lipid molecules.

241—J (Chapter 15) The sodium/potassium ATPase, or sodium pump, catalyzes the translocation of two sodium ions out of the cell and three potassium ions into the cell during the hydrolysis of one ATP. This enzyme (ATP) is the target of digitalis.

242—A (Chapter 18) Osteoporosis is associated with the demineralization of bone. Calcium salts, which are essential, are used in the prevention and management of this common disorder. Fluoride occurs in bone and may be beneficial, but fluoride is not essential.

243—D (Chapter 18) An excess of copper leads to both liver and brain pathology (hepatolenticular degeneration, or Wilson's disease).

244—F (Chapter 18) Thyroid cells actively take up iodine for incorporation into thyroxine. In people with Graves' disease, radioiodine is administered to inhibit thyroid cell function.

245—D (Chapter 16) Glycine is the only amino acid residue small enough to fit in its unique position in the triple helix of collagen; mutant forms of collagen containing other amino acids in this position lead to Ehlers-Danlos syndrome.

246—C (Chapter 18) Protein-glutamate residues are found in factors II, VII, IX, X, and XIII; the carboxylation of specific glutamate residues in these proteins is mediated by vitamin K.

247—J (Chapter 15) Tryptophan is converted into 5-hydroxytryptophan, which undergoes a decarboxylation to yield serotonin.

248—A (Chapter 6) Alanine is liberated from muscle into the circulation and is taken up by the liver, where it serves as a substrate for gluconeogenesis as part of the Cori cycle.

249—E (Chapter 6) Fructose 1,6-bisphosphatase, an enzyme important in gluconeogenesis, is inhibited by fructose 2,6-bisphosphate, which is formed in response to activation of the cyclic AMP second-messenger system.

250—M (Chapter 6) Isocitrate dehydrogenase, which is activated allosterically by ADP, is the rate-limiting enzyme of the Krebs citric acid cycle.

251—L (Chapter 7) Hormone-sensitive lipase is activated after phosphorylation as catalyzed by protein kinase A, or cyclic AMP-dependent protein kinase. Glycogen synthase and liver pyruvate kinase are also substrates for protein kinase A, but these two enzymes are inhibited by phosphorylation.

252—A (Chapter 7) Acetyl-CoA carboxylase, which catalyzes the conversion of acetyl-CoA to malonyl-CoA, is the rate-limiting enzyme in fatty acid biosynthesis.

253—J (Chapter 15) Thyrotropin-releasing hormone is a tripeptide produced by the hypothalamus and leads to increased secretion of thyroid stimulating hormone, or thyrotropin.

254—A (Chapter 15) Adrenocorticotropic hormone (ACTH) contains the sequence corresponding to melanocyte-stimulating hormone. When there is a deficiency of adrenal steroids, which occurs in Addison's disease, there is increased production of ACTH, which leads to increased skin pigmentation.

255—H (Chapter 15) Renin is a protein that is secreted by the juxtaglomerular cells of the kidney in response to norepinephrine release of postganglionic sympathetic neurons. It operates on angiotensinogen, a plasma protein that is produced by the liver.

256—F (Chapter 15) The insulin receptor, which consists of two α-chains and two β-chains, is an integral membrane protein whose β-chains possess protein-tyrosine kinase activity.

257—D (Chapter 8) Besides phenylalanine hydroxylase deficiency, the deficiency of dihydropteridine reductase produces hyperphenylalaninemia.

258—E (Chapter 8) Tyrosine aminotransferase, which is a hepatic enzyme, is the rate-limiting enzyme for tyrosine catabolism; its deficiency leads to type I tyrosinemia.

259—A (Chapter 8) Alkaptonuria is due to a deficiency of homogentisate oxidase. Tyrosine metabolites are excreted in urine and form dark-colored products after reacting with dissolved oxygen.

260—A (Chapter 6) Acetyl-CoA is required for the expression of pyruvate carboxylase activity, an enzyme that participates in gluconeogenesis.

261—E (Chapter 7) Farnesyl pyrophosphate is a 15-carbon intermediate in the pathway for cholesterol biosynthesis.

262—G (Chapter 9) N^5-Methyltetrahydrofolate reacts with homocysteine to produce methionine; a failure to produce N^5-methyltetrahydrofolate results in homocystinuria.

263—I (Chapter 7) Palmitoyl-CoA reacts with serine to initiate the synthesis of sphinganine biosynthesis.

264—C (Chapter 5) The k_{cat}/K_m is the specificity constant.

265—E (Chapter 5) The K_m is the substrate concentration that corresponds to $V_{max}/2$, or half-maximal velocity.

266—H (Chapter 5) In noncompetitive inhibition, the V_{max} is decreased without a change in the K_m.

267—G (Chapter 5) Increasing the substrate concentration can overcome competitive inhibition (no change in the V_{max}).

268—D (Chapter 18) Rickets is due to a deficiency of vitamin D, or dehydrocholesterol.

269—A (Chapter 18) Vitamin C, or ascorbate, is a substrate for the dopamine β-hydroxylase, an enzyme that converts dopamine to norepinephrine.

270—H (Chapter 18) Kwashiorkor is due to a deficiency of protein and calories.

271—F (Chapter 18) Pellagra is due to a deficiency of niacin.

272—K (Chapter 18) Beriberi is due to a deficiency of thiamine.

273—I (Chapter 18) Night blindness is due to a deficiency of vitamin A, or retinol.

274—M (Chapter 18) Abnormal bleeding is due to a deficiency of vitamin K.

275—I (Chapter 6) A deficiency of glucose-6-phosphatase, which catalyzes the hydrolysis of glucose 6-phosphate, leads to von Gierke's disease, or type I glycogen storage disease.

276—E (Chapter 6) The α-1,6-glycosidic linkages occur at branch points in glycogen; a deficiency of branching enzyme (amylo-1,4-1,6-transglycosylase) results in glycogen with abnormally long straight chains and decreased branching.

277—C (Chapter 6) UDP-glucose pyrophosphorylase catalyzes the reaction of uridine triphosphate and glucose 1-phosphate to produce UDP-glucose and inorganic pyrophosphate. This reaction is a reversible process with an equilibrium in favor of the reactants and not products. The hydrolysis of inorganic pyrophosphate helps to pull the reaction forward.

278—B (Chapter 8) A deficiency of ornithine transcarbamoylase decreases urea production (and leads to hyperammonemia). The carbamoyl phosphate that is synthesized in mitochondria traverses to the cytosol and serves as substrate for pyrimidine synthesis, leading to orotic aciduria.

279—H (Chapter 8) The cofactor for aspartate aminotransferase and all other mammalian transaminases is pyridoxal phosphate.

280—C (Chapter 8) ATP furnishes the chemical energy for the synthesis of argininosuccinate as ATP is converted to AMP and PP_i.

281—E (Chapter 8) Arginase catalyzes the hydrolysis of arginine to form urea and ornithine. As a result, there is an accumulation of argininosuccinate and citrulline.

282—J (Chapter 8) A deficiency of pyruvate carboxylase leads to acidosis (lactic acid accumulates) and a deficiency of aspartate. The aspartate deficiency is the postulated mechanism for decreased urea cycle activity and hyperammonemia.

283—L (Chapter 8) Approximately half of the individuals with homocystinuria respond to pyridoxine and the homocystinuria disappears.

284—E (Chapter 6) Isocitrate dehydrogenase, the rate-limiting enzyme of the Krebs tricarboxylic acid cycle, is activated allosterically by ADP.

285—C (Chapter 8) Fumarylacetoacetate hydrolase deficiency leads to hepatorenal tyrosinemia (tyrosinemia type I). Fumarylacetoacetate undergoes a nonenzymatic decarboxylation that leads to succinylacetone.

286—B (Chapter 8) A defect in homogentisate oxidase, which results in alkaptonuria, is accompanied by the excretion of homogentisate, a substance that turns black on standing or when treated with alkali.

287—A (Chapter 8) Tetrahydrobiopterin is a required cofactor in the phenylalanine hydroxylase reaction.

288—J (Chapter 7) The carboxylation of propionate to form methylmalonyl-CoA is an ATP-dependent process that involves a biotin prosthetic group.

289—H (Chapter 6) Succinate dehydrogenase is inhibited by malonate, a succinate analogue.

290—G (Chapter 6) The conversion of succinyl-CoA to succinate is accompanied by the conversion of GDP and P_i to GTP, a substrate-level phosphorylation.

291—A (Chapter 6) β-D-glucose, which occurs in the urine of people with diabetes mellitus, has hydroxyl groups that alternate up, down, up, down, and up in the standard Fischer projection formula.

292—D (Chapter 6) β-D-galactose differs from glucose at C^4 where the hydroxyl group is up in the standard Fischer projection formula.

293—E (Chapter 6) Sucrose is a nonreducing sugar (not oxidized by alkaline copper) because its aldehyde functional group occurs in a glycosidic linkage and is not available for an oxidation-reduction reaction.

294—D (Chapter 2) The amino group of proline is attached to the α- and δ-carbon atoms; this cyclic amino acid is an imino acid.

295—C (Chapter 8) Glutamate can be synthesized from α-ketoglutarate in one step via transamination or via reduction catalyzed by glutamate dehydrogenase.

296—F (Chapter 2) Tryptophan contains the aromatic indole ring.

297—G (Chapter 16) Myosin is the protein that is the chief ATPase of muscle that transduces chemical energy into mechanical energy.

298—C (Chapter 17) Factor IIa, or prothrombin, is a serine protease. Factor Ia is fibrin, formed by the action of factor IIa. Factor IIIa is thromboplastin.

299—F (Chapter 16) Lysyl oxidase, procollagen amino-terminal protease, and procollagen carboxyterminal protease function extracellularly during the synthesis of mature collagen.

Biochemistry
Must-Know Topics

The following are must-know topics discussed in this review. It would be useful to formulate outlines on these subjects because knowledge of the related material will be key to understanding the subject and material and for performing well on the examination.

The Big Picture: Blueprint of Metabolism

- The interrelationship of the mainstream pathways of metabolism
- The anatomy of the cell and the main function of the plasma membrane, nucleus, mitochondrion, lysosome, peroxisome, endoplasmic reticulum, Golgi, and secretory vesicles
- The fluid-mosaic model of biological membranes
- The stages of the cell cycle

Structure of Biomolecules

- Important elements that make up the body and the structure of important organic functional groups
- The functions of calcium, fluoride, cobalt, and iodine ions
- The classes of amino acids
- The structure of proteins and the type of bonds that contribute to the stability of proteins

pH and the Henderson-Hasselbalch Equation

- Use the Henderson-Hasselbalch equation to calculate pH, measure of acid strength, and the log of salt/acid when any two are given
- The titration curve of alanine
- The main physiologic buffers

Bioenergetics

- The relationship of the equilibrium constant and standard free energy change
- The advantage of energy-rich compounds in biosynthesis

Enzymes

- The role of enzymes in catalyzing reactions
- The Michaelis constant (K_m), the maximal velocity, the specificity constant
- Use the Michaelis-Menten equation to calculate the velocity, maximum velocity (V_{max}), K_m, and substrate concentration when any three of the four quantities are given
- Competitive, noncompetitive, and uncompetitive enzyme inhibitors and their corresponding diagnostic Lineweaver-Burk (double reciprocal) plots
- Irreversible enzyme inhibitors
- Cooperativity and how to diagnose it
- Mechanisms of enzyme regulation by covalent modification and by allosteric mechanisms
- Use the molecular logic of the cell to deduce the effect of allosteric agents

Carbohydrate Metabolism

- The structure of glucose, galactose, mannose, glycogen, and starch
- Glycolysis and its regulation
- Galactose metabolism and its aberrations
- The enzymology of pyruvate dehydrogenase and its regulation
- The Krebs tricarboxylic acid cycle and its regulation
- Electron transport and adenosine triphosphate (ATP) synthesis by the chemiosmotic mechanism
- Inhibitors of electron transport
- Calculation of ATP yields from metabolism of glucose
- Glycogen metabolism, its regulation, and its metabolic aberrations
- Gluconeogenesis and its regulation
- Regulation of carbohydrate metabolism by insulin and by glucagon
- Functions and vitamins of the pentose phosphate pathway

Lipid Metabolism

- Fatty acid activation, oxidation, biosynthesis, and regulation and aberrations in fatty acid oxidation
- Metabolism of propionyl-coenzyme A and the role of vitamin B_{12} in this pathway
- Phospholipid biosynthesis and catabolism and aberrations in phospholipid catabolism
- Ketone body metabolism
- Cholesterol biosynthesis, metabolism, and transport in circulation

Amino Acid Metabolism and Urea Biosynthesis

- Transamination reactions
- Ammonium ion production
- Urea cycle and its metabolic aberrations
- Catabolism of amino acids including the branched chain amino acids and phenylalanine
- Inborn errors in the metabolism of phenylalanine, tyrosine, leucine, valine, isoleucine, and urea
- Nitrogen balance
- Glycogenic, ketogenic, and essential amino acids

One-Carbon and Nucleotide Metabolism

- Role of *S*-adenosylmethionine, biotin, tetrahydrofolate, and vitamin B_{12} in metabolism
- Methyl trap hypothesis and folate and vitamin B_{12} actions
- Pyrimidine biosynthesis from named precursors and regulation of synthesis in humans and *Escherichia coli*
- Purine biosynthesis from named precursors and regulation of synthesis in humans
- The salvage pathways of purine metabolism
- Biochemistry of gout

Principles of Molecular Biology and Nucleic Acid Structure

- Central dogma of molecular biology
- DNA structure
- RNA structures, classes, and functions

Replication—DNA Synthesis

- Enzymes required for replication and DNA repair in humans and in bacteria
- Leading and lagging DNA strand biosynthesis
- Reverse transcription
- Mutations and cancer
- DNA repair mechanisms
- Mechanisms of cancer production
- Targets of cancer chemotherapy

Transcription—RNA Synthesis

- Classes of human and bacterial RNA polymerases
- Promoters and transcription signals
- The RNA polymerase reaction

Translation—Protein Synthesis

- General properties of the genetic code
- Amino acid activation
- The ribosome
- Initiation, elongation, and termination reactions
- Signal peptide hypothesis
- Biochemistry of glycosylation
- Steps in bacterial protein synthesis that are inhibited by antibiotics

Recombinant DNA Technology

- Polymerase chain reaction
- Restriction enzymes
- Using recombinant DNA for diagnosis and for preparing proteins

Cell Signaling

- Neurotransmitter biosynthesis, inactivation, and receptors
- Pathophysiology of myasthenia gravis and parkinsonism
- Steroid hormone action
- Cyclic adenosine monophosphate, cyclic guanosine monophosphate, calcium, and diglyceride second-messenger systems
- Protein serine/threonine kinases and protein tyrosine kinases

Muscle and Connective Tissue

- Role of ATP in muscle contraction
- Creatine phosphate metabolism and the clinical usefulness of creatine phosphokinase measurements
- Proteoglycan structure
- Collagen structure and biosynthesis
- Collagenopathies

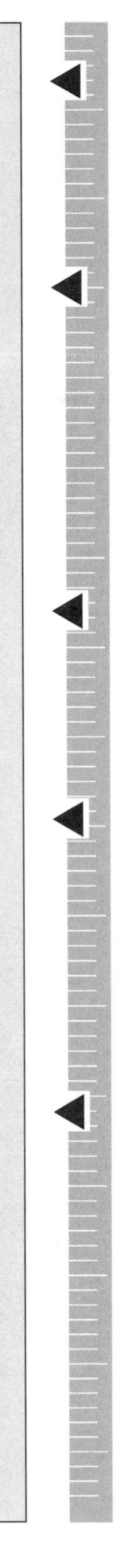

Hemoglobin, Heme, and Blood Coagulation

- Hemoglobin structure, function, properties, and disorders
- Heme biosynthesis and degradation
- The blood clotting cascade and intrinsic and extrinsic pathways
- Mechanisms of fibrinolysis and drugs used to produce fibrinolysis
- Genetics of hemophilia and von Willebrand's disease

Nutrition

- Energy content of fats, carbohydrates, proteins, and ethanol
- Role of water-soluble vitamins in metabolism
- Role of fat-soluble vitamins in metabolism
- Vitamin deficiency diseases
- Essential fatty acids
- Essential amino acids

Index

Note: Page numbers followed by *f* refer to figures; page numbers followed by *t* refer to tables.